S0-BKP-349

TURNING AND BORING
ANGLES AND APPLICATIONS

Donald O. Wood
Editor

Robert E. King
Manager

Published by

Society of Manufacturing Engineers
Publications Development Department
Marketing Division
One SME Drive
P.O. Box 930
Dearborn, Michigan 48121

TURNING AND BORING
ANGLES AND APPLICATIONS

First Edition

First Printing

Library of Congress Catalog Card Number: 85-50271

International Standard Book Number: 0-87263-169-9

Manufactured in the United States of America

SME wishes to acknowledge and express its appreciation to the following contributors for supplying the various articles reprinted within the contents of this book. Appreciation is also extended to the authors of papers presented at SME conferences or programs as well as to the authors who generously allowed publication of their private work.

The Carbide and Tool Journal
The Society of Carbide and Tool Engineers
P.O. Box 437
Bridgeville, Pennsylvania 15017

Cutting Tool Engineering
CTE Publications Incorporated
464 Central Avenue
Northfield, Illinois 60093

Machine and Tool BLUE BOOK
Hitchcock Publishing Company
Hitchcock Building
Wheaton, Illinois 60488

Manufacturing Engineering
Society of Manufacturing Engineers
One SME Drive
P.O. Box 930
Dearborn, Michigan 48121

Production
Production Publishing Company, Incorporated
Box 101
Bloomfield Hills, Michigan 48303

Sandvik, Incorporated
Coromant Division
1702 Nevins Road
Fair Lawn, New Jersey 07410

Tooling & Production
Huebner Publications, Incorporated
6521 Davis Industrial Parkway
Solon, Ohio 44139

Cover photo courtesy of The Lodge & Shipley Company

PREFACE

Essential to the success and production of any manufacturing metalcutting operation, is an understanding and basic knowledge of the various types of cutting tools available in today's advancing technological world. Equally important is the comprehension of the basic turning and boring operations.

Throughout the years of the industrial revolution and before, turning and boring operations have been fundamental metalcutting operations. Reshaping a raw forging or casting into a functional context is to provide manufacturing personnel with the knowledge and practical know-how to better enable them to select proper tooling in the basic turning and boring operations.

Troubleshooting and updating current metalcutting operations is also a vital part of this text. Application data of all types of cutting tool materials is clearly defined. Attention is directed to the workpiece, machine tool considerations and practical threading operations also are addressed.

Chip control plays a vital role when consideration is directed to the overall success of a turning or boring operation. One chapter of the text directs attention to the methods of obtaining proper chip control.

With the advent of the goal of the unmanned factory of the future, much additional responsibility will be placed upon manufacturing engineers to provide the most trouble-free turning operations as possible. The information contained in this text will provide assistance in attaining this goal.

Special thanks is due to Paul Neumann, Manager, Technical Services, The Lodge & Shipley Company, Cincinnati, Ohio, who researched with me, the various publications which were made available to us. Mr. Neumann has been long recognized as an authority in the metalworking industry.

I wish to thank all of the companies, organizations, publishers, and authors who gave permission to have their articles reprinted in this volume. Thanks also to the Publications Development Department staff at SME for their assistance in the research and development required in making this book possible.

Donald O. Wood
The Lodge & Shipley Company

ABOUT THE EDITOR

Donald O. Wood has been associated with the metal-working industry for the past 17 years. Mr. Wood is employed at The Lodge & Shipley Company as their Mid-Central Regional Sales Manager. He has served in this position for the past seven years. Prior to his joining The Lodge & Shipley Company in 1978, he served in various capacities with one of the leading ceramic cutting tool manufacturers throughout the world. His final position was that of Director of Marketing.

Mr. Wood's basic manufacturing experience was obtained in his years at Dana Corporation, Toledo, Ohio. While serving Dana, Mr. Wood served in various management positions including Cutting Tool Coordinator.

Mr. Wood obtained his degree in Industrial Engineering from the University of Toledo, majoring in Industrial Engineering. Additionally, he earned an Associate Degree in Industrial Management.

A member of the Society of Manufacturing Engineers, The Society of Cutting Tool Engineers and The Association for Integrated Manufacturing Technology (formerly the Numerical Control Society), Mr. Wood has authored numerous papers on ceramic cutting tools. He has chaired numerous programs for SME including the Cutting Tool Clinic.

SME

The informative volumes of the Manufacturing Update Series are part of the Society of Manufacturing Engineers' many faceted efforts to provide the latest information and developments in engineering.

Technology is constantly evolving. To be successful, today's engineers must keep pace with the torrent of information that appears each day. To meet this need, SME provides, in addition to the Manufacturing Update Series, many opportunities in continuing education for its members.

These opportunities provide:

- Monthly meetings through five associations and their more than 300 chapters and 130 student chapters worldwide to provide a forum for membership participation and involvement.
- Educational programs including seminars, clinics, programmed learning courses, as well as videotapes and films.
- Conferences and expositions which enable engineers and managers to examine the latest manufacturing concepts and technology.
- Publications including the periodicals *Manufacturing Engineering*, *Robotics Today*, and *CIM Technology*, the *SME Newsletter*, the *Technical Digest*, *Journal of Manufacturing Systems*, and a wide variety of text and reference books covering everything from the basics to manufacturing trends.
- Information on Technology in Manufacturing Engineering database containing technical papers and publication articles in abstracted form. Other databases are also accessible through SME.

The SME Manufacturing Engineering Certification Institute formally recognizes manufacturing engineers and technologists for their technical expertise and knowledge acquired through experience and education.

The Manufacturing Engineering Education Foundation was created by SME to improve productivity through education. The foundation provides financial support for equipment development, laboratory instruction, fellowships, library expansion, and research.

SME is an international technical society dedicated to advancing scientific knowledge in the field of manufacturing. SME has more than 80,000 members in 70 countries and serves as a forum for engineers and managers to share ideas, information, and accomplishments.

The society works continuously with organizations such as the American National Standards Institute, the International Organization for Standardization, and others, to establish and maintain the highest professional standards.

As a leader among professional societies, SME assesses industry trends, then interprets and disseminates the information. SME members have discovered that their membership broadens their knowledge and experience throughout their careers. The Society of Manufacturing Engineers is truly industry's partner in productivity.

MANUFACTURING UPDATE SERIES

Published by the Society of Manufacturing Engineers and its affiliated societies, the Manufacturing Update Series provides significant up-to-date information on a variety of topics relating to manufacturing. This series is intended for engineers working in the field, technical and research libraries, and also as reference material for educational institutions.

The information contained in this volume doesn't stop at merely providing the basic data to solve practical shop problems. It also can provide the fundamental concepts for engineers who are reviewing a subject for the first time to discover the state of the art before undertaking new research or applications. Each volume of this series is a gathering of journal articles, technical papers and reports that have been reprinted with expressed permission from the various authors, publishers, or companies identified within the book. Educators, engineers, and managers working within industry are responsible for the selection of material in this series.

We sincerely hope that the information collected in this publication will be of value to you and your company. If you feel there is a shortage of technical information on a specific manufacturing area, please let us know. Send your thoughts to the Manager, Publications Development Department, Marketing Division at SME. Your request will be considered for possible publication by SME or its affiliated societies.

TABLE OF CONTENTS

CHAPTERS

1 CHARACTERISTICS INFLUENCING METAL CUTTING

2 TURNING

3 BORING

CHAPTER 1

CHARACTERISTICS INFLUENCING METAL CUTTING

Reprinted courtesy of Sandvik, Incorporated, Coromant Division
Fair Lawn, New Jersey

Modern Metal Cutting: Turning Theory

Introduction

In part 1 we talked generally about metal cutting, machinability and the main components of a modern cutting tool: tool material, cutting edge geometry and edge clamping method. We discussed these points irrespective of the metal cutting method. Now the time has come to apply these points to various metal cutting methods. In the following parts of Modern Metal Cutting we will discuss the three principal metal cutting methods: turning, milling and drilling as well as the application of tools for the respective methods.

This part deals with the theory of turning, by far the most common metal cutting method. Nearly a quarter of all metal cutting machines are for turning. It is also the least complex of the metal cutting methods. Turning is the cutting off of chips from a workpiece to obtain a specific circular shape in most cases using a single point tool.

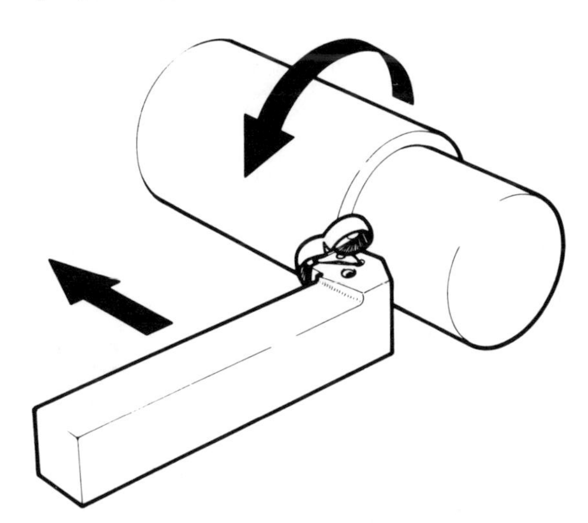

The Turning Tool

In turning the metal workpiece rotates in a machine and the cutting tool is forced in against it. (In some special machines the workpiece is stationary and the tool rotates round it). Turning is made up of two main movements: the rotation of the workpiece and a linear movement of the tool along or across the workpiece. The workpiece rotates at a certain speed while the tool is moved along at a certain feed and at a certain depth into the workpiece.

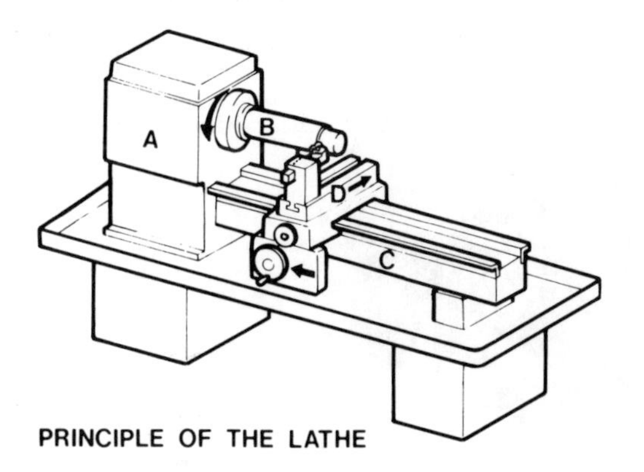

PRINCIPLE OF THE LATHE

Principal parts of a turning lathe:

A - the headstock which provides the power and motion for the workpiece to rotate.
B - Workpiece held and rotated.
C - Lathe bed with slides for carriage, performing longitudinal movement.
D - Carriage with toolpost.

Later on in a part of Modern Metal Cutting we shall describe the machines known as turning machines, normally called lathes, which rotate the workpiece and move the cutting tools linearly against the workpiece through various mechanical devices. While the above drawing shows the principal parts of a lathe the below photographs show a manually operated lathe (left) and a numerically controlled lathe (right) in action in machine shops. These machines and others for turning will be described in more detail in a later part.

The Cutting Edge

It does not take a lot of imagination to see that there is a lot more than what meets the eye to a cutting edge. As we indicated in part 1 there are three main fields in the design of any cutting tool. We have talked about tool materials so now let us take a general look at geometries.

It is essential for the efficiency of machining that the correct cutting edge condition prevails. We all know from everyday experience that a sharper edge cuts much easier than a blunt one. With added reflection we also know that with more sharpness comes a weaker edge--the cross-section of the edge is smaller. In metal cutting this difference is quite straightforward.

Let us imagine that we are looking at a section of the workpiece and tool and that we have a line across the center of the workpiece. Generally it can be said that the cutting tool is always placed with the edge tip at this line so as to obtain the correct cutting conditions. The front side of the tool, below the edge that cuts, should taper away somewhat so as not to rub against the rotating workpiece. The tool is said to have a certain amount of clearance. This clearance angle of the tool does not vary much--it amounts to a few degrees for most tools. Keeping in mind that the front of the tool more or less always slants away at a few degrees, we can look at what happens to the top face of the tool with a varying cross-section.

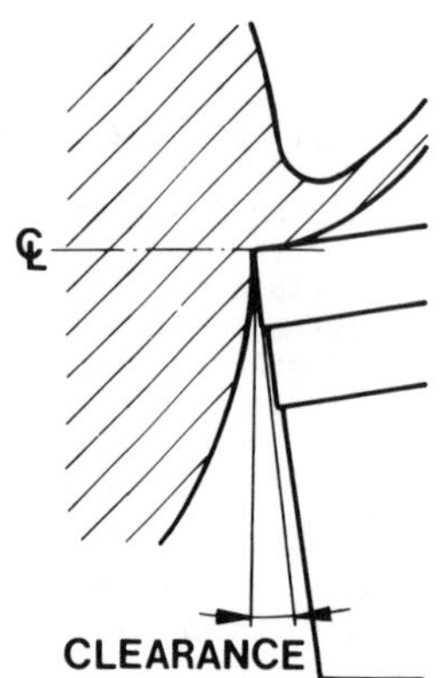

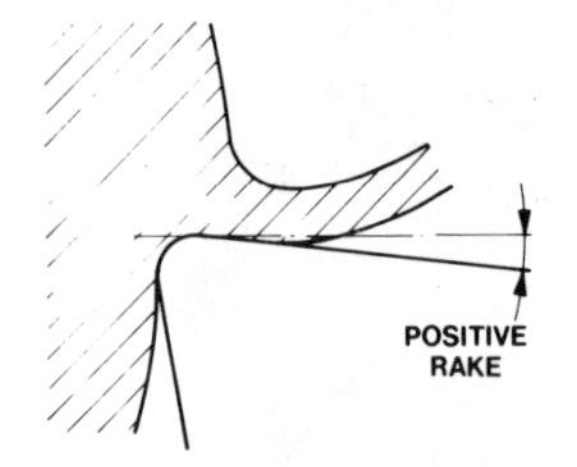

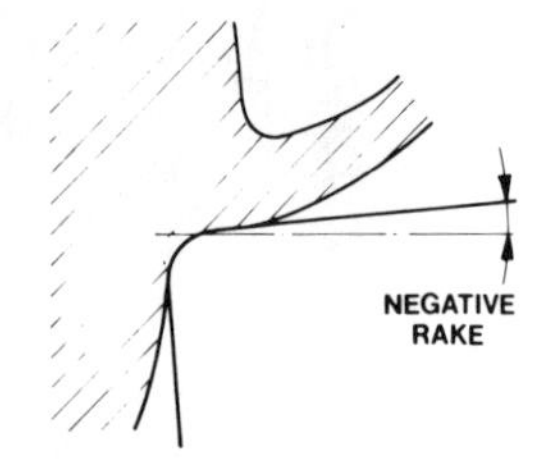

The left figure shows a sharper cutting edge than that of the right. (Look at the top of the tool nearest the workpiece). On the other hand the left tool has a weaker cross-section than the right one, and in metal cutting edge-strength is a vital asset. The left edge is a positive type cutting edge (the angle at the front of the tool cross-section is less than 90°). The right edge is a negative type cutting edge (the angle is 90° or more). The choice of angle is closely associated with the workpiece material.

In this connection we talk about the positive or negative rake angle. If we go further back from the edge we may find that although a tool may have a positive rake at the front it may have a negative inclination further back, as can be seen in principle in the next figure. This is often the case with modern indexable insert tools.

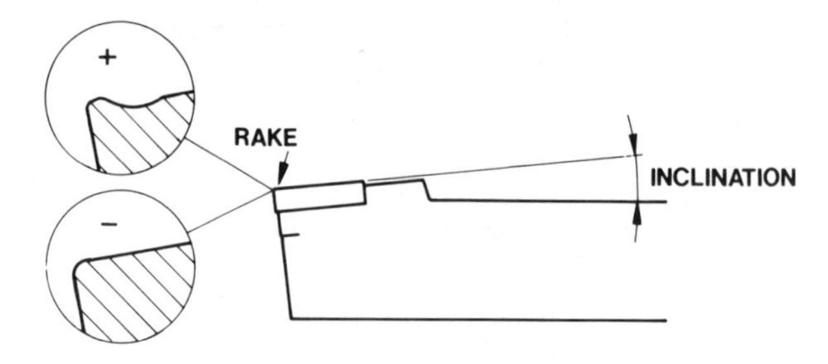

It is also important to note the difference between basic insert shapes. A negative basic shape is characterized by the various faces being perpendicular in relation to each other. The basic shape is positive when the sides are less than 90° in relation to the top face of the insert. In the toolholder, this means that the negative shape insert must be given a negative angle of inclination so as to provide the tool with the necessary clearance. The positive shaped insert, on the other hand, can be positioned with a positive angle of inclination.

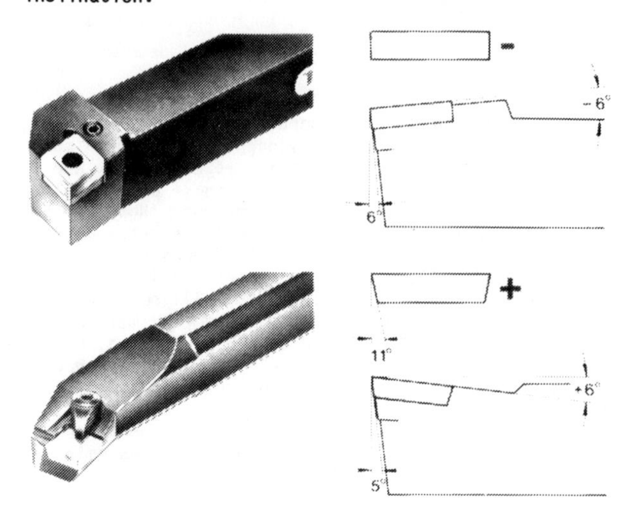

The cutting edge can be analyzed into a number of other angles but here we will mention only the important ones from a practical point of view. Looking down on top of various tools we find that the cutting edge point is different. The angle between the cutting edge sides varies. This is called the point angle. As with the previously mentioned angle of the side cross-section of the edge, a large angle means a stronger edge than that obtained with a small one.

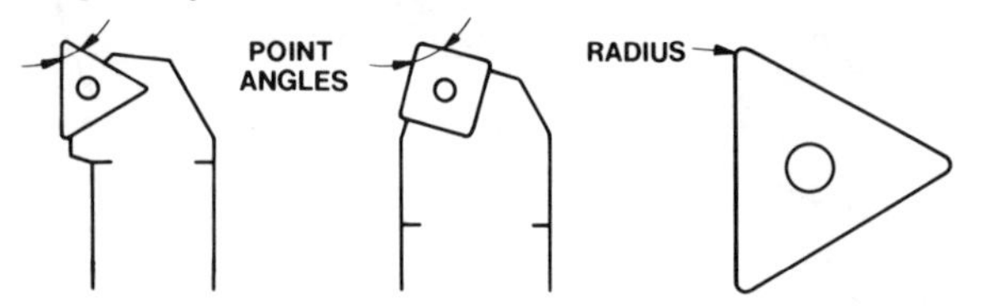

Also important in this context--especially in turning--is the strength of the cutting edge as affected by the configuration of the edge corner. The corner, when looking down on the tool, is always rounded to various radii. Part of the cutting tool salesman's job is to select a correct edge radius on the tool he is to apply. The point radius is important in that it has some direct effect on performance and result of the machining. A large radius means a strong cutting edge but it may also lead to vibration tendency. The finish of the turned surface and the accuracy is affected by the combination: point radius and the feed rate at which the tool is moved along at.

Finally, we have the most important angle as regards the choice of a tool for a certain turning operation--the entering angle or the approach angle. This is the angle between the main cutting edge of the tool and the face of the workpiece along which the tool is cutting. (This angle is also expressed through the lead angle or side cutting angle which is the angle

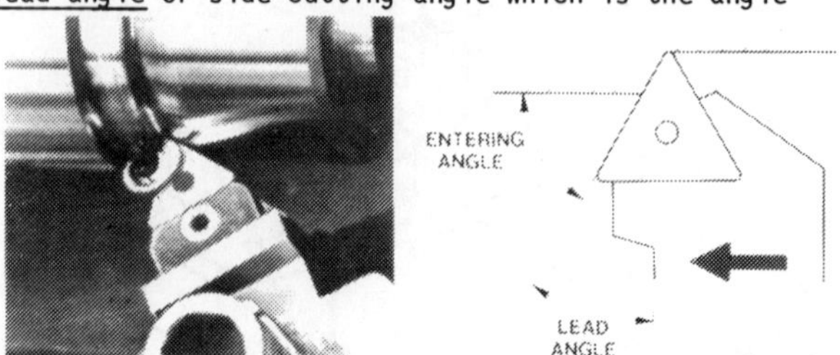

between the main cutting edge and the longitudinal axis of the tool). When the cutting tool engineer chooses a tool for an operation it is this angle he determines first of all and for this reason catalogue material on turning tools lists the tool according to entering angles. It is often the configuration of the workpiece which determines the entering angle.

The entering angle is important in that it directly affects the thickness and direction of the metal chips cut off. It also affects the forces involved in the cutting action. As with most other factors in machining, one value is advantageous in one way but disadvantageous in another and so the working result usually turns out to be the most suitable compromise. Small entering angles such as 45° are advantageous in that with the thinner chip the cutting forces are distributed over a longer part of the cutting edge. This means that the tool can cope better with heavier and difficult cutting and that the increase and decrease of thrust on the tool at the start of cuts and at end of cuts will be more gradual. However, the tool presses very heavily against the workpiece which can give rise to vibrations and deflections if the workpiece is long in relation to its diameter.

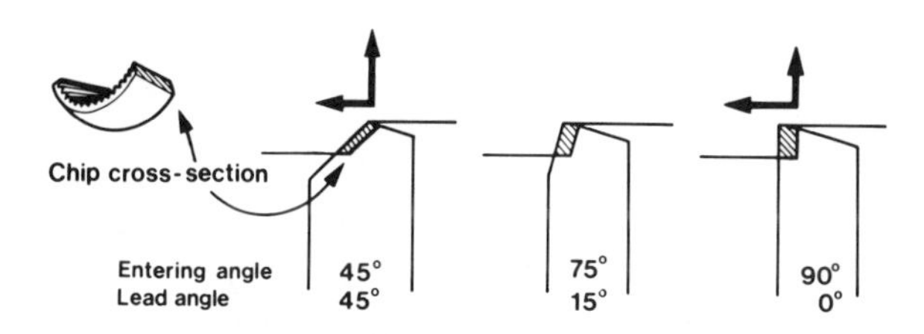

Large entering angles involve smaller forces acting radially, towards the center of the workpiece. Instead the tool presses more in the direction of the length of workpiece which mean less vibration tendencies especially when long and weak workpieces are machined. The disadvantage with a 90° entering angle (0° lead angle) is the sudden loads imposed on the cutting edge at the start of the cut and the sudden release of loads at the end of the cut. Unless cuts are very heavy and difficult or the workpiece is long and weak and the shape of the workpiece demands a very large angle, entering angles of 60-70° (lead angles of 15-20°) are useful compromises.

The Chipbreaker

If the cutting edge is made to cut as a plain wedge forced into workpieces, which are of ductile materials such as most steel, the metal that comes off will be long continuous strands. The more brittle metals like most cast-irons break up into small flakes when cut. Consequently, chipbreakers for such metals are not as vital. The long strands will be very difficult to handle and be not only a safety hazard but become tangled with the workpiece and cutting tool. Today's modern lathes, working at very high removal rates would not be able to function if cutting tools did not have some means with which to break these long strands into shorter chips. Good chipbreaking is absolutely essential in certain materials for trouble-free production machining.

The consequences of poor or non-existant chipbreaking.

Most of today's turning tools have chipbreakers included in the actual indexable insert. A modern carbide insert is a carefully balanced unit of cutting angles and chipbreaker functions. The angles in the grooves give a positive rake and act as a chipbreaker. Combinations of angles and radii are form-sintered onto the insert during manufacturing. Some inserts are more complicated than others ranging from just one simple chipbreaker groove to an insert face made up of ridges, valleys and flats which break chips of different materials at various conditions. The latter can cope with heavy to fine cutting as it has chipbreakers in several dimensional steps.

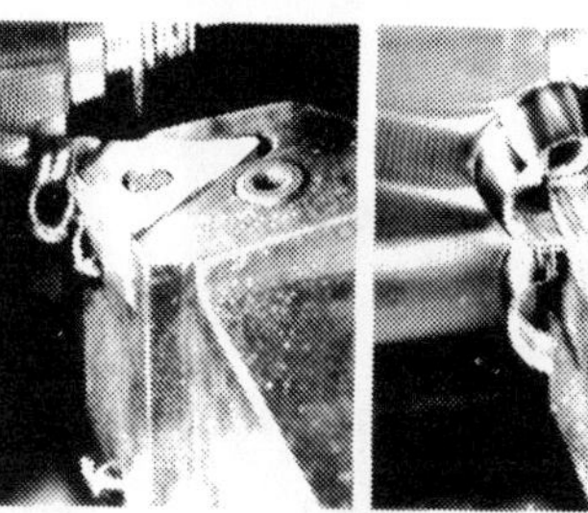
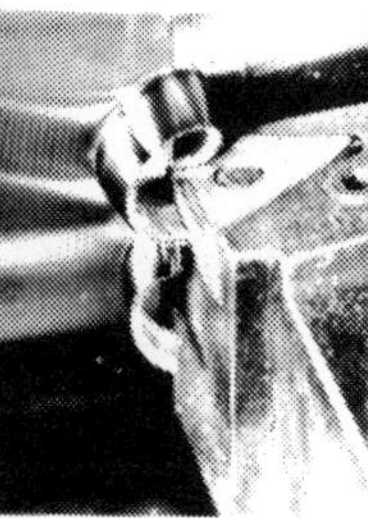

At small depths of cut the point of the insert is in action and the chip often comes off as helically formed chips in various lengths. When the depth of cut is increased a larger part of the edge and chipbreaker comes into action and the chip adopts a shorter form. The chip will then break off either as a result of the deformation it is subjected to or because of the stresses built up in the chip as it comes off against the tool or workpiece.

Also how rapidly the tool is fed along the workpiece will affect how the chip passes the chipbreakers. The depth of cut will influence the width of the chip, while the feed influences the thickness of the chip. The first chipbreaker stage takes care of chips created by smaller feed rates. By stepping up the feed rate of the tool, the chip will become thicker and come off against another stage. Thicker chips, from higher feed rates are usually easier to break.

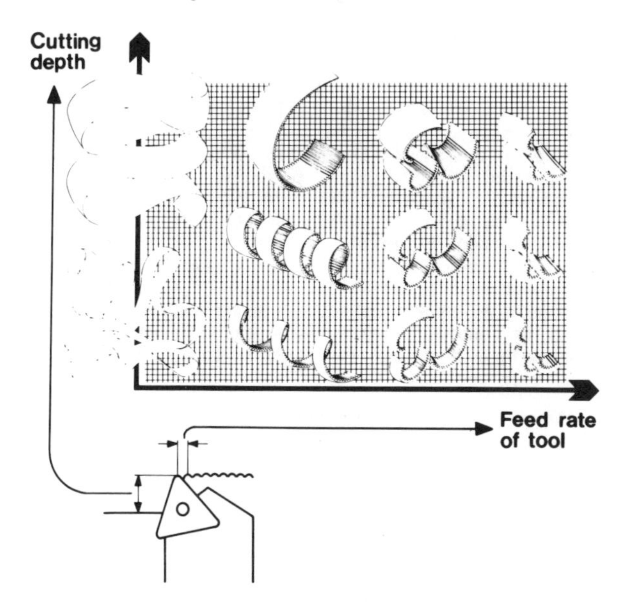

The chipbreaking ability of an insert is often the decisive factor for which operation it will be applied for. The application range of an insert is thus usually indicated through a simple diagram with cutting depths plotted against feed rates. A practical diagram can be made up for an insert in the machine shop by running various combinations of depths and feeds of the tool and placing the chips obtained on a large drawn diagram. When a certain amount of values have been covered the various types of chips are studied to see which are acceptable. The part of the diagram enclosing acceptable chips forms the application range of the insert.

Some tools do not have the chipbreaker integrated in the insert. Older tools and certain heavy duty tools with plain inserts have a loose piece clamped on top of the insert against which the chips are broken. One advantage of these are that they can be adjusted to suit specific cutting conditions. There exist also special purpose tools which have chipbreakers that are ground into the insert.

Chipbreaker on clamp

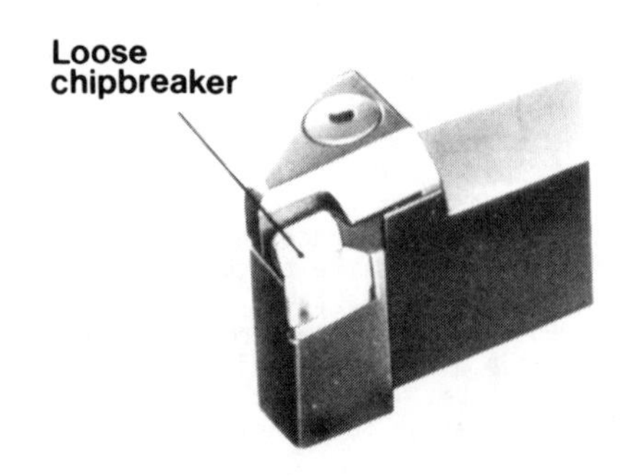

Edge preparation

In addition to the basic geometry of the cutting edge, made up of cutting angles and chipbreakers, the edge of the insert itself can be modified to have important effects on the durability and performance of the tool. These modifications may sometimes only be microscopic in size but, especially in milling and more brittle tool materials, substantial modifications play a vital role. Above all, edge strength is improved through edge preparation. As a rule, modern carbide indexable inserts have these modifications incorporated during manufacturing but some basic knowledge of these techniques may come in useful to solve problems at one time or another.

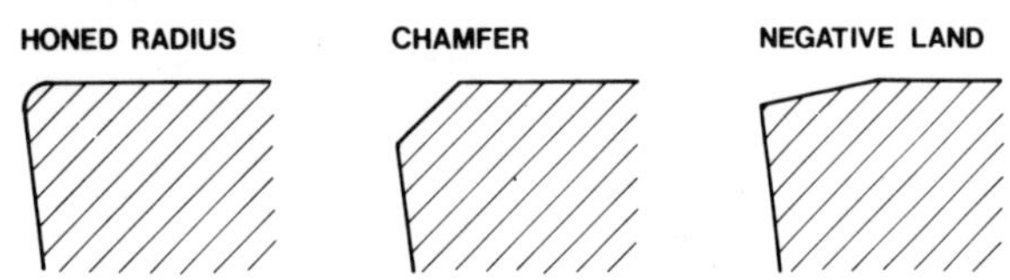

There are three ways in which the edge can be prepared: honing, chamfering and the application of a negative land. Most turning inserts are honed as part of the manufacturing process. It entails a rounding off of the edge: a small smooth radius is applied on all of the insert edges. This is a mechanical process in which the inserts are tumbled for a certain time with an abrasive mixture. Honing is an important preliminary step performed prior to coating inserts.

When an insert is chamfered it is provided with a bevel generally ranging from 20-45°. Honing and chamfering can be used to take away the sharp delicate edge. When an insert is provided with a negative land it is given a bevel ranging from a few to about 15°. This method is especially useful for heavy duty or turning involving shock loads. The negative land modifies the rake angle to be more negative thus strengthening the edge and redirecting cutting forces acting on the cutting edge to a more advantageous direction. However, the wrong application of a negative land may lead to excessive loads which deteriorate the life of the edge and raise the power consumption.

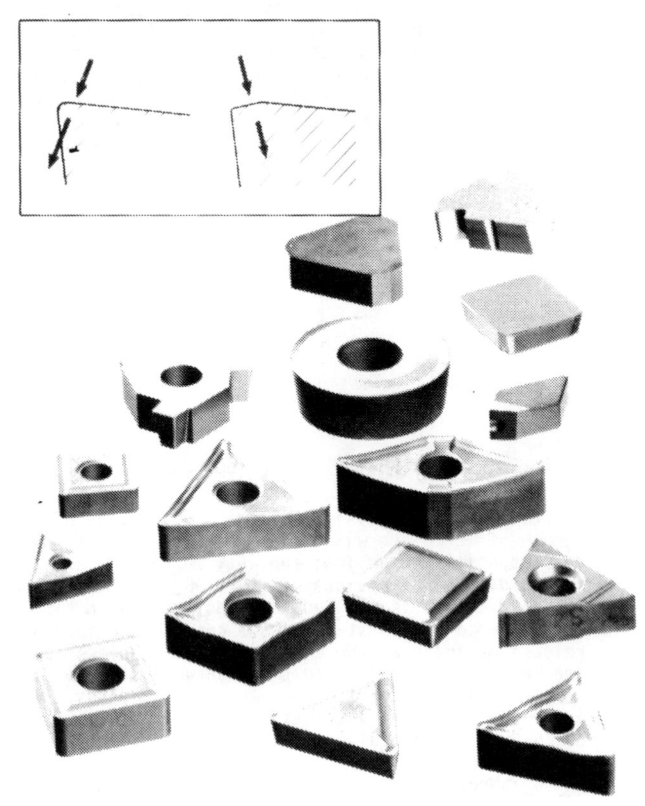

The Turning Operation

As mentioned, turning is made up of two main movements: the rotation of the workpiece and a linear movement of the tool, along or across the workpiece. This is expressed through the workpiece rotation at a certain spindle speed and the tool being fed along at a

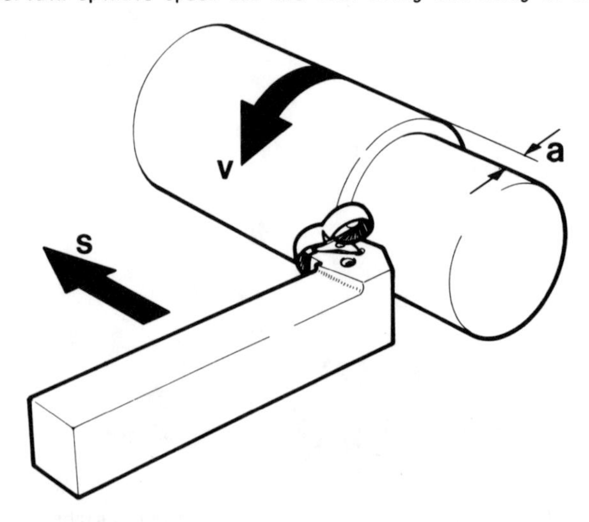

certain feed rate. Moreover, so as to cut material off, there has to be a certain depth of cut. The turning operation is practically defined through various values: cutting data.

CUTTING DATA

Spindle speed, symbolized by the letter n, is the number of revolutions that the machine turns the workpiece during one minute. Consequently, it is denoted in rev./min. This value is machine tool oriented and is found on the machine control panels.

Cutting speed, is symbolized by the letter v, and expressed in surface feet per minute (sfm) or m/min in metric values. This is a value for the velocity at which the periphery of the workpiece being cut passes the cutting edge. This is obviously a tool oriented value related to what diameter is being machined. Often, a workpiece consists of several different diameters. If the workpiece rotates at a constant spindle speed the cutting speed will vary with the diameters--a fact which can have consequences for the outcome of the machining. Some modern machines, however, can be programmed to run at constant cutting speeds--the spindle speed increases when the smaller diameters are being machined and decreases when larger diameters are turned.

The cutting speed is related to and can be calculated through the spindle speed and workpiece diameter:

$$= \frac{n \times \pi \times D}{12} \quad \text{(sfm)}$$

The diameter D is in inches, π is a constant 3.14 and 12 converts the value into feet.

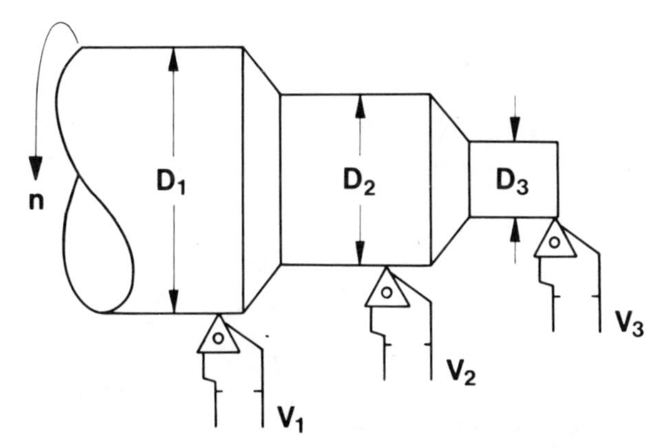

In metric, the calculation is as follows:

$$= \frac{n \times \pi \times D}{1000} \quad \text{(m/min)}$$

where D is in mm.

The cutting speed is very important for performance in turning. It influences the metal removal rate, tool-life, power consumption, etc. Nominal cutting speeds are available for tool materials and designs to machine specific workpiece materials--with important consideration taken to the hardness. The selection of speeds is also made on the basis of experience and apart from being largely influenced by the tool and workpiece material, the type of operation, workpiece shape, machine condition and above all the tool-life desired will affect the cutting data.

In part two of this series we mentioned a phenomenon known as built-up edge. This is a problem which can arise during the machining of ductile materials, stainless steel, etc. Material from the workpiece welds onto the cutting edge. A good way to avoid built-up edges is to have a cutting edge of a certain design and to run it at suitable cutting data.

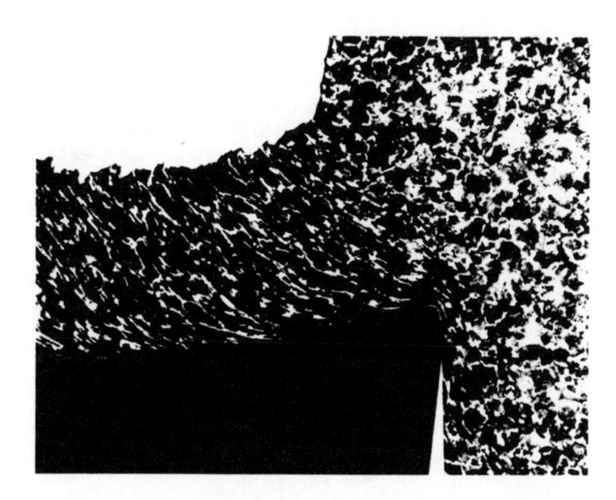

With higher cutting speeds come higher temperatures and higher temperatures can improve machinability through a layer effect. The contact zone between the chip and the cutting edge experiences the formation of workpiece-metal-layers which act as lubrication films, improving the passage of the chip over the edge. Modern indexable inserts have been designed to incorporate this effect and come to their full advantage when used at high cutting data.

Feed, is symbolized by s and expressed in inches per revolution (ipr) or mm/rev. This is the distance the tool moves for each revolution of the workpiece. The feed can also be expressed as the length moved per minute. As indicated in the chipbreaking diagram, the feed determines not only the thickness of the chip being machined off but also the quality of chip control.

Roughing

Finishing

The feed rate is also an important influencing factor on the metal removal rate for roughing operations (where the object is to remove material) and surface finish for finishing operations (where the object is a satisfactory finish). It does not influence tool-life or power consumption as much as the cutting speed does.

The two figures below show the feed and cutting depth when the turning tool is fed longitudinally, along the axis of rotation in the lathe, and across towards the center of the workpiece.

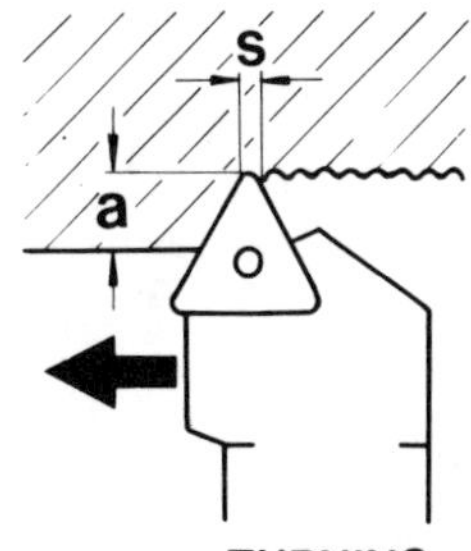

TURNING

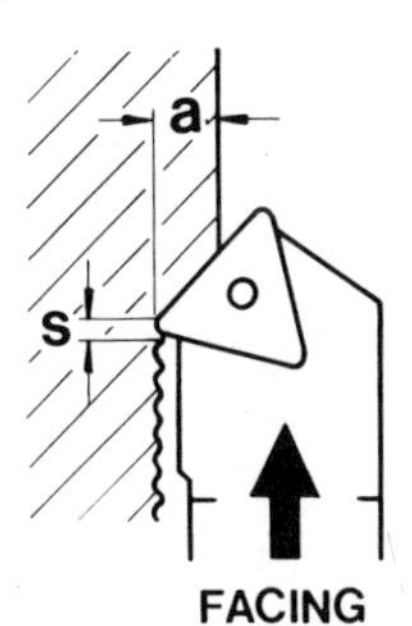

FACING

Cutting depth, is the distance (a) between the uncut and cut surface of the workpiece. The depth of cut is usually determined by factors such as the working allowance and configuration of workpiece, etc. To achieve the best possible performance when performing roughing cuts the largest cutting depth possible should be chosen. The cutting depth does not influence tool-life to any mentionable extent but to some extent does affect the power consumption.

Another important point is that the whole cutting edge length of the insert cannot be utilized in one cut. Selection of inset size is to a considerable extent based on cutting depths to be taken. The high

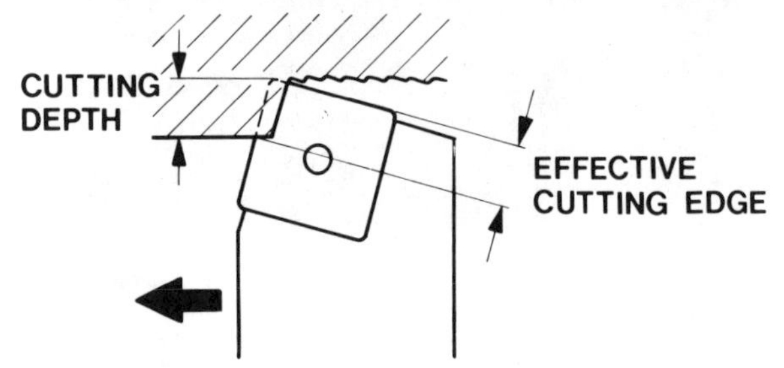

efficiency of modern indexable inserts and other varying factors make it difficult to establish any set rule stating how much of the cutting edge length can be utilized in a cut. However, a rule which has been applied in practice has been that one should avoid using more than 2/3 of the edge as cutting depth. Some inserts cannot, however, use as much as this.

POWER

To remove metal efficiently demands sufficient power for the operation in question. The machine tool is powered by an electric motor from which the rotational movement is transmitted through a gearbox to the spindle which rotates the workpiece. In machining, the power is consumed for the cutting of the chip, friction at various points between the tool, chip and workpiece and efficiency losses as the power is transmitted from the motor to the cutting edge.

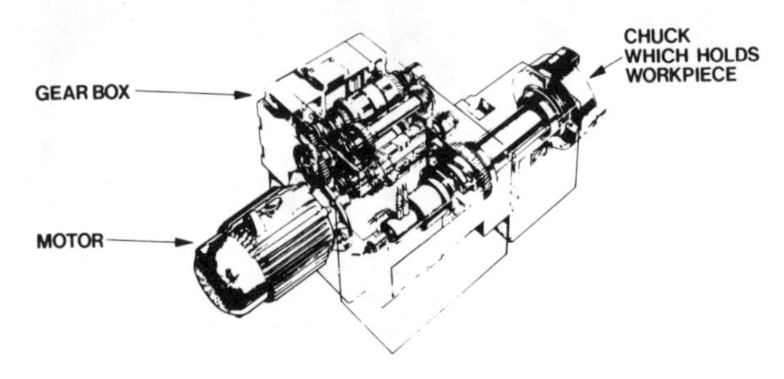

The power needed for metal cutting is mainly affected by the following factors: cutting speed, feed rate, cutting depth and the workpiece material. Generally speaking, the higher the first three factors are and the stronger and harder the workpiece material is, the higher the power requirement. Most modern lathes have sufficiently large motors to provide the power required for turning operations with indexable insert tools.

The two photos below show two extremes: the one on the left is on a heavy duty lathe where very large workpieces are machined. The cutting depths that have to be taken are sometimes so large that they have to be divided so as to be taken by several cutting edges. The picture on the right shows a small workpiece being machined with comparatively small cutting depths. The difference in power required to turn at various cutting data is indicated by the large lathe needing some 150 hp and the small one about 8 hp.

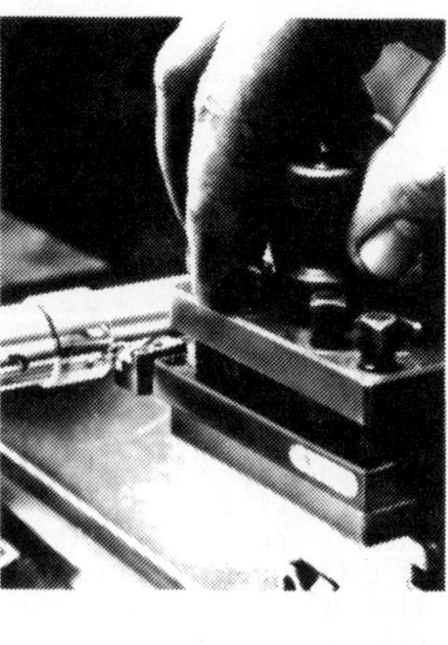

OPERATION TYPES

A lot of turning operations consist of straightforward longitudinal machining of a round bar--simply removing a rough outer surface, turning it to another diameter. But there are also several other operation types in turning. The outer part of the bar may need a smaller diameter part of the way along. The step down to this smaller diameter may be via a chamfer or a perpendicular shoulder. This change in the configuration may result in just a groove or recess or it may be that the rest of the workpiece will continue at this diameter or even come down again. Some part of the workpiece might require a screw thread or some other special form cut. The same operations might have to be performed internally--in a hole through the workpiece.

When a workpiece is to be machined, the machining is broken down into simpler sequences. The manufacturing engineer will analyze the machining necessary for the workpiece, decide which machine tool is most suitable and plan the operations required. Machining operations have to be specified for operating, setting or programming of the machine tool. It is here, also, that the cutting tool salesman can be of assistance in advising suitable tools.

Longitudinal turning is the most common turning operation where the tool moves along the workpiece axis, turning the diameter down.

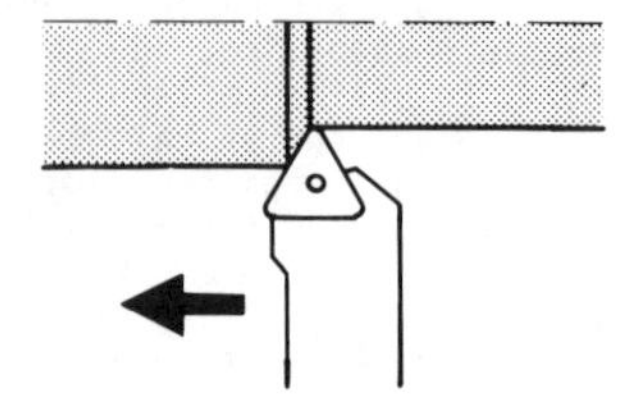

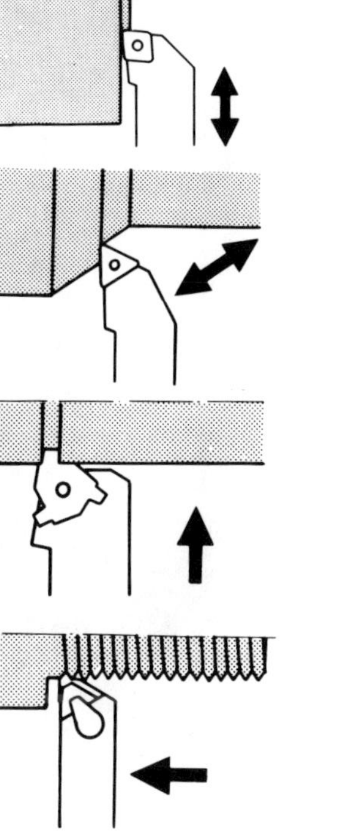

Facing is also a common operation where the tool turns a face perpendicular to the workpiece axis, either out away from center or in towards center.

Out-copying and in-copying can be performed at various angles. Some workpieces consist of combinations of these cuts and acute angles which make demands on the accessibility that can be obtained with the tool.

Form cuts are performed with tools which have been shaped to the specific form that has to be cut. The most common are various types of grooves, recesses or chamfers.

Threading is performed when the workpiece needs a screw head. This can be performed with a tool having the same form as the thread which is then fed along at a rate which corresponds to the pitch of the thread. (Screw threads can, however, also be produced using taps and dies).

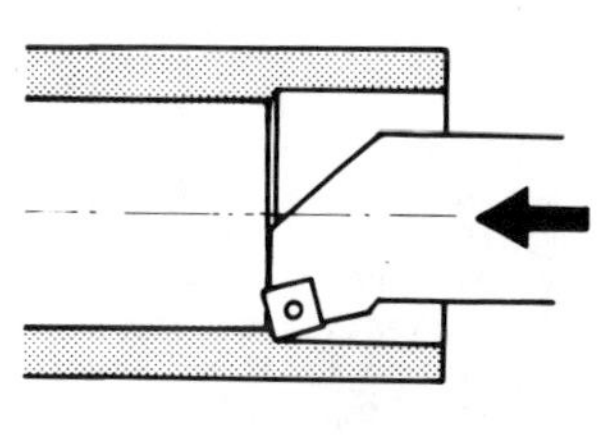

Boring is the internal turning of a workpiece--the machining of a hole already drilled. Most of the above mentioned external operations are performed internally as well with tools specially developed for this purpose.

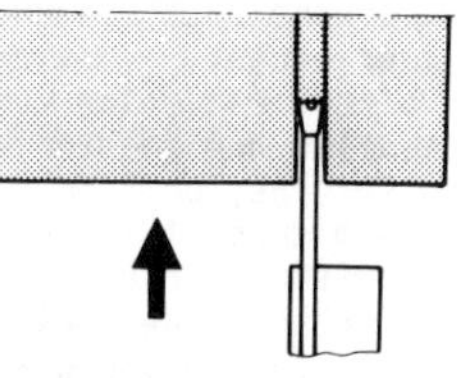

Cutting off is performed when the workpiece has been ready turned, at least at one end. Instead of sawing off the component a special parting tool is used to cut off the component from the workpiece bar.

These are the basic operations of turning. Most workpieces are a combination of a couple, few or all of these thus requiring a set of tools to complete their machining. Many modern automatic or numerically controlled machines have sufficient tool positions to cope with all the turning operations necessary for many complicated components.

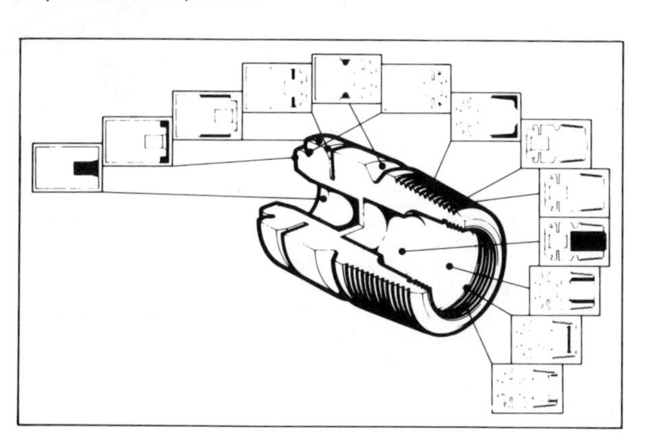

The next part in this series will deal with the practical part of turning, choosing cutting tools to suit workpiece configurations and operative demands.

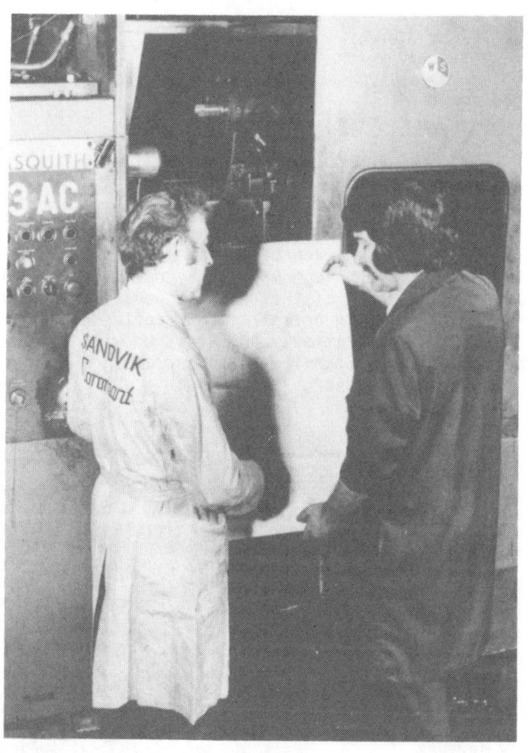

Reprinted courtesy of Sandvik, Incorporated, Coromant Division
Fair Lawn, New Jersey

Modern Metal Cutting: Turning Tools

Introduction

This part will deal with the practical part of turning, actually selecting the tool types for specific operations. Part 4, Turning Theory finished off with Operation Types, where the machining of a workpiece was broken down into seven basic operations. We will carry on in this way throughout part 5, reviewing the choice of tools for each of these basic operations and bringing up the most important factors influencing the choice and application. But first we will review the Sandvik Coromant turning tool program.

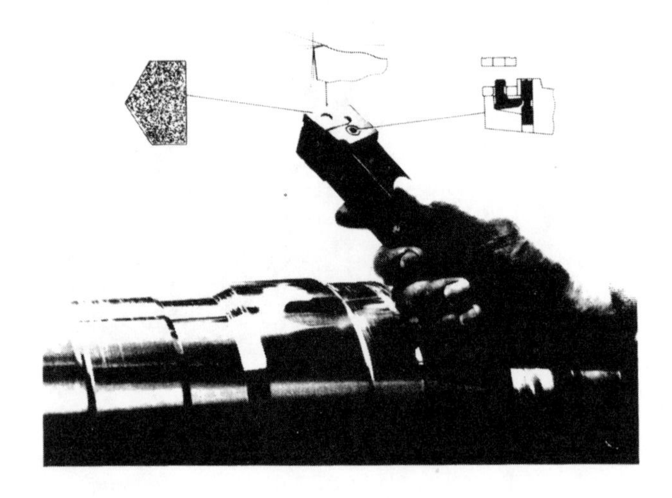

Again let us remember that basically a cutting tool is made up of three main components that have to be taken into account: the tool material, cutting geometry and edge clamping method. The selection of these components have by far the greatest influence on the successful outcome of the machining. There are, however, also a number of additional machining factors which are important for getting the best utilization out of the available machinery and achieving good machining economy.

The instrument for selecting a tool for an application is the catalog. The Sandvik Coromant Turning Tool Catalog groups tools according to their fields of application. It shows a current standard program of turning tools, a total of well over 2000 items, allowing the selection of toolholder, insert, tool material and spare parts. The program is made up of various tool systems, each one optimized for various uses. Each system complements the others and simplifies the job of choosing the correct tool.

Turning Program

The Sandvik Coromant turning program is based upon a few basic tool systems: T-MAX P, T-MAX S, T-MAX and T-MAX U. All the systems make use of indexable inserts which means a number of advantages for the user:

- no regrinding or brazing of tool edges
- better tool materials and geometries can be used
- uniform performance throughout machining
- simpler and safer handling of tools

T-MAX P

The T-MAX P system is intended mainly for external turning. (The first three holders on the left in the picture above). It is a versatile, general purpose system for light to heavy machining. It holds a number of different inserts, which covers most applications. All inserts are clamped through a center hole. There are three types of T-MAX P tools: lever, wedge and wedge clamp. All three types use shims: supports made of carbide that lie underneath the inserts in the holder pocket. They are fixed in the pocket by means of a shim pin. The support provided by the shims is essential and adopted wherever possible in cutting tools. They provide clearance to the effect that maximum support is achieved for the insert to absorb the cutting forces. In the event of insert fracture, they protect the toolholder from damage and in this way provide the toolholder with a life that will see hundreds of indexable insert changes.

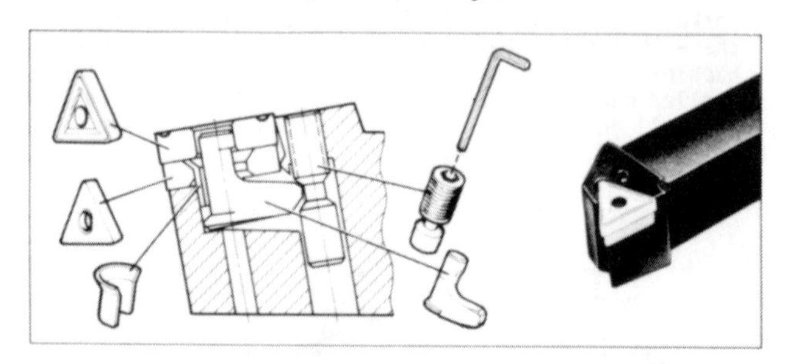

The lever design has a pivoted lever which tilts by an adjustment of the clamping screw. The lever forces the insert backwards into a pocket, locating the insert firmly against two sides. Excellent stability and locating accuracy is achieved for the insert in the holder. The design ensures unobstructed chip flow through the clear top face. The lever is easy to operate for quick indexing.

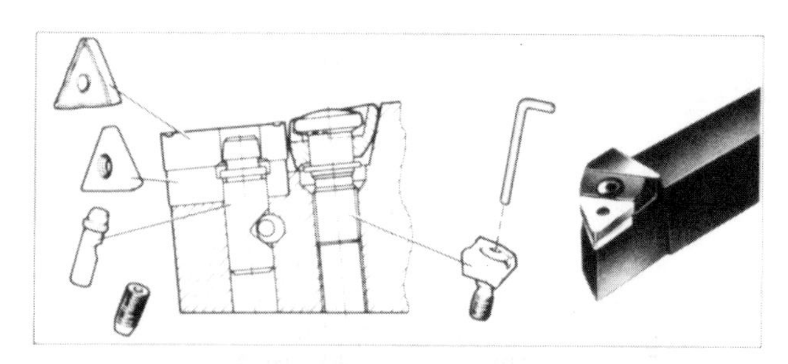

The wedge and wedge clamp design are intended for operations that require more tool accessibility than provided with the lever design, in which the insert is clamped into a pocket. The design utilizes a wedge which forces the insert forward against a fixed pin and at the same time clamps it down. When the wedge is forced down by the screw, it makes contact against the angular back of the toolholder and the back of the insert. The insert is held very securely but released through just a slight turn of the screw. The design also provides a clear top face. Obviously, this design does not provide the same clamping stability as the two-sided pocket design of the lever design. But this aspect has to be balanced against that of the accessibility provided through this design and the toolholder type should be selected for operations accordingly.

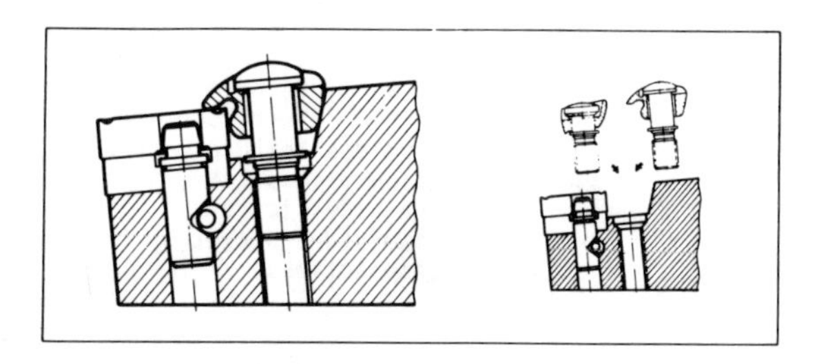

The wedge clamp design works on a similar principle, the difference here being that the wedge also has a top clamp. In addition to the wedge forcing the insert against the pin and down, the top clamp makes sure that the insert is retained down against the shim. This is especially useful in operations involving conditions that can affect the position of the insert. For instance, when intermittent cuts are involved or outfacing, some double-side inserts have a tendency to wriggle up the pin. The clamp set and wedge clamp set are interchangeable in the toolholder.

T-MAX P Inserts

T-MAX P inserts all have a center hole for clamping. The program includes a number of types, shapes and sizes: double and single-sided inserts; light duty, general purpose and heavy duty inserts; plain and with form-sintered chipbreakers; triangular, square, diamond and round. Inserts are usually described in

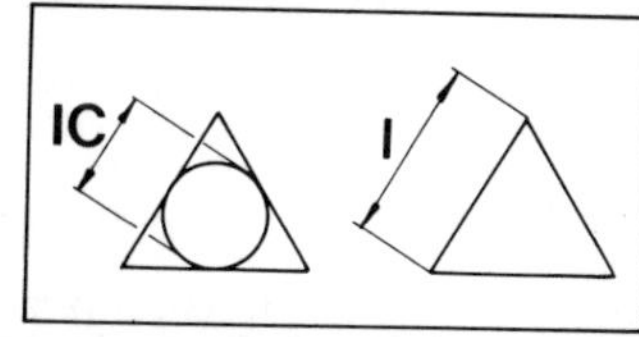

size by the inscribed circle or the cutting edge length which for T-MAX P inserts usually vary from 1/4 (6.35 mm) to 3/4-1 inch (19-25.4 mm). Standardization has meant that inserts of the same basic shape and size are interchangeable in one toolholder.

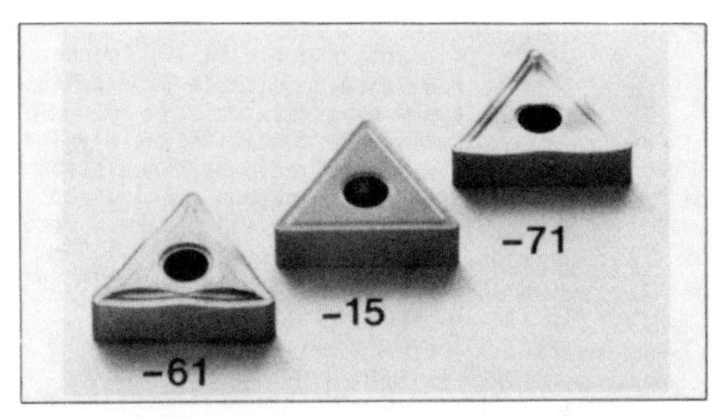

For general purpose turning, light to heavy duty, there is a modern range of inserts selected for optimum performance called P+. There are three cutting geometry types denoted 61, 15 and 71. They complement each other to cover a large, frequently occurring, application range. The P+ cutting edges have been developed for good chip control at high cutting data.

Double-sided insert with wavy cutting edge and positive top rake angle. Outstanding for light machining and finishing and chip control at low cutting data. Feed range approx. .004-.016 in/rev (0.1-0.4 mm/rev). Cutting depth approx. .04-0.16 in (1-4 mm). The ideal choice where good chip control is a necessity, e.g. automatics especially for low carbon content material, cold-extruded components, some free cutting steels and rolled stainless materials.

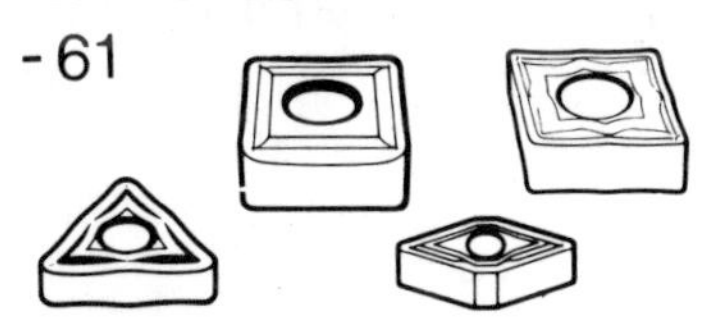

A double-sided insert for use over a large application area with a straight cutting edge and only the one-step chipbreaker being positive. Offers production security even where the feed and cutting depth vary on the same machining operation, e.g. during copy-turning. Highly suitable for turning of large production series such as in the automotive industry. Feed approx. .008-.02 in/rev (0.2-0.5 mm/rev), cutting depth .06-.24 in. (1.5-6 mm).

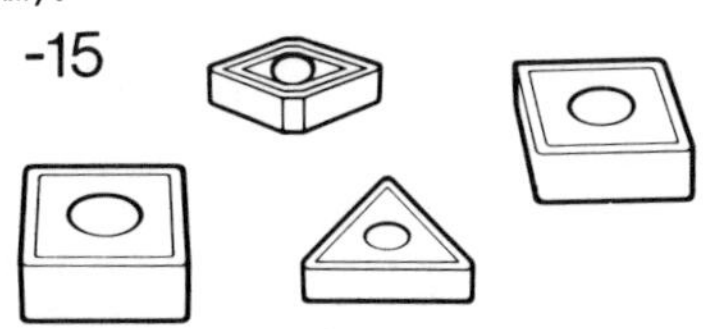

A single-sided insert with a large range application area having wavy cutting edges and a positive top rake. For fine to medium roughing. Should be applied with raised cutting speeds. Feed range approx. .012-.04 in/rev (0.3-1 mm/rev). Often the best solution for vibration-sensitive workpieces and in limited-power machines. Good chip control--even in tough material where chips are guided away from the workpiece. Ideal for most types of steel.

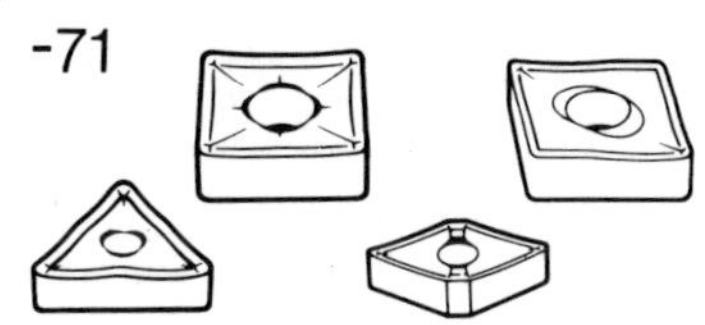

In addition to the 61, 15 and 71 type inserts, there are a number of T-MAX P inserts for more specific applications.

Complementary to 71 for roughing purposes with wavy edge and positive rake. Feed range approx. .012-.06 in/rev (0.3-1.5 mm/rev). Range also suitable for intermittent machining. Under severe conditions this diamond insert with 100° corner should be used. Suitable for machining most types of steel.

-41

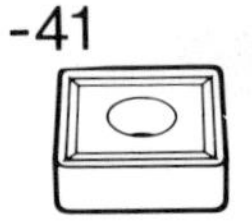

For severe conditions mainly involving stainless steel, acid heat resistant materials, and low carbon steels. Straight cutting edge and positive rake. Feed range approx. .016-.048 in/rev (0.4-1.2 mm/rev). High edge reliability--especially suitable for intermittent machining.

-31

Complementary to 71 for medium to rough machining of components with difficult-to-machine surfaces, as on castings and forgings. Straight cutting edge and positive rake. Extremely broad range of application. Feed range approx. .02-.07 in/rev (0.5-1.8 mm/rev). Cutting depth approx. .16-.59 in (4-15 mm). Also useful for intermittent machining alloyed, non-alloyed and stainless steels.

There are also a number of other T-MAX P inserts which complement the P+ range for more special applications.

First there are the double-sided inserts with form-sintered chipbreakers. These have straight cutting edges and single-step chipbreakers--just one groove behind the edge. They have smaller top rake angles. Suitable for light to medium rough machining with moderate feeds.

The round double-sided insert has a strong cutting edge, offering many cutting edges through the circular indexing of the insert. Being round, the insert is useful for intermittent machining.

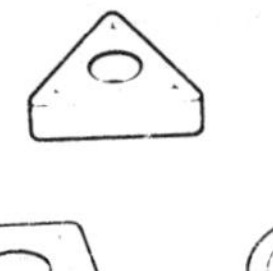

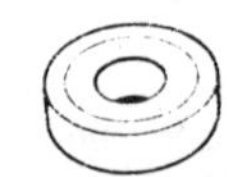

These are single-sided inserts with multi-step chipbreakers--several distinct grooves behind the edge. Again the edges are straight and the top rake angle considerably smaller. Suitable for light to heavy machining over a large range: .01-.03 in/rev (0.25-0.76 mm/rev). Often a solution for when a more negative cutting edge is required for certain carbon, hardened and stainless steels.

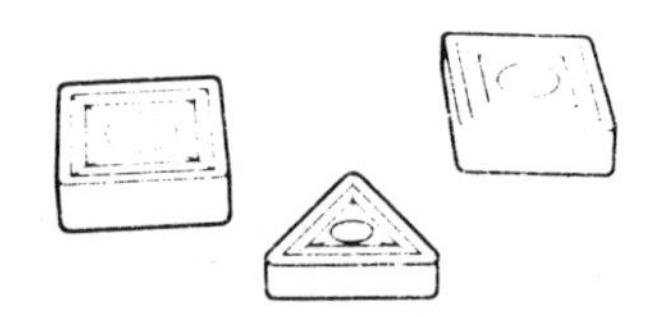

Another special solution type of single-sided insert, these are straight cutting edge inserts. They have a special geometry for light to medium roughing of stainless and certain heat resistant materials.

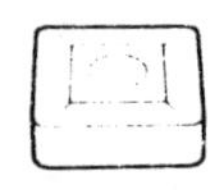

Plain double-sided inserts are used for machining cast-iron, castings and for intermittent machining of steel, where a negative top rake is useful. Through not having a form-sintered chipbreaker a large, strong edge cross-section is obtained. Also a useful insert for other short-chipping materials not requiring a chipbreaker function.

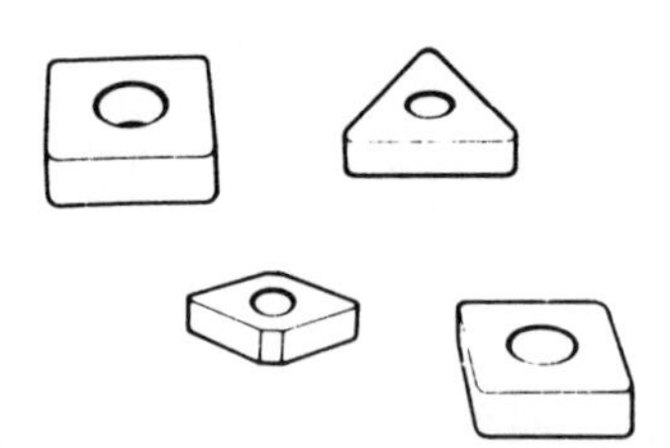

The single-sided button type insert has a considerable positive top rake combined with a strong cutting edge. Provides many cutting edges and produces good surface finish at heavy feeds. Useful for some copy-turning operations.

T-MAX S

This system is intended for boring and light external turning. T-MAX S tools do not clamp inserts through a center hole but by way of a top clamp. The clamp is brought down onto the insert by a screw through the clamp. The clamp is designed and positioned so as not to hamper chip flow. The clamp and screw come in a set held together by a circlip. When the screw is loosened, the circlip lifts the clamp. A tenon at the back of the clamp keeps the clamp in position. The T-MAX S system also uses shims in most of the holders. In some, the shim pin hole would reduce the supporting face in the toolholder too much.

The T-MAX S design provides features especially suitable for boring operations: positive rakes with fine chipbreaking and extra side clearance. The insert is securely fixed in the two-sided pocket by the positive, direct screw clamp action.

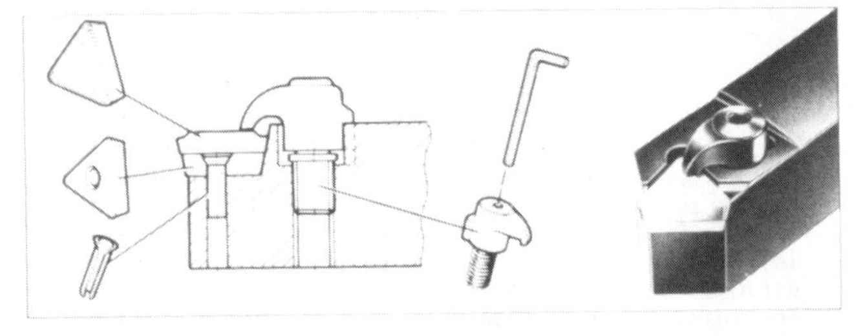

T-MAX S Inserts

Inserts for use in the T-MAX S system do not have any holes as they are all held by a top clamp. They do, however, all have a positive basic shape whereas most T-MAX P inserts have a negative basic shape. This means that the inserts have some built-in clearance and can be inclined positively in the toolholder. The T-MAX S inserts are available in two versions: utility and precision. This means that the insert manufacturing tolerances are different. The utility version has a M-tolerance, where the tolerance on the diameter of the inscribed circle of the insert is ±.002 in (±0.05 mm). The precision version has a G-tolerance: ±.001 in (±0.025 mm). T-MAX S inserts are often used for precision finishes on workpieces and the selection of insert version makes an important difference.

Form-sintered T-MAX S inserts have positive rakes that are specially suitable for light boring operations and other light duty operations. Fine chipbreaking and low cutting forces are obtained. There are two chipbreaker versions, one with a narrower chipbreaker groove for chipbreaking at very fine cutting data. There are also plain T-MAX S inserts for machining aluminum alloys, cast-iron and other short-chipping materials. These have extra large clearances to enable extra inclination in specially adapted toolholders.

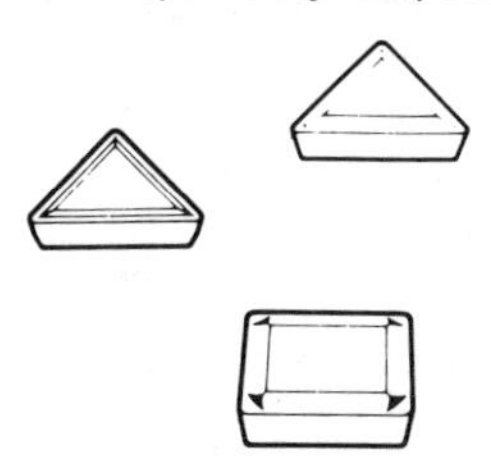

T-MAX

The T-MAX system of turning tools contains a varied assortment of toolholder and insert types. Today, the system cannot be said to be characterized by any specific type of edge clamping method as in T-MAX P, T-MAX S and, as we shall see, T-MAX U. The T-MAX system contains top-clamp holders with loose chip-breaker for plain inserts, top-clamp holders for copying with form-sintered inserts, thread-cutting toolholders, groove-cutting toolholders and cutting off toolholders--all with different clamping mechanisms. The T-MAX system contains mostly special-purpose tools.

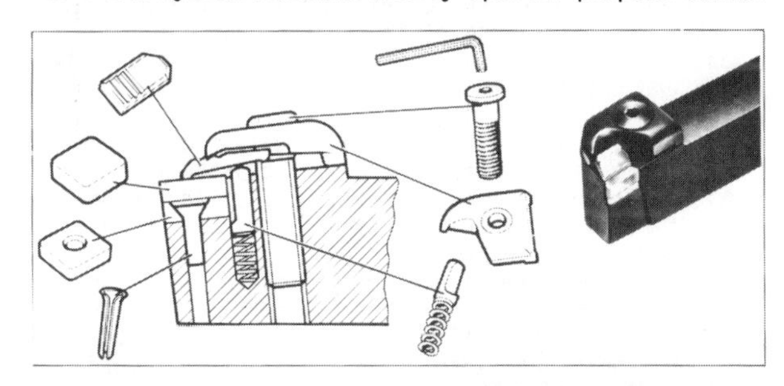

The original top-clamp toolholder for plain inserts is a tool with several years of service behind it. The design dates back to the early days of indexable inserts during the 1950's. Since it was then not feasible to mass-produce form-sintered inserts, the tool was equipped with a loose chipbreaker positioned between the insert and clamp. The steps of the chip-breaker have to be adjusted manually through the clamp. This T-MAX toolholder is still popular in some machine shops. However, it is considerably out-dated as regards performance and handling by the T-MAX P tools. It can be used for some copying operations and others requiring a negative or a positive rake or for inserts in other tool materials, for instance ceramics.

The T-MAX copying tool with top-clamp is also a well established tool. The insert has form-sintered chipbreakers and a basic positive form. It has a positive cutting geometry which means low cutting forces. The thick insert has high cutting edge strength with good stability and the top clamp holds the insert fixed at two points--one in the chipbreaker and one on the angular flat. The flat is inclined and perpendiclar to the chipbreaker groove so as to prevent the insert from moving.

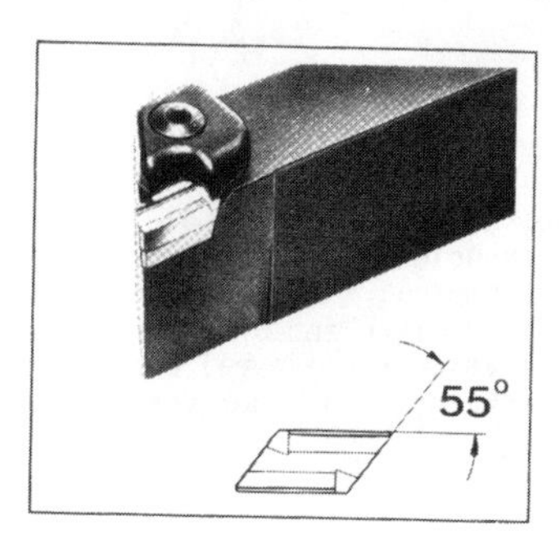

Threading tools are used for turning various screw-threads. The toolholder is equipped with an insert producing the required thread form. It is then made to turn the thread by passing along the workpiece at a feed which corresponds to the pitch of the thread. It is usually a question of the thread being cut with several passes. This is because the large cutting depth of the V-groove in the thread would be far too much for the pointed cutting edge to take in one pass. Instead, the full thread depth is divided up into smaller cutting depths. Correct, full-profile threads are produced.

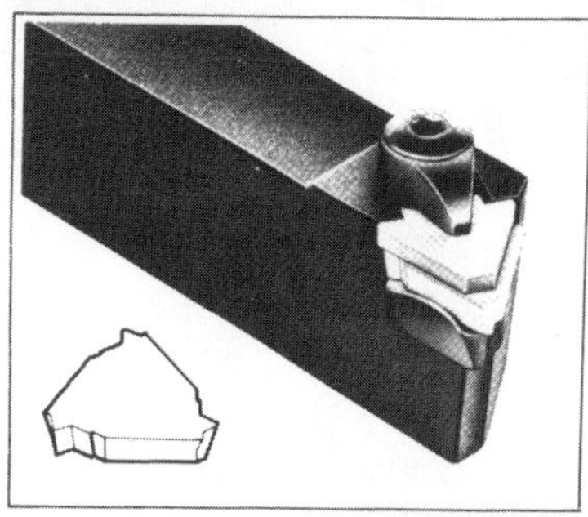

The T-MAX toolholder for threading holds the inserts by way of a top clamp--the same type as used in the T-MAX S system. The inserts are plain and being triangular and of positive basic shape have three cutting edges. A loose chipformer on top of the insert is used on some of the tools are shims. Three different loose chipformers are available and can be used either way up. This means that six variants are available for setting the width for the best chip flow. Inserts are accurately located and indexing does not involve any resetting in the machine.

Form cutting tools

This heading includes the three main groups of groove cutting tools. Grooves for O-ring seals are turned with inserts shaped to produce the correct form and size. This grooving tool uses the lever type clamping as in the T-MAX P system to hold a basically triangular insert. Like the threading insert, it has a positive basic insert form and is accurately and easily indexed.

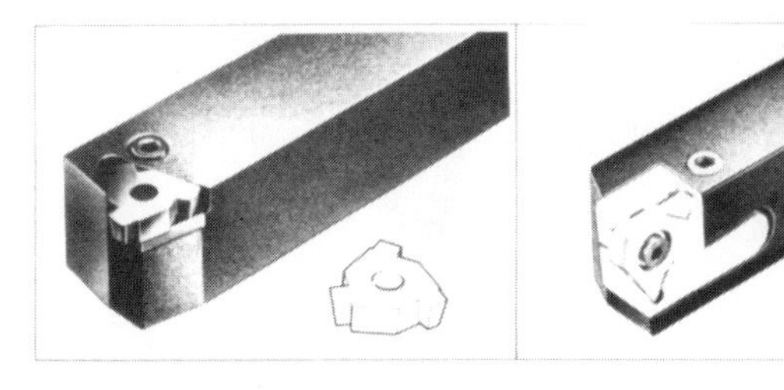

Grooves for retaining rings (circlips) on shafts or in holes are a common requirement. For this purpose there are T-MAX grooving tools with inserts shaped to produce suitable grooves for standard rings to be fitted. These toolholders are different in that inserts are held tangentially, up-right in a pocket. Quick indexing can be performed by changing the insert seat unit in the holder. This unit is held in the holder pocket by one screw.

There are T-MAX tools, employing a T-MAX S type top clamp, for general purpose grooving operations. The rectangular inserts are positioned by a pin in the insert pocket and no shim is employed. These tools are particularly suited for special grooving forms and even special-profile threads. Inserts are available as

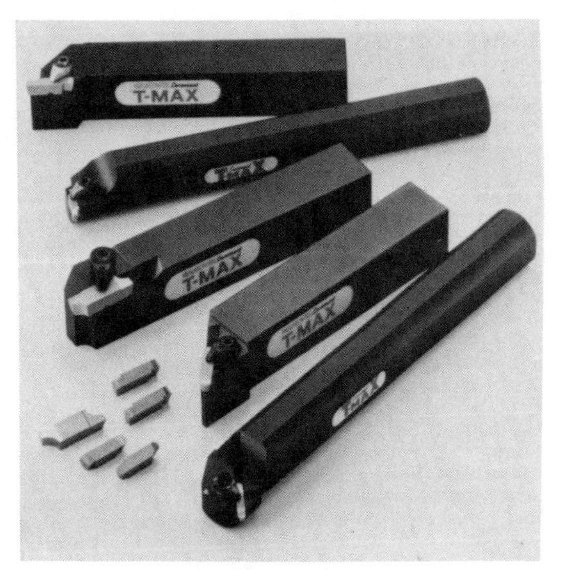

fully formed or as blanks that are ground to special profiles. When grinding insert, the toolholder can be used as a fixture.

Parting tools are used to cut off components or parts of components. There are two T-MAX parting tools: the adjustable-blade type and the shank-blade type. The first type is a toolholder in the form of a blade which is held in a block. The blade is adjustable in that the overhang--the length the blade protrudes beyond the block--can be varied. The blade toolholder holds the insert in a rather unconventional way: the insert is clamped by a spring arm, which is part of the blade.

The shank blade type clamps the insert through a loose top clamp, which is tightened by a screw. The front end of the toolholder has been made into a parting blade, thus limiting the diameter that can be cut off. However, there are gains in the way of stability and the tool is suitable for more demanding operations. The blade being an integral part of the toolholder with firmer clamping makes the tool useful not only for parting off but for other turning adjustments.

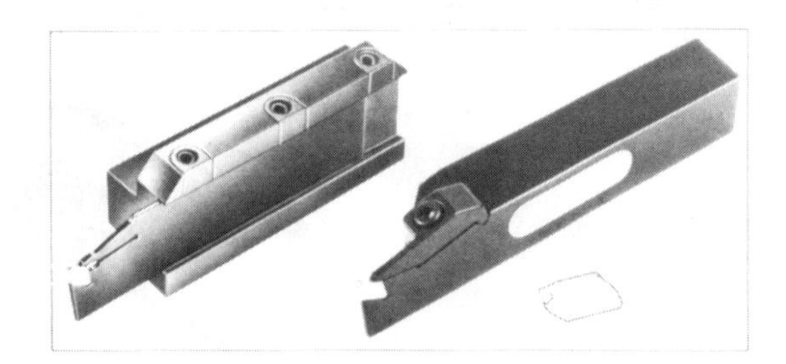

T-MAX U

The T-MAX U system uses a screw to secure inserts with a center hole. The system is advantageous in that

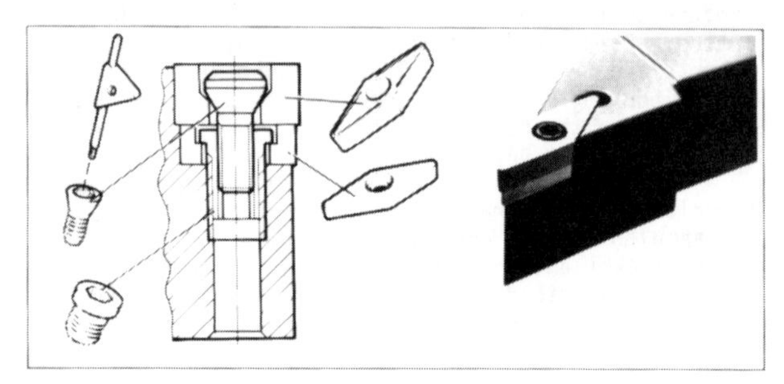

much less space is needed to clamp the insert when compared to lever-type clamping and wedges. This can be made use of in tools for boring operations and for copy-turning. In some cases, advantage can also be made of the fact that it is less complicated to include the T-MAX U insert location in a tool than other methods.

The T-MAX U toolholder has a sleeve which screws into a hole in the insert pocket. Into this sleeve goes the screw which holds the insert. The sleeve also retains the shim in place. Instead of the insert being inclined 6° negatively as in the T-MAX P holders, the inserts are held neutrally at 0°.

The T-MAX U system is excellent for copy-turning tools where accessibility is important. A 35° point angle is achieved while securing the insert into a two-sided pocket. The T-MAX U copy-turning tool is especially suited for operations where very acute in- and out-copying is demanded. A single step chipbreaker and a sharp, positive basic form provides an extensive application range, where the point radius is an important factor.

Qualified Toolholders

When machining is performed in numerically controlled machines, where most functions are programmed, the positioning of the cutting edge is an important factor. For the machine to function automatically it must have the cutting edge at the same point all the time and not have to be checked and compensated each time the cutting tool is changed. For this reason, NC machines require either pre-set tools or qualified tools. The pre-set tools are set accurately in a block which fits on the machine tool. Qualified toolholders do not require any pre-setting as they have been manufactured more accurately.

When they are placed in the machine tool-location, the cutting edge is positioned to sufficient accuracy. The toolholder is qualified to $\pm$.003 in. ($\pm$0.08 mm) on length and width/length and side in accordance with ANSI standards.

Boring Tools

Boring covers a wide range of hole size and length ratios and a number of operations. It is generally performed with a single edge tool but can also employ a multi-edge tool. Machining in a hole means limits on the size of the holder, strict demands for satisfactory chip control, tool rigidity and controlled cutting forces for the edge to reach in and perform satisfactorily.

This requires a comprehensive program of tools adapted for boring which can employ all the toolholder systems. The single-edge tools are in the form of boring bars which are either integral-shank bars with the insert clamped at the end or with the cutting head being exchangeable.

The integral boring bars employ T-MAX S for light duty in shorter holes and the other systems for heavier work and copy turning. The T-MAX positive copy-turning insert on the end of an integral bar is often a suitable application for general boring and internal copying where the holes are not very deep.

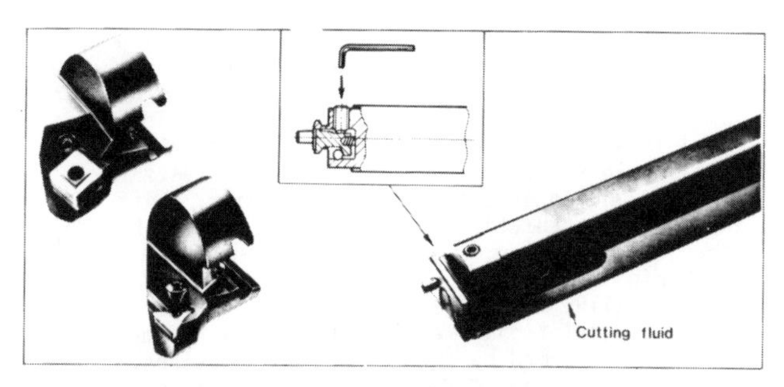

Flexibility is achieved with boring bars having exchangeable cutting heads. The same boring bar can take heads suitable for varying operations. The heads slide off sideways and are locked securely by a screw. The cutting heads can also be adjusted somewhat radially (sideways). The boring bars are also available with cutting fluid supply. A hole is provided in the bar just behind the head. During machining, fluid is sprayed out onto the cutting edge to cool, lubricate and facilitate chip removal out of the hole.

A more sophisticated type of bar is the anti-vibration TNS boring bar. These use the same exchangeable cutting heads as described above but the actual bar has a built-in arrangement for moderating vibrations. There is a long version coping with overhangs of up to ten times the diameter which can be adjusted

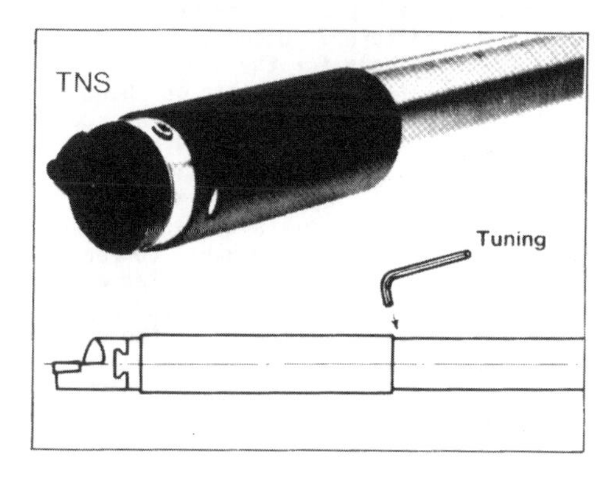

or tuned to achieve the best performance for each individual case. A shorter version, for up to seven times the diameter does not have this adjustment.

Many boring operations require a number of cuts which are often possible to perform during the same in-feed of the boring bar into the hole. In some operations it is wise to distribute the depth of cut on several cutting edges. For this purpose, special boring bars are often designed based on standard cutting units. The small units available for this are several: T-MAX S round shank boring tools are mounted in holes in the boring bar and can be accurately adjusted. At the back of the shank there is a screw for radial adjustment of the cutting edge in the boring bar. T-MAX S seating units are fitted in an open pocket in the boring bar. The unit has a screw hole for mounting. The bottom surface has a tenon which slides in a groove when being adjusted. Cartridges--a type of insert seating unit--have a more extensive use with adjustment screws for radial and axial fine adjusting. They are mounted with flat surfaces in the bar pocket and held in place by a screw through the unit. Cartridges are ideal tool solutions for multi-edge boring bars through the flexibility obtained with the extensive range, mounting method and adjustment possibilities.

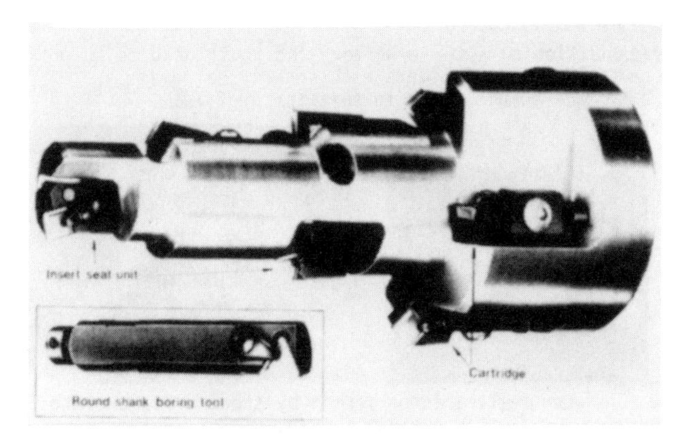

Tools for Specific Machining Ranges

HEAVY DUTY TURNING is the turning of usually large workpieces at high feeds and large cutting depths. This type of extreme heavy duty machining is applied during the manufacturing of steel-mill rolls, marine axles, shafts, etc. Specially designed machine tools are used which are very powerful, rigid and large enough to rotate the workpiece.

The Coromant program includes various systems for heavy duty machining. The T-MAX HD system is a series of toolholders which, with a top clamp, holds long-edge inserts. The clamp supports and retains the insert by resting against a shelf on the back of the insert. These tools cope with most of the general purpose turning, facing and grooving operations that occur.

The T-MAX P stepped toolholder has tangentially mounted inserts. The tool consists of a holder and a number of insert units, each with a clamping mechanism for the on-edge inserts. The large cutting depths that occur in these type of applications can thereby be distributed over a couple of stepped edges.

The Cassette system for heavy duty turning simplifies and speeds up the handling of the large and heavy tools in these operations. The whole cutting end of the large toolholder comes off with the release of two screws while the toolholder is left secured in the machine. The cassette units can be equipped with various cutting tools which provides flexibility and un-complicated changes from one type of cut to another.

RAILROAD WHEEL CONDITIONING, BAR PEELING and OIL-TOOL MACHINING are machining which occurs in a relatively few specialized industries. For these, Coromant have tool systems which have been developed to perform the operations in the best way. Railroad wheels are worn and damaged during their service and need reconditioning after a while. The tools used to turn the wheels are toolholders with insert seat units in various combinations to suit the profile.

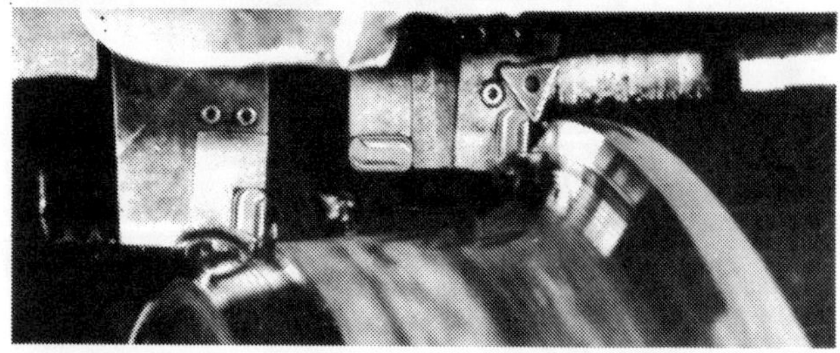

Bar peeling is performed by steel mills and other companies. This entails turning off the outer layer or skin of bars and rods. The bars are fed through a specially designed machine equipped with several toolholders, which combine together to peel the bar. The toolholders and inserts are designed specfically for the purpose.

There is also a program of threading tools for the oil tool industry. The drill stems employed in the prospecting and extraction of oil are made up of a series of pipes and connections which are screwed together. Special oil industry threads are used for this purpose. For the threading of these parts there is a program of different toolholder systems using flat-mounted lever-clamp inserts, flat-mounted top-clamp inserts and tangentially mounted screw-clamp inserts.

For the concept of unmanned machining in the mass-production industry with automatic machining and work-piece handling, there is T-MAX Automatic. This is a tool in which the cutting edge is changed automatically. In the toolholder is built in a magazine of inserts and an insert-changing mechanism. With hydraulic, pneumatic and electric functions the tool changes cutting edges after a pre-determined amount of machining.

Finally, there are numerous turning operations for which special tools are required--where the machining is best performed by special design solutions. The most rational way of designing and manufacturing special tools is the use of standard tool elements and inserts.

Application Technique

Solving machining problems is a matter of making sure the cutting conditions are right. The problems have to be solved by the people involved wih metal cutting production in a machine shop. These include, in addition to the machine operator, supervisor, foreman and shop manager, also the cutting tool salesman. The later is usually called in to put right what is a question of going through the basics of metal cutting.

The ideal way for a metal cutting operation to be optimized is for the cutting tool supplier to be present at the planning stage of the manufacture of a component. Studying the drawing he can then suggest the cutting data which provide the best machining economy for the manufacturer. Based on this data he will select the correct tools for the operation. Then, based on this, the manufacturer can go ahead in buying the machine tool which will fulfill these requirements--that is to say, a machine tool having sufficient power, rigidity, etc., to fully utilize modern cutting tools.

However, the above ideal situations are rare. Usually the cutting tool salesman has to recommend tools and cutting data on the basis of an existing machine tool, whatever the condition and capability, along with the established drawing of the component to be manufactured. Usually, with considerably limited parameters, he has to select the tools which will do the best job under the circumstances.

The selection of tools is based on the available program and the tool systems therein. The first half of this part described the tool systems, T-MAX P, T-MAX

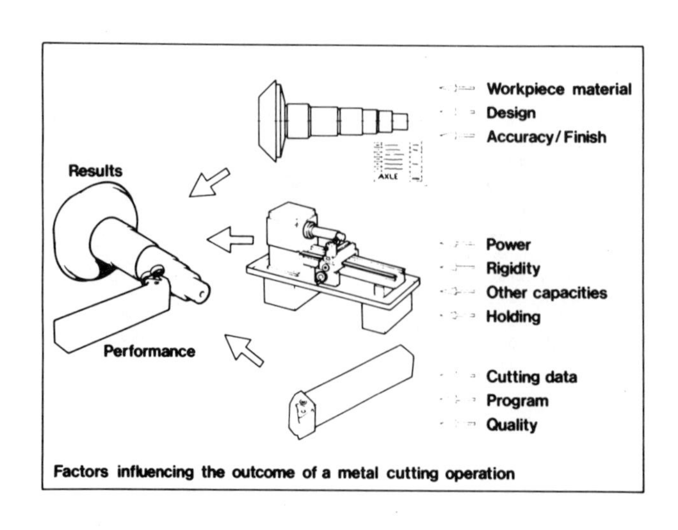

Factors influencing the outcome of a metal cutting operation

S, T-MAX and T-MAX U, and the main purpose of each one. When faced with the choice of tools for machining the first thing to do then is to establish the best system for each operation. For instance, the normal rough and finish turning down of a workpiece would lead to tools from the T-MAX P part of the program while groove cutting and threading would lead us to T-MAX.

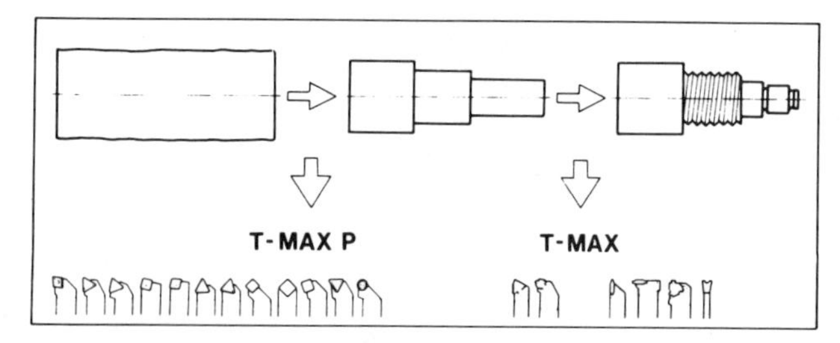

EXTERNAL TURNING AND FACING

In these the most commonly occurring turning operations the tool either moves along the workpiece axis turning down the diameter of the workpiece or in facing the tool moves perpendicular to the workpiece axis, shortening a length dimension on the workpiece. This can be either as a roughing operation where the object is to, as efficiently as possible, remove workpiece material or as a finishing operation where a certain accuracy and surface finish has to be obtained. These two operations are different as regards demands, results and tools used and are best discussed separately.

ROUGH TURNING AND FACING is usually a task for the T-MAX P system. These tools cover most applications that come under this heading with the three ways of clamping the insert, various insert types and tool sizes. If the operation is a straightforward bar turning operation, use a toolholder having a normal lead angle (entering angle) of 15 or 30° (75° or 60°). If the workpiece is long and thin, a lead angle of 0° (entering 90°) is the best choice to direct cutting forces in a way that reduces vibration tendencies. The same angle has to be selected if the operation involves turning up against a shoulder. If the machining includes severely interrupted cuts or hard skin, a 45° angle is often a reliable choice. This is because the edge centers and leaves the cut more

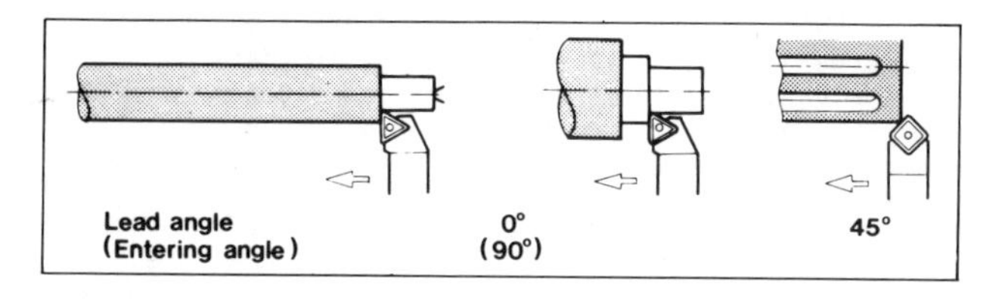

gradually, the workpiece makes contact away from the weaker point of the edge and the cutting forces are distributed over a longer part of the edge. Generally then, it can be said that an intermediate lead angle is

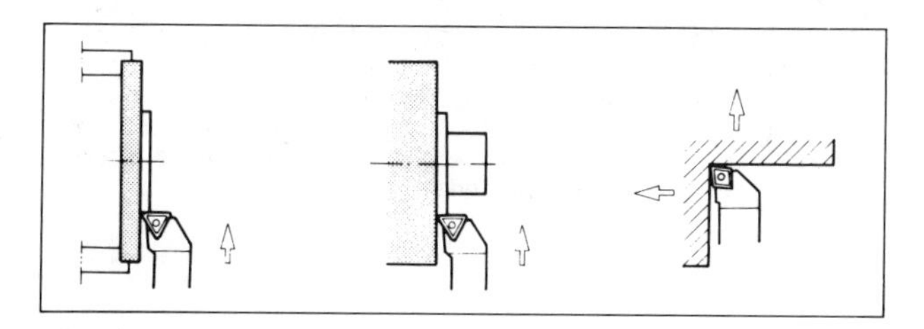

the best compromise for plain turning because higher feed rates can be applied without increasing the chip thickness. A smaller or larger lead angle should be applied if any of the mentioned complicating factors are involved.

Obviously, for facing the same principles apply. It is here a matter of selecting the lead angle with the radial direction of cut in mind instead and that vibrations might appear when machining across thin, disc-type workpieces instead of long, slender ones. It should also be noted that in many machines the cutting speed becomes lower along the cut towards the center. This may affect the performance and result of the machining. A very useful tool is the 95° diamond insert tool which can turn and face in the same set-up. The insert has a strong 80° point angle. More about this tool in copy turning.

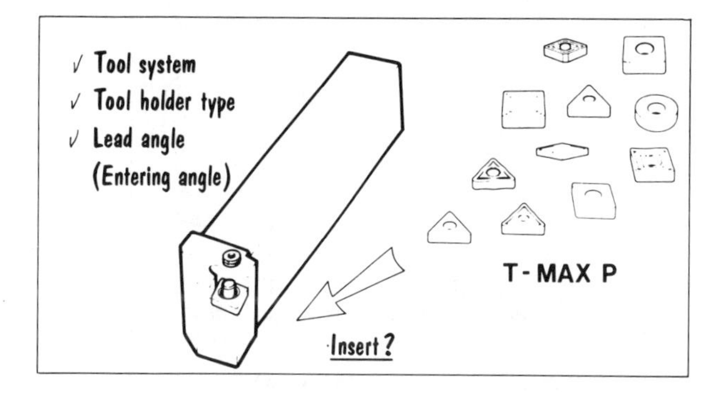

Now we have selected the tool system, toolholder type (although you still do not know the insert size and shape) and the suitable lead angle. Next, the insert should be chosen and the first to be determined is the cutting depth. This should be as large as possible in roughing. The cutting depth and entering angle results in an effective cutting edge length and from this you can select the suitable insert edge length. A well established rule-of-thumb is that the effective cutting edge length should be about 2/3 of the length of the insert edge. (With some modern inserts this has been subjected to re-consideration.)

Next, you should consider the type, shape and size of the indexable insert to be used in the toolholder. Having selected the tool material (the cemented carbide grade) for the machining, consider the workpiece material as this makes demands on the strength, geometry and chipbreaking. Here also, machining conditions and workpiece rigidity affect the choice. Generally, you should select the largest possible point angle and nose radius to obtain a strong tool point. Inserts are available with various nose radii. A large nose radius permits large feeds but an excessively large radius lead to vibrations. Square-shaped inserts have large strong points and round inserts have extremely strong cutting edges. A general rule-of-thumb is that for strength reasons, the feed per rev. should not be set any higher than 2/3 of the insert radius.

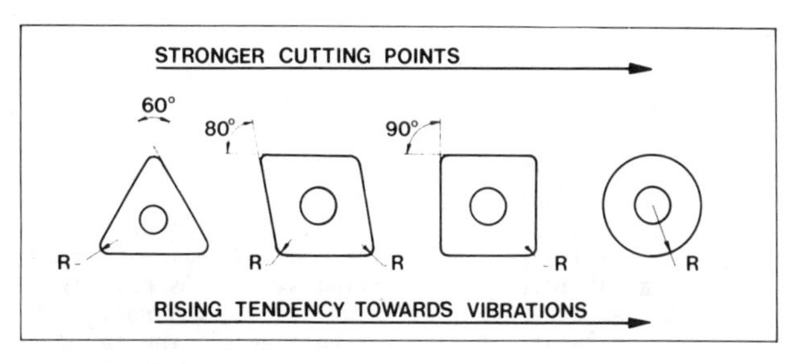

Another important point in roughing is that of the choice between a single-sided insert and a double-sided one. A single-sided insert has many advantages over the double-sided when it comes to operations rated as medium to heavy machining. They can be provided with a fully optimized cutting geometry and chipbreaking capability, without having to compromise with the demand for supporting faces because both sides have to act as cutting sides and contact side in the holder. A much stronger and stable cutting edge is achieved with a single-sided insert. The available machine power is sometimes a limiting factor in rough machining. Here

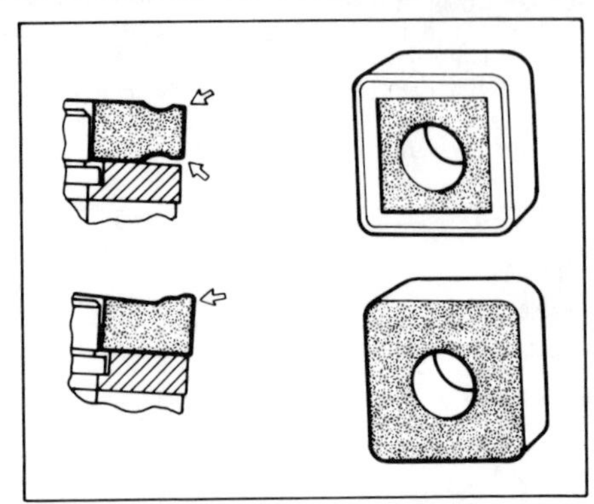

again the single-sided insert is at an advantage because the cutting geometry can be formed so as to have a positive rake, giving rise to lower cutting forces.

FINISH TURNING AND FACING is also a task for T-MAX P. T-MAX P inserts for finishing, medium and heavy machining are interchangeable in the same holder. Here the prerequisites are quite different to those in roughing. Of course, the principles of cutting tool selection as described for roughing apply to finish turning as well but, finishing being much lighter, the limits are not approached as close as in medium or rough machining.

Generally, the surface finish and accuracy can be improved with higher cutting speeds and positive rakes. The largest influence on the finishing result is the combination of insert nose radius and feed rate. The surface generated by an insert is a series of valleys and ridges formed by the nose as it moves along. The higher the feed, the higher the ridges. The lower the feed, the more ridges will be machined. The following figure showing the basic profile and a formula, describing that the height (H) of the ridges are related to the square value of the feed divided by eight times the nose radius (r), tells us a lot about the importance of selecting the right combination. The largest possible nose radius should be selected and the feed rate then set so as to give rise to a satisfactory finish.

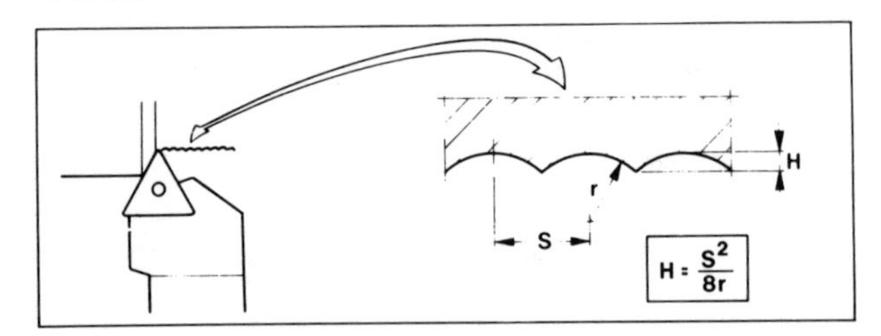

The chipbreaking and rake of the cutting edge affect the surface finish. For this reason, it is important to select an insert with the right chipbreaking ability at light feeds and cutting depths and to have a sufficient high cutting speed. As regards the obtained accuracy (having to machine the workpiece diameter or length to within a tolerance), the surface finish may play a role. Otherwise accuracy is often a result dependent upon various other factors in the machining such as that the more rapid tool wear on a pointed tool gives rise to greater dimensional fluctuations than does the wear on a more blunt tool.

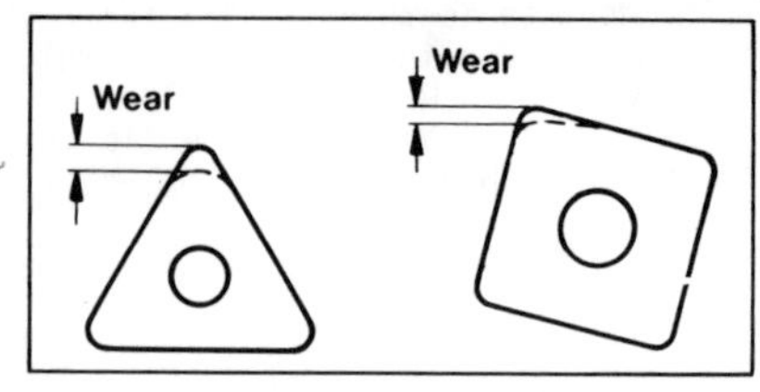

COPY TURNING

Copy turning is today given a wider meaning. Previously, copy turning was a process by which given a sample component a copy lathe could reproduce this shape with a suitable turning tool. This is still the case and for a lot of mass-production manufacturing copy or profiling lathes are the best choice of method. However, copy turning as defined by type of cutting operation is today also performed in numerically controlled lathes.

Copy-turning is a composition of cuts that vary in direction. The adjoining figure illustrates the basic operations of copy-turning. In addition, these workpieces usually require threading, grooving and cutting off. Apart from general operations like longitudinal turning and facing we have in-copying, out-copying and undercutting which can be performed at various angles. In and out in this context refers to towards and away from the center of the workpiece while undercutting speaks for itself as being a type of in- or out-copying under the outer material.

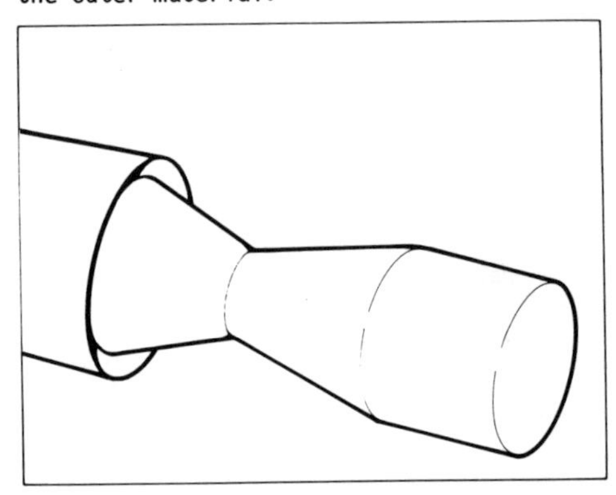

The tools used in copy-turning are subjected to large variations in stresses and cutting data due to the varying cutting directions and diameter differences. Because of the point angle often being smaller than those used in more general turning, the edge strength is reduced. For this reason it is important to see that the cutting depths and feeds are kept to a reasonable level. In some extreme cases, it might be a good idea to perform a general rough turning operations prior to the copying. The largest possible point angle should be used for the operation. The clearance, at each cut is important to check as is the active entering angle, to ensure satisfactory chip formation.

It is quite apparent that one of the most important characteristics of a copy-turning tool must be the accessibility obtained with the cutting edge. The point angle of the copying tool is an ultimate factor. Let us start by looking at a rough copying operation which, while being characterized by large dimensions, requires the tool to machine nearly all of a half-circle. These operations are often to be found among heavy duty applications.

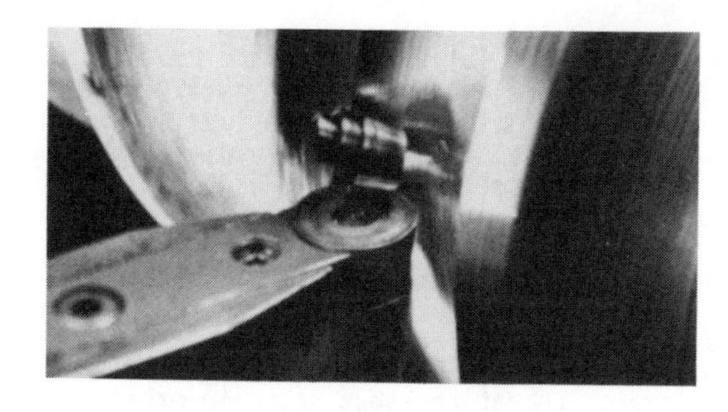

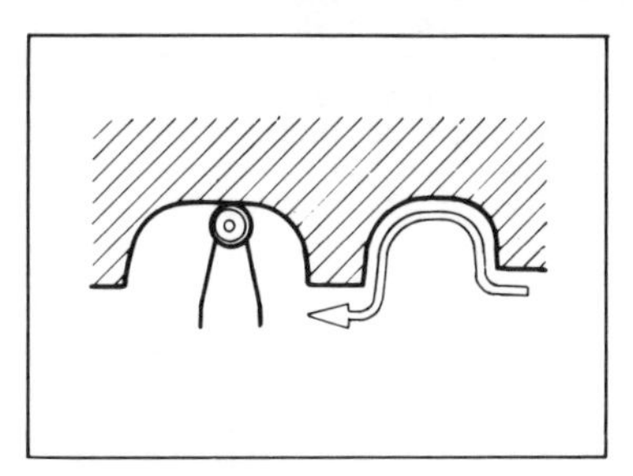

This is a task suitable for a T-MAX P tool having a positive button insert. It is a tool with a 180° copying range with the insert not being inclined in the holder. As was previously pointed out, it has a very strong cutting edge. Although the positive cutting geometry gives rise to smaller cutting forces, it is important to be wary of vibration tendencies because of the large cutting edge engagement. Machine rigidity, power and workpiece clamping should be checked. Round negative inserts are also available for similar purposes.

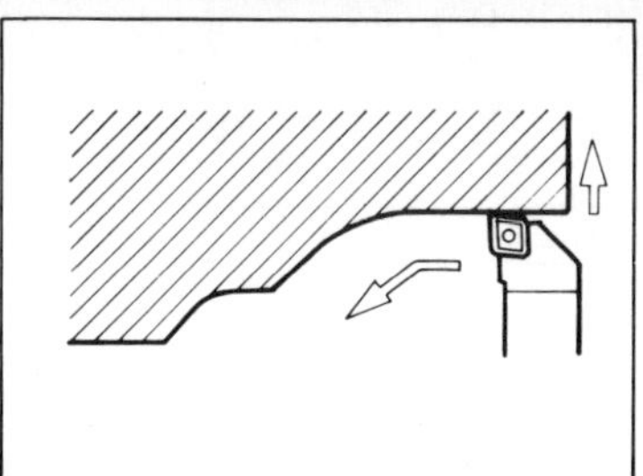

Next, we have the T-MAX P tool with the 80° point angle insert. With a 95° lead angle for turning and facing, this tool is extremely universal for many general copying operations. Various diamond inserts are available to suit different machining demands, conditions and workpiece materials.

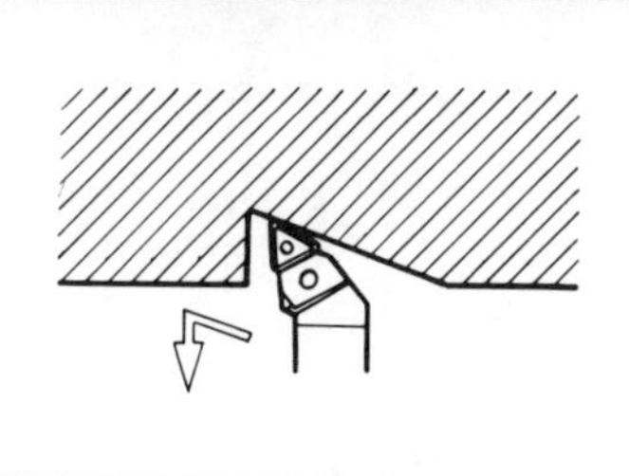

Triangular inserts held either by the T-MAX P wedge or wedge clamp offer wide copying capability. With a 60° point angle the tool is capable of working into acute corners.

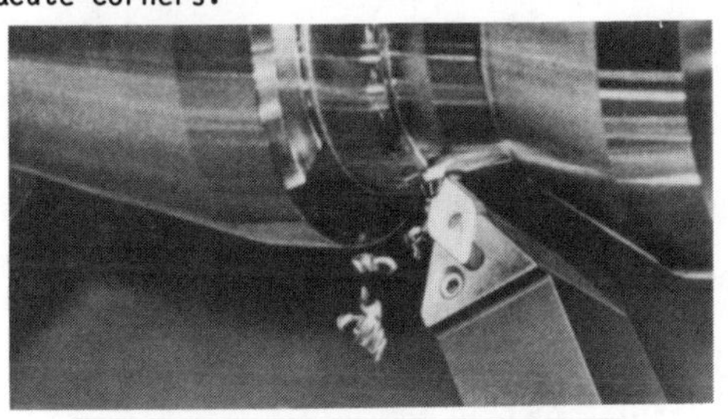

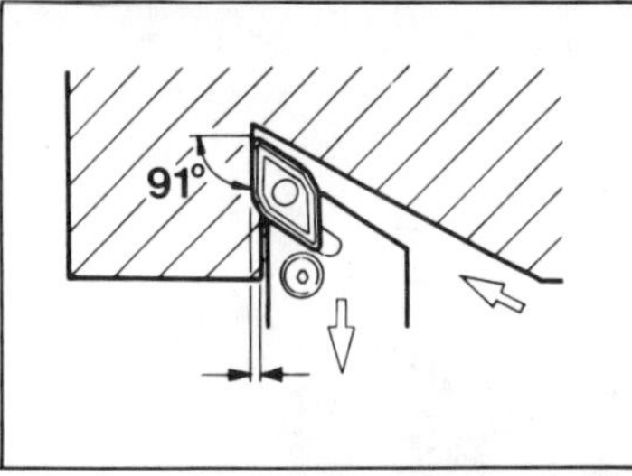

The T-MAX P tool with inserts having a 55° point angle has an even greater copying capability. This is a tool developed especially for copy-turning. The insert has a rhombic basic shape and, by having additional cutting edges on the sides, has an excellent

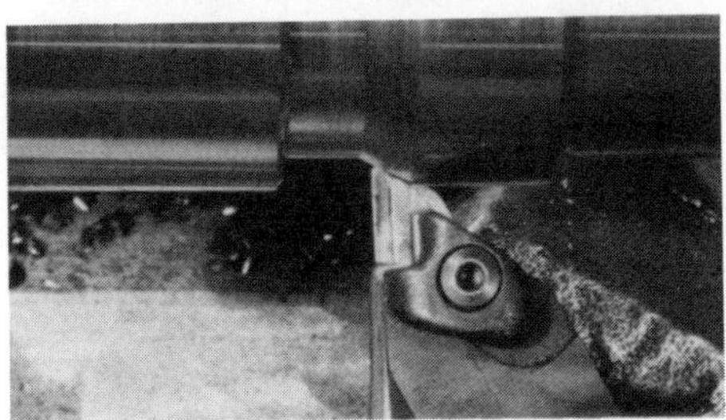

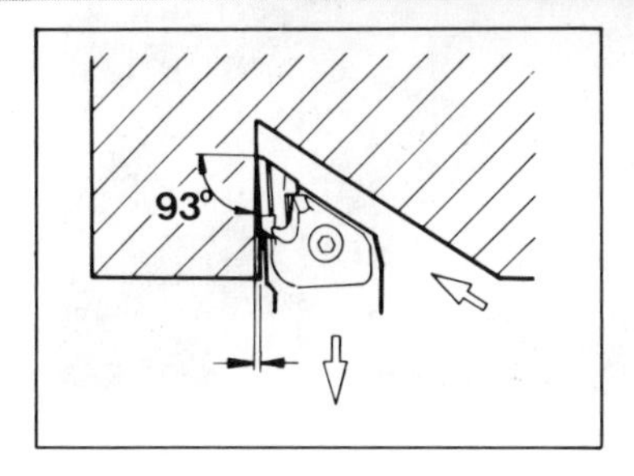

capacity for out-copying. There are various lead angles (entering angles) and insert types available so as to provide an extensive application area.

Here the T-MAX copy-turning tool should also be mentioned. It also has a 55° point angle. The rhomboidal inserts have a basic positive shape and are excellent for applications where cutting forces play an important role. This tool does not have the same out-copying capacity as the above described T-MAX P tool.

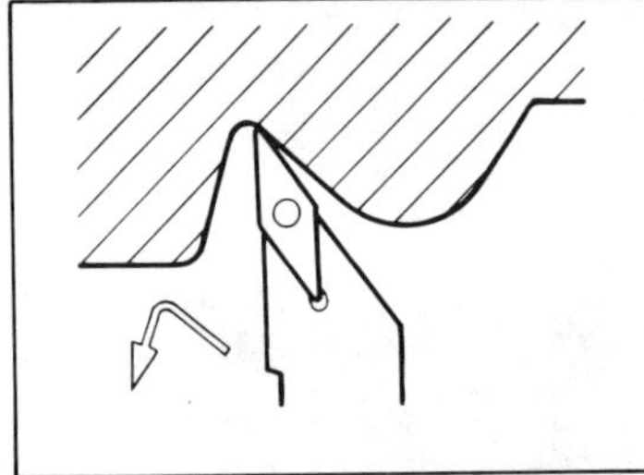

Finally, we have the T-MAX U copy turning tool with a 35° point angle. This tool is intended for applications where the extremely sharp point angle is necessary to copy in and out to form complex contours. The top and sides of the inserts are ground to give it a sharp cutting edge. This provides a good finish and lower cutting forces in machining.

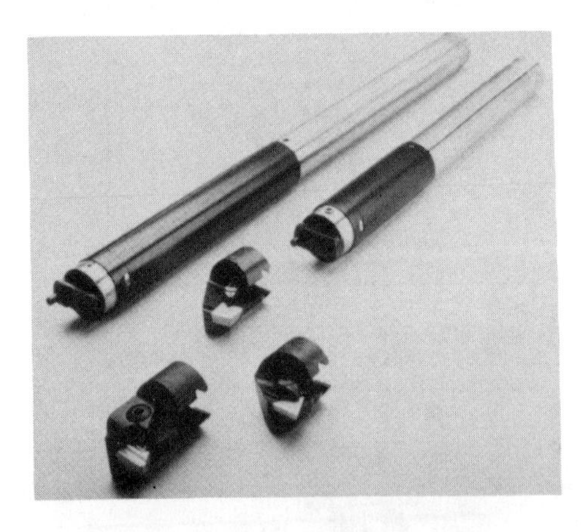

BORING

Boring makes special demands on the tools to be used. With tools for external machining it is far easier to use a toolholder with dimensions that will stand up to the forces and stresses involved in the cut and to minimize the tool overhang. In boring, the tool has to go into a hole of a certain diameter which limits the size of the tool shank. It has to reach the length of the cut which means plenty of overhang. Moreover, machining in a hole means that the removal of chips is not going to be as easy as in external machining.

The type of operations that occur in external machining are also performed in boring. Apart from tools for the more general type of internal turning operations, there are boring bars especially for copy-turning, threading and grooving. The heavier roughing cuts are not so common as the holes are usually pre-machined through the drilling operation. The longitudinal turning operation in boring should be performed with a very small or negative lead angle (large entering angle) for the same reasons as discussed for external turning of keeping cutting forces to a minimum. Most boring operations are light so the risks of cutting edge engagement as discussed before are not as apparent. The vital object is to keep the vibration tendencies to a minimum. This is best done by using the anti-vibration TNS boring bars with exchangeable cutting heads. This also provides the flexibility of with the same bar set-up in the machine to change to the most suitable cutting head for the operation.

There are heads suitable for light to heavy turning, facing, copy-turning, etc., which can take most of the available inserts. The anti-vibration bars, however, only dampen the vibrations which arise in the bar. To achieve a good result, the clamping of the workpiece tool and the stability of the machine must be satisfactory. Chipbreaking is also important to facilitate the removal of swarf. Sufficient space must be left between bar and hole walls to allow efficient removal of swarf. Boring bars with cutting fluid supply can be used to assist the removal out of the hole.

Multi-edge boring bars designed with insert seating units, cartridges and round shank boring tools should be regarded as more advanced cutting tools in that they often have to be made to fulfill combined demands of finish, accuracy, etc. They are usually special tools involving either the customer tool and methods department or the local special tools department within Sandvik Coromant.

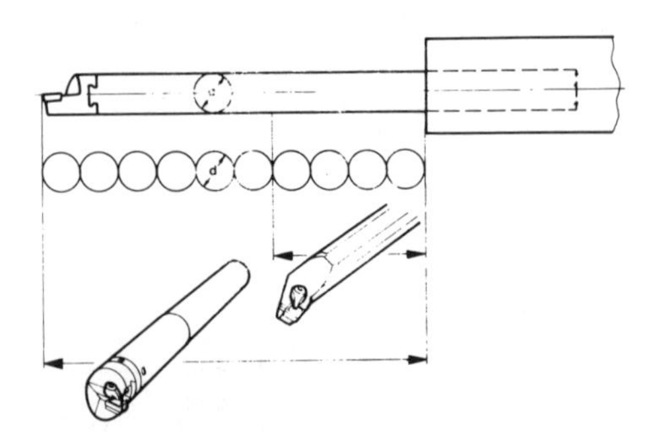

The diagram indicates the metal removing capacity of the various types of boring bars in relation to each other. It clearly shows that for overhangs of up to four times the diameter, the best solution for most cases is the solid steel boring bar. Above 4 x d overhang the removal rate must be severely reduced with this bar because of vibration tendencies. The tuned boring bars are then the only efficient solution. The short TNS bar is suitable for machining with an overhang of 4-7 x d and the long one 7-10 x d. Bar number four is a special tool for coping with very large overhangs: 10-12 x d. It is a tuned bar that has been reinforced with cemented carbide rings.

The tuned boring bar only reduces vibrations that generate in the bar. In order to obtain a satisfactory result, the clamping arrangements must be very stable. A clamped part of the bar of at least 4 x d is always recommended.

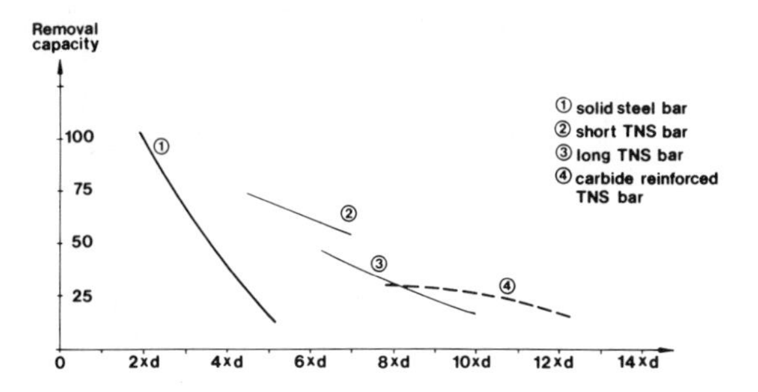

Threading

A screw-thread is a helical ridge of uniform section formed on the outside of a bar or inside a hole. If a right-angled triangle is wound round a cylinder in a way that the base corresponds to the circumference of the cylinder the hypotenuse, or long side opposite the right angle, forms a screw-thread line round the cylinder. The height of the triangle is then called the pitch of the screw-thread. The angle opposite the pitch is called the pitch angle or helix angle.

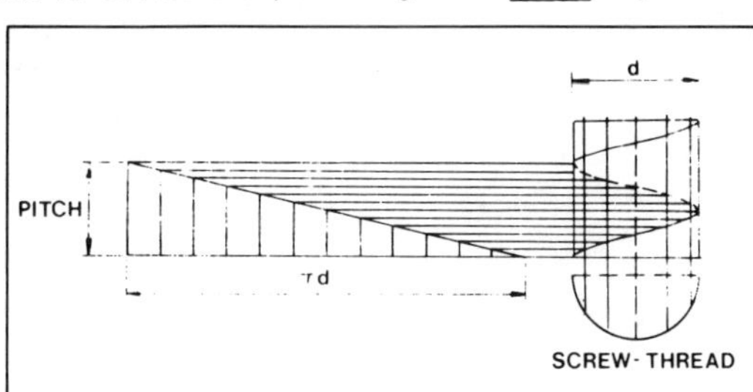

To start with let us describe the thread generally. There are the major and minor diameters which are the largest and smallest diameters. The pitch diameter is that of an imaginary cylinder which passes through the thread at a point where the groove and thread widths are equal. The pitch is the distance from a point on one thread to a corresponding point on the next thread, measured parallel to the axis. Number of threads (t.p.i.) is an important identification in the selection of inserts for inch threading. It states the number of threads or pitches contained in one inch. For metric threads the pitch is stated in mm/rev. Lead is the distance a screw thread advances in one revolution. (On a single start thread the pitch and lead are equal.) Root is the bottom joining the sides of adjacent threads. (The root of an external thread is on its minor diameter while the root of an internal thread is on its major diameter). Crest is the top joining two sides of a thread. (The crest of an internal diameter is on the minor diameter.) Flank is the side which connects the crest with the root. The angle of the thread is the angle between the flanks measured in the axial plane. Thread depth is the distance perpendicular to the axis between the crest and the root.

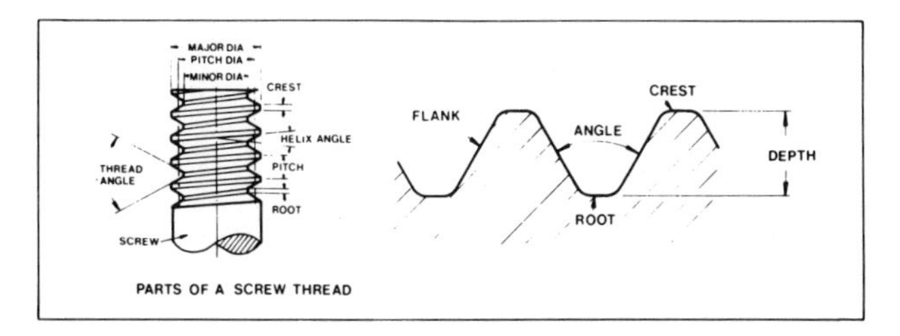

PARTS OF A SCREW THREAD

There are right-hand and left-hand threads--the right-hand being such that the nut or component screwed on to the external thread is turned in a clockwise direction. There are also differently shaped threads. Threads have been standardized for some time now and there have been a number of standard threads deriving from various countries. The International Organization for Standardization (ISO) drew up an agreement a few years ago to cover standard metric threads and standardized profiles, sizes and pitches for a range of diameters for various threads in the ISO Metric Thread Standard. These metric threads are identified by the letter M and have gained widespread use. For instance an ISO thread with an outside diameter of 10 mm and a pitch of 1.50 mm would be denoted: M10 x 1.50. Other important threads are Whitworth (British Standards Whitworth) and the ISO Inch which has got the same basic profile as the ISO metric.

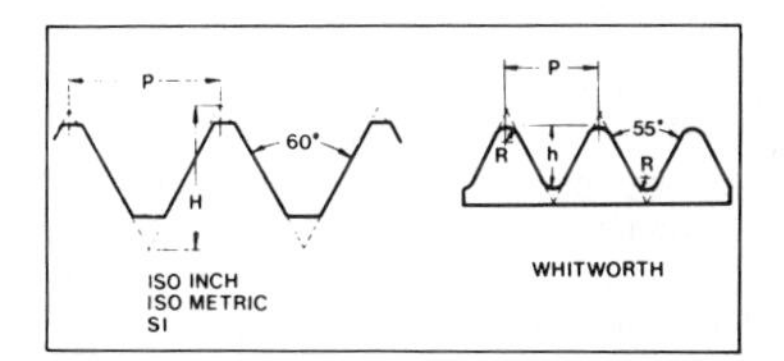

Also important in the production of threads is the fit. This is regulated by the tolerances--the permitted variation in size from the nominal dimension. The thread tolerances are classified in various grades or classes. For ISO inch class 2A, for ISO metric class 6, for SI and Whitworth Medium are frequently occurring general purpose tolerances.

As mentioned, the production of screw threads with turning tools is a matter of feeding the profiled cutting edge along in a number of passes. The indexable insert has the thread profile accurately ground which means that the screw-thread produced is finished off correctly. A full profile is produced by the insert which means that the complete thread with crest is shaped. There are inserts for ISO metric, ISO inch, Whitworth and SI threads and one toolholder can be used for different threads. (It is important to note that inserts are different for external and internal threads--this is marked on the inserts by a line and circle, respectively.) Accurate setting of the tool in the machine is vital for screw thread production so as to obtain the right profile.

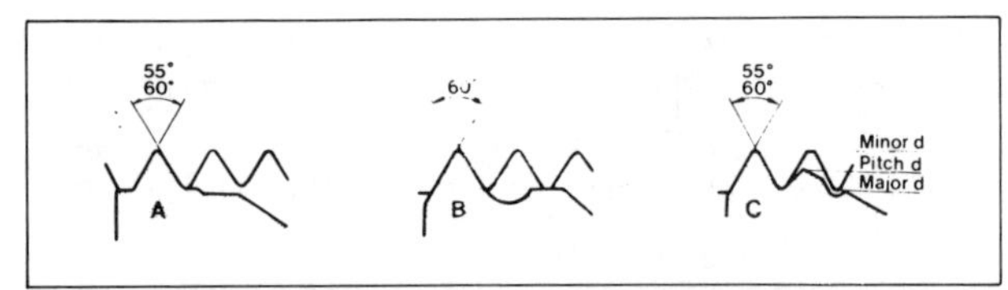

There are three insert types available as regards the different pitches. Type A is for moderately coarse pitches forming the flat crest in front of the thread being cut. Type B is for smaller pitches and forms the crest behind the thread being cut. Type C forms the crest immediately behind. Right- and left-hand threads can be cut by combining suitable inserts and rotational directions. The largest possible toolholder should always be chosen with consideration to the pitch size and workpiece material. It is suitable to choose an infeed which decreases successively during the threading. The following table will provide an indication of the number and sizes of in-feeds for different pitch sizes.

Form Cuts

By form cuts in this context we mean the cutting of specific grooves or recesses. These may have to be added in the form of a groove to take a fastener or a seal. Recess cutting is extremely common to provide relief in corners, end of threads, etc. Up to quite recently it was common to take these types of cuts with tools that had brazed and ground carbide cutting edges or ground high speed steel edges. Whereas these tools still have an application area for which sharp, positive, often involving small, thin tools are required, indexable insert tools are today also here the means with which to obtain rational machining. Much of what has been said generally about cutting edges also applies here. Characteristic for form cuts of the type mentioned here is that the duration of cut is

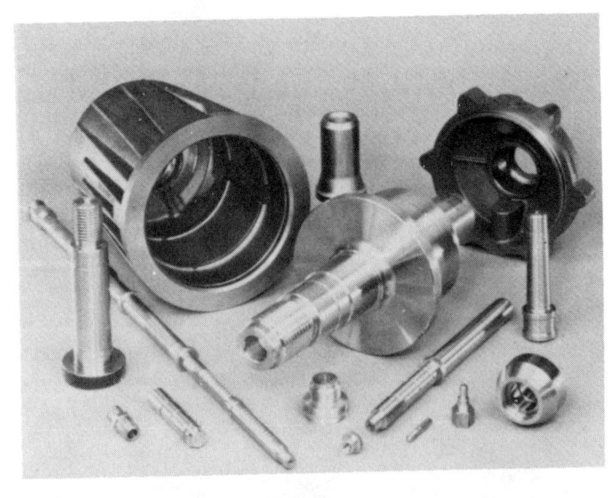

relatively short. The cutting edge is usually shaped in the final form of the groove and thus needs only a plunge into the workpiece.

To start with let us look at cutting grooves for O-ring seals. The rubber O-rings are the most common industrial sealing and used mainly for static purposes but also to some extent for dynamic. The basic O-ring has a circular cross-section. It is placed in a specific groove on one of the fitting components and compressed against a mating face on the other. In this way it forms a seal to hold fluid at pressure.

T-MAX P tools are used to turn the grooves for O-rings. The clamping and positioning of the insert corresponds to the lever holding where allowances have been made for the shape of the insert. The form-sintered insert is available to finish-cut a range of standard O-ring grooves. With three cutting edges per insert it is the efficient means with which a machine shop obtains the ready form tool for machining with high machining rates.

IN-FEEDS FOR ISO mm EXTERNAL THREADS

(For approx. inch values multiply the metric values by .04)

No of in-feeds	Pitch mm 0.75	1.0	1.25	1.5	1.75	2.0	2.5	3.0	3.5	4.0	4.5	5.0
1	0.20	0.20	0.23	0.25	0.25	0.25	0.28	0.30	0.35	0.35	0.40	0.45
2	0.15	0.16	0.18	0.20	0.20	0.22	0.25	0.25	0.30	0.35	0.40	0.45
3	0.10	0.14	0.14	0.18	0.17	0.20	0.22	0.22	0.25	0.30	0.35	0.40
4	0.05	0.10	0.10	0.15	0.14	0.18	0.19	0.20	0.20	0.25	0.30	0.35
5		0.06	0.10	0.12	0.12	0.15	0.16	0.18	0.18	0.20	0.25	0.30
6			0.06	0.06	0.10	0.12	0.14	0.16	0.18	0.18	0.20	0.25
7					0.08	0.10	0.12	0.14	0.15	0.15	0.15	0.20
8					0.06	0.06	0.10	0.12	0.15	0.12	0.15	0.15
9							0.08	0.10	0.15	0.12	0.10	0.10
10							0.06	0.08	0.12	0.12	0.10	0.10
11								0.08	0.10	0.10	0.10	0.10
12								0.06	0.08	0.10	0.10	0.10
13										0.10	0.10	0.10
14										0.08	0.10	0.10

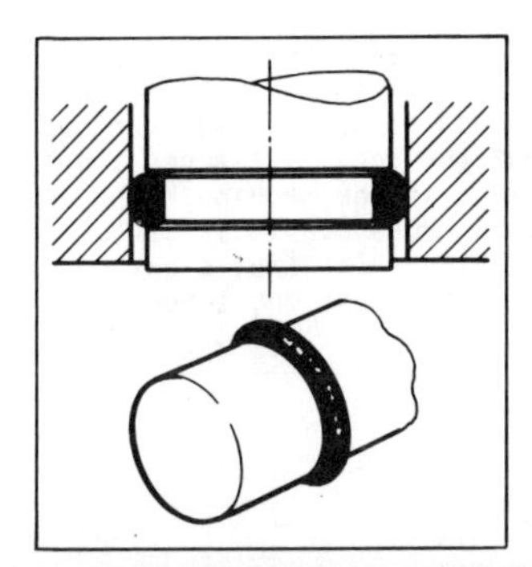

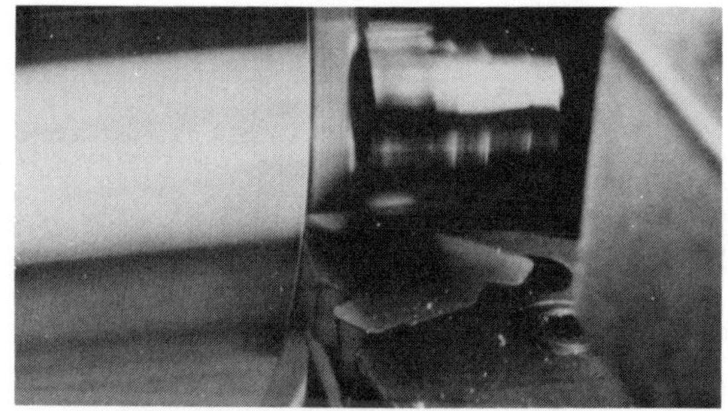

A common fastener on shafts and in holes is the retaining or snap ring (circlip ring). This fits in a groove to locate, retain or lock components on the shaft or in the hole. The ring is made of spring metal and put on with special pliers. Partly in the groove and partly out, the retaining ring forms a strong, accurate axial stop.

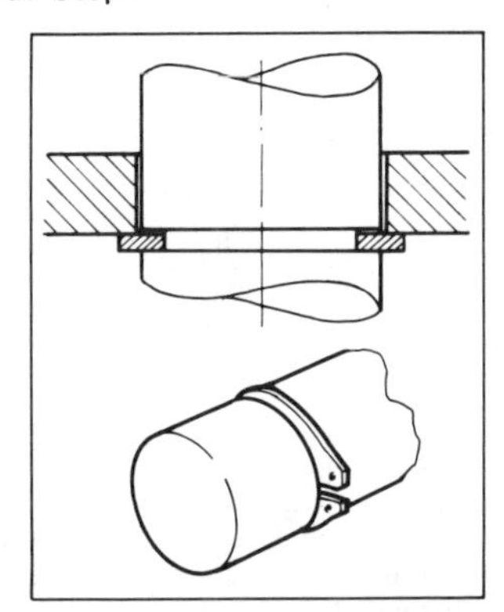

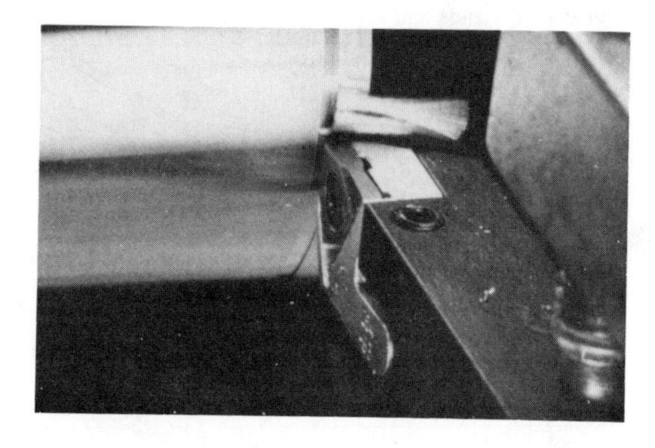

A T-MAX tool has been especially developed for machining retaining ring grooves. It is available as an external or internal tool. The insert is clamped tangentially by a center screw in a unit, which fits into the toolholder. Through retaining ring grooves being relatively thin, it has been found the best way to place the three-edge insert vertically. Also here inserts are available to cut a range of grooves to suit standard ring sizes.

These T-MAX grooving tools have a wider application range in that they can be used for efficient cutting off of thin-wall tubes, up to .118 in. (3mm) deep. They can also be used to cut grooves of larger sizes and chamfers.

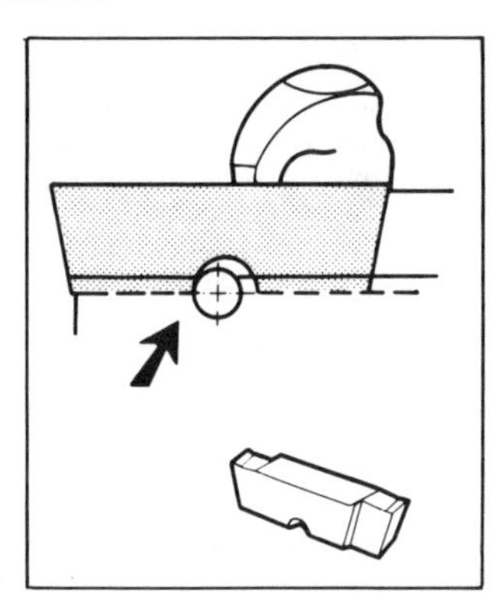

For more general groove and recess cutting there is another range of T-MAX tools. These tools are intended for the machining of grooves and reliefs for

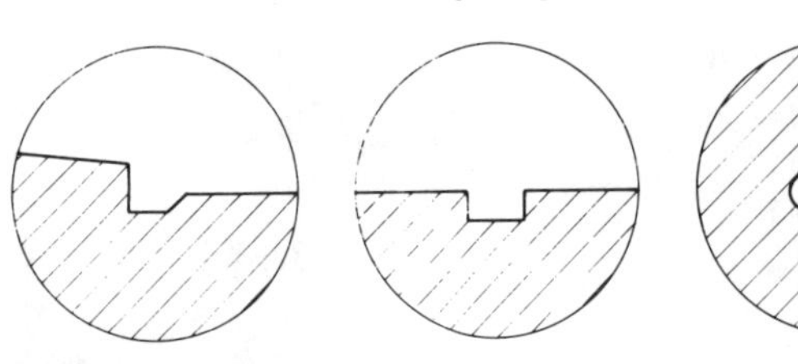

various purposes of up to .315 in (8 mm) widths and .354 in (9 mm) depths and can also be used for special profile threading. A square or rectangular cross-section insert is located in a V-shaped seat in the toolholder, which also has a pin locating in a corresponding recess in the insert. This ensures correct location of the insert when indexed and acts as suport during machining. The inserts are retained by a T-MAX S clamp.

These tools are available for external and internal machining and have either straight or angular mounting of the insert. The tools have been designed strong enough to permit not only the plunging operations of the actual groove cut but also some longitudinal turning and facing. Apart from the ready form inserts--square-shaped grooves, convex radii and round profiles for reliefs--insert blanks are available which can be ground to specific shapes in the toolholder. This provides the machine shop with an extremely versatile form cutting system for machining with high cutting data.

Cutting Off

When a component is machined from bar or tube stock it normally has to be cut off from the part held in the machine chuck and is then either ready or has to be machined on the cut-off end. Moreover, some manufacturing industries produce components that need few or no other turning operations apart from cutting off.

Again this is an operation previously dominated by high speed steel or brazed carbide blade tools. Cutting off is, however, demanding machining because of considerable cutting force acting on a thin tool with long overhang, chipforming and evacuating presents more of a problem and the cutting speed is often reduced to next to zero just before the center of the workpiece is reached. (Modern machines often compensate by speeding up as the tool is fed across.) Modern indexable insert tools can do a lot to boost productivity and improve reliability.

The T-MAX adjustable-blade and the shank-blade parting tools have been developed to optimize cutting off operations. One is adjustable to different overhangs with a changeable blade and the fixed-blade has limited cutting depth capability but can stand up to more unfavourable conditions. It can also be used to perform some turning, facing, chamfering and grooving operations.

The same type of insert is used in both tools. The insert fits into the V-groove in the holder and has a form-sintered chip-former. Conventional chipbreaking is difficult to achieve in threading, grooving and parting off. In order to get the chip out of the deep groove being cut in parting off operations, the insert has been provided with a dimple (rill) on the rake face. This has the effect of reinforcing the chip, making it narrower than groove. The chip rolls up and ensures smooth chipflow.

In order to obtain the best possible performance with the parting tool there are a few but important points that should be noted. To start with it is imperative that the tool is clamped and supported sufficiently well in the machine. The least possible overhang of tool cross carriages or turrets should be the first objective. The tool block or holder must have a stable rest and strong screws to clamp it in the machine.

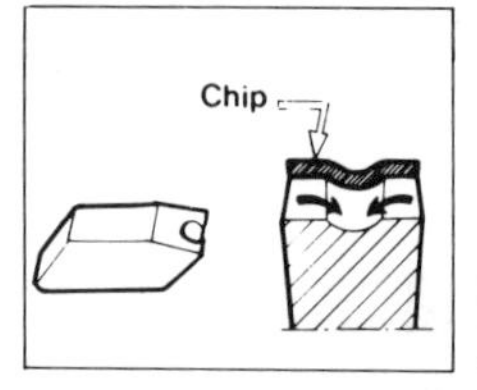

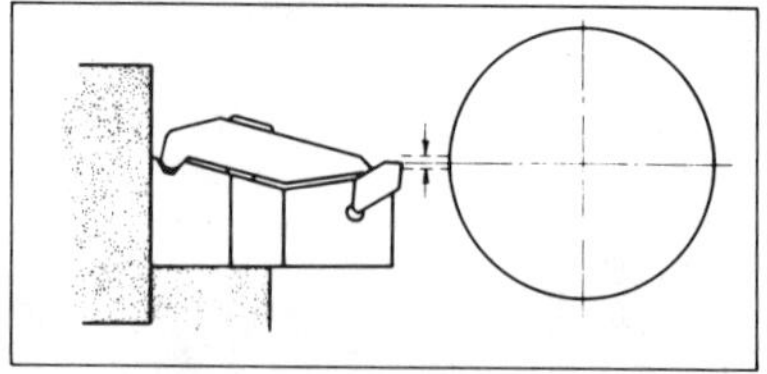

The parting tools should be carefully positioned as regards height to the center line. The adjustable blade holder to within -004 in. (0.1 mm) and shank blade holder to within ±.004 in. (±0.1 mm). The widest possible insert should be used in order to obtain the best insert strength and stability. Due to the pressures involved in the machining, the insert forms a depression in the blade seat. For this reason, the cutting edge may after a while be found to be somewhat below the center line. Plenty of cutting fluid should be used before and during the cut. This will help towards a better surface finish, reduce sudden temperature variations (which can cause cutting edge cracks) and evacuate chips. Feed movements should be even throughout the cut.

The shank blade parting tool can be used for facing operations performed with the same cutting data as parting off. The adjustable blade tool cannot perform any planing if less than 3/4 of the cutting edge width is engaged. Moreover it must not be used for turning adjustments.

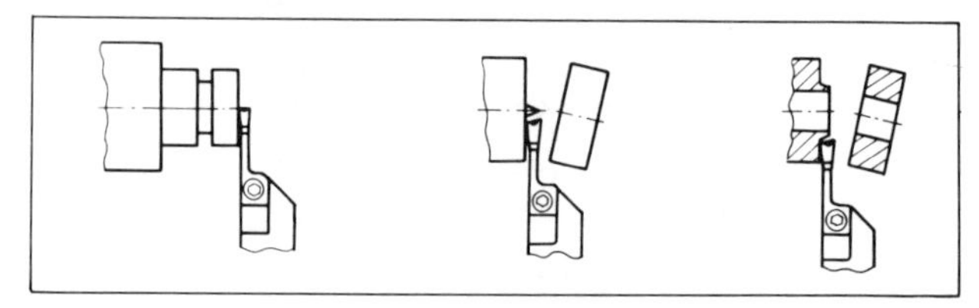

The final part of the cutting off is as mentioned a critical point in the operation. Not only because of the cutting speed dropping towards the center, but also because of the last pip or ring which remains just before center is reached. These two breakthroughs make considerable demands on the cutting edge and tool life

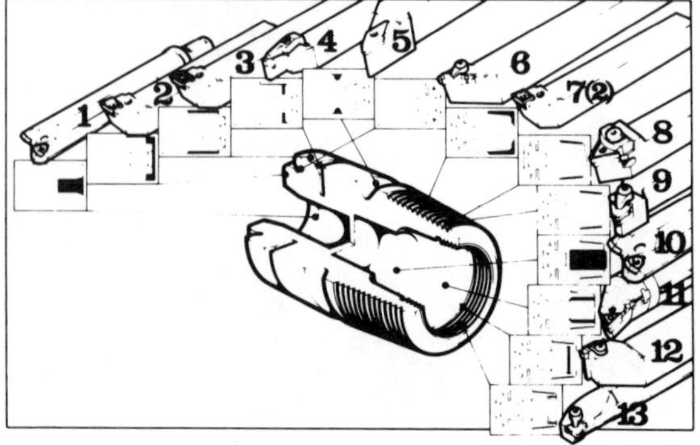

can often be much improved through carefully regulating the feed rates before this part is reached and stopping the tool before it crosses either the center line or breaks past into the tube hole.

A Modern Turning Example

Let us finally look at the machining of a complete and somewhat complicated component which includes most of the turning tool applications described. The finished component is not a typical example from the manufacturing industry, rather a composition of operations that often occur a few at a time. This workpiece represents machining with a modern and efficient selection of cutting tools. It took four minutes to machine from solid alloy-steel bar in an NC lathe. Its finished length is 6.90 in. (175 mm) and the outer diameter 4.53 in. (115 mm).

Operation 1 involves drilling which will be treated separately in a later part of Modern Metal Cutting. At this stage it will only be mentioned that a T-MAX U short hole drill is used at the machining rates of turning.

Operation 2 is roughing and a combination of three facing and turning passes on the end of the piece for which a T-MAX P toolholder is selected. The 80° diamond insert enables the tool to be fed in the two directions required. This is an ideal application for the 71-type insert geometry in combination with GC 015, a double-coated carbide grade as the tool material.

v : 660 fpm (200 m/min) at outer diameter
s : .018 ipr (0.45 mm/rev)
a : .18 in (4.5 mm)

Operation 3 is again a T-MAX P tool with an 80° diamond insert. This is turning down the rough outer diameter and although it is roughing, involves only a light cut which removes the outer bar material. This is performed in one pass and the tool is also utilized to finish off the round contour of the shoulder previously roughed by the tool in operation 1. A 61-geometry insert also in GC 015 is selected.

v : 990 fpm (300 m/min)
s : .008 ipr (0.2 mm/rev)
a : .0984 in. (2.5 mm)

Operation 4 involves the cutting of a deep groove which includes a chamfer on one side. The groove width is .158 in (4 mm) and 1 in. (25 mm) deep. The T-MAX parting tool, shank blade type, is a natural choice here because of the depth of the groove and the subsequent chamfering. GC 135 grade is used.

v : 660 fpm (200 m/min)
s : .007 ipr (0.18 mm/rev)
a : 4 x 1 in. (4 x 25 mm)

Operation 5 is a copying operation producing a groove with sides at 45°. The T-MAX U 35° copy turning tool is chosen because of the in and out copying required. The groove is roughed out in four longitudinal passes and finished off with a light cut that follows the contour. Again GC 015 is used.

v : 820 fpm (250 m/min)
s : .012 ipr (0.3 mm/rev)
a : 4 x .12 in. (4 x 3 mm)

Operation 6 is the cutting of a relief. The T-MAX grooving tool with an insert having the round cutting edge is used. The recess made is .079 in. (2 mm) wide. Grade S1P is used.

v : 990 fpm (300 m/min)
s : .0032 ipr (0.08 mm/rev)
a : .079 in. (2 mm)

The workpiece which is a bar somewhat longer than the length of the component to be produced is turned around in the machine.

Operation 7 uses the same tool as in operation 2: T-MAX P, 71-geometry and GC 015. A series of four turning and facing passes is performed producing a taper.

v : 660 fpm (200 m/min)
s : .018 ipr (0.45 mm/rev)
a : .24 in (6 mm)

Operation 8 is somewhat of a specialized application: the roughing of a large oil screw thread on the tapered bar. Instead of using the threading insert to perform all the passes necessary to produce the full, deep and the correctly formed thread, a T-MAX P tool for general turning is used for the preliminary machining of the thread. Modern NC machines allow roughing and finishing of threads to be performed with separate tools. This can often be advantageous as regards the tool life of the tools involved as the lives of threading cutting edges are often limited by accuracy of the thread produced. A 15-geometry insert is used in GC 015.

v : 500 fpm (150 m/min)
s : 0.25 ipr (6.35 mm/rev)
a : 7 passes (7 passes)

Operation 9 is the actual finishing of the API thread in three passes with the full form insert in a T-MAX threading tool. Grade S1P is used.

v : 500 fpm (150 m/min)
s : .25 ipr (6.35 mm)
a : 3 passes (3 passes)

Operation 10 is again drilling with a T-MAX U short hole drill and will be treated later.

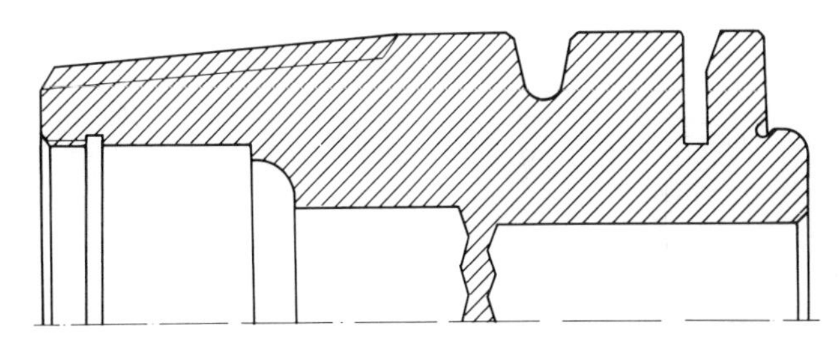

Operation 11 is boring the drilled hole from 2 in. (52 mm) to 3.15 in. (80 mm). The depth of the part to be bored is only 2.36 in (60 mm). A 1.575 in. (40 mm) diameter boring bar is used. The tool selected is a T-MAX boring bar consisting of an exchangeable head and the short version TNS tuned bar. In this operation it was argued that although the tool overhang was not very long (about twice the diameter of the bar) the TNS bar would offer the capability to with higher machining rates ensure a good surface finish. A T-MAX P type cutting head with a 15-geometry insert in grade GC 015 is used.

v : 860 fpm (260 m/min)
s : .01 ipr (0.25 mm/rev)
a : .12 in. (3 mm)

Operation 12 is the internal cutting of a groove .163 in. (4.15 mm) wide and deep to act as a relief after the screw thread. This is a task for the T-MAX grooving tools with its wide range of grooving sizes. Grade S1P is used.

v : 990 fpm (300 m/min)
s : .0032 ipr (0.08 mm/rev)
a : .163 in. (4.15 mm)

Operation 13 is the cutting of a short internal ISO screw thread along part of a bored hole. This type of thread does not require any preliminary roughing as did the larger external oil-tool thread. An internal T-MAX threading tool produces a fully formed thread in six passes.

v : 620 fpm (180 m/min)
s : .079 ipr (2 mm/rev)
a : 6 passes (6 passes)

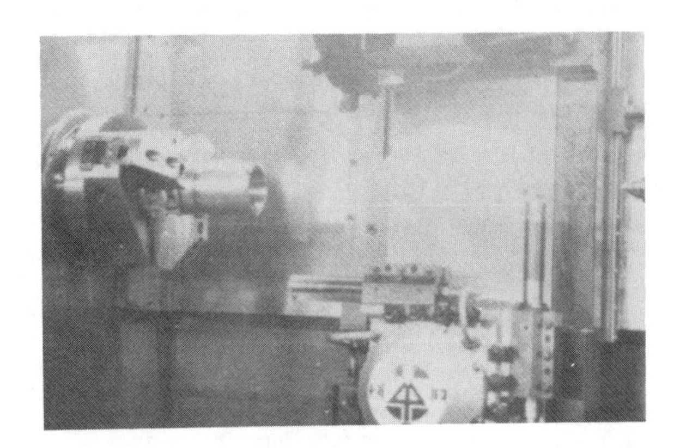

Reprinted from *Tooling & Production*, April 1983

Coated carbides for the common job

Coated-carbide cutting tools have gained widespread acceptance since their introduction in this country about 12 years ago. They account for at least 40 percent of all carbide cutting tools used today. Recent advancements will make them even more useful.

The popularity of coated tools lies in their increased application range, longer tool life and higher speed capability. These characteristics result from the presence of an adherent, hard and chemically stable compound on the surface.

As an example, consider a typical uncoated insert. Under given conditions, at a constant tool life of 10 min, the presence of a TiC (titanium carbide) coating allows a cutting speed increase of 50 percent, and the presence of an Al_2O_3 (aluminum oxide) coating allows a speed increase of 90 percent.

Although coated inserts can increase tool life up to five times in some areas, there is usually much more to be gained by taking advantage of higher machining speeds. In cases where the machine has the horsepower, capacity and rigidity to handle higher metal-removal rates, and the workholding device and workpiece are capable of withstanding the increased forces, coated carbides can provide substantial reductions in per-piece costs. **Table 1** shows the productivity gains in a metal-cutting operation where machining parameters were adjusted to take advantage of carbide inserts coated with TiC and Al_2O_3. The operation involved turning of a hot-rolled 8620 steel shaft using a triangular-shaped cemented-carbide insert.

by Keith H McKee
Manager, Engineering
Carboloy Systems Dept
General Electric Co
Detroit, MI

How coated carbides work

An advantage unique to coated inserts is that the coating and substrate can each be tailored to contribute as a system to maximum tool performance. Thus, coated-carbide cutting tools represent a composite approach to solving the contradiction between the need for high wear and deformation resistance on the one hand and high breakage resistance on the other hand.

The role of the *substrate* is to provide resistance to breakage and thermal deformation, enabling the tool to withstand increased chip loads from higher cutting speeds. The role of the *thin coating* is to provide the necessary resistance to wear from abrasion or chemical interaction with the workpiece. Abrasive wear resistance is basically a function of hardness and determines the rate at which flank wear progresses (**A** in **Figure 1**). Very hard materials such as TiC have very high abrasion resistance and therefore flank-wear resistance.

Chemical wear, or dissolution of the components of the tool material, can occur when the cutting temperature is high enough to promote diffusion between the tool material and the workpiece chips. The maximum diffusion rate occurs at the chip-contact zone, since this is where the highest temperatures are obtained, as evidenced by the formation of a crater (**B** in **Figure 1**).

Chemical wear resistance—chemical stability—depends primarily on how tightly the atoms in a compound are bonded together. This bonding energy of a coating compound is determined by measuring the thermodynamic free energy of formation. Of a number of coating candidates evaluated, Al_2O_3 has been found to have very high chemical wear resistance.

Because the demands on tool materials vary with machining conditions, no one coating works over the wide range of conditions used in industry. For example, when cutting ferrous materials at high speeds (from 500 sfm to 1200 sfm), where crater wear is often the primary failure mechanism, Al_2O_3 is the superior coating. Ceramic-coated inserts provide the greatest tool life at these elevated rates. At lower speeds where abrasive wear, or flank wear, is the limiting mechanism, harder coatings such as TiC provide the greatest tool life. You can generally obtain the best productivity gains under heavy roughing conditions with a plain-TiC-coated grade.

The medium range

Metal-cutting operations that fall within the medium-duty or general-purpose steel machining range are particularly challenging because a high demand is placed on all of the tool-failure mechanisms: wear, breakage and deformation. These are the operations that run at speeds from 300 sfm to 800 sfm, which comprise about 70 percent of all steel machining today. Here, several problems have long been common.

On the high side of this range, the trouble spot is usually deformation of the insert caused by softening from the heat developed during cutting—or cratering caused by chips rubbing the top of the insert. On tougher jobs run at lower speeds, edge chipping and breakage are common difficulties. These problems mean that two or more grades have long been required for general-purpose steel machining; even using several grades, tool life has normally been disappointing.

The solution to these problems has now been found with the development of several highly engineered material systems that employ a multilayer combination coating, improved bonding and a special substrate. The new-generation coated carbides, such as Carboloy Systems' ProMax™ 550 inserts, provide two key benefits. They offer tool-life gains throughout the general-purpose machining spectrum, and they frequently can replace several conventional coated grades because they provide optimum performance throughout the entire spectrum. For example:

- A 8622 steel differential case is rough turned at 719 sfm, 0.014 ipr feed and 0.100″ depth of cut. The conventional general-purpose coated inserts used previously produced 25 pcs/edge. ProMax 550 inserts produce 39 pcs/edge.
- A 4340 steel spindle is rough turned at 500 sfm, 0.025 ipr feed and 0.125″ depth of cut. Conventional multicoated inserts produced 17 pcs/edge. The new-generation inserts produce 30 pcs/edge.

The substrate developed for ProMax 550 inserts provides substantial improvement in the strength and deformation combination, **Figure 2**. This has been accomplished by forming a graded carbide structure having a thin, tough surface zone over a bulk structure with high deformation resistance. The tough zone blunts surface cracks that form in rough cutting applications, preventing edge failure by breakage. The heat-resistant bulk structure prevents softening and loss of cutting-edge sharpness caused by thermal deformation. The extra edge strength of the substrate in fact allows thicker coatings, thus improving cratering and deformation resistance.

Flank-wear resistance is also higher because the multicoating system has a layer of TiC that provides high resistance to abrasive wear. The outer layer of TiN provides high lubricity to minimize built-up edge, thus improving tool life at slow speeds and surface finish on soft steels. This coating layer also resists grooving or notching at the depth-of-cut line (**C** and **D** in **Figure 1**).

The new inserts will machine all carbon and free-machining steels, alloy steels, tool steels, stainless steels, and high-temperature and high-strength alloys and hardened steels. The wide ranges allow them to replace several different conventional-coated steel-cutting grades. This in turn minimizes application mistakes and allows smaller, simpler inventories. With one grade replacing several, inserts may be purchased in generally larger quantities with proportionately higher discounts. ■

How coatings cut costs

	Uncoated	TiC coated	Ceramic coated
Speed (sfm)	400	640	1100
Feed (ipr)	0.020	0.022	0.024
Cutting-tool cost/edge ($)	0.80	0.92	1.12
Tool life (pcs)	40	40	40
Tool life (min)	192	108	60
Nonproductive time/pc (min)	0.5	0.5	0.5
Machining time/pc (min)	4.8	2.7	1.5
Tool-change time/pc (min)	0.075	0.075	0.075
Nonproductive cost/pc ($)	0.50	0.50	0.50
Machining cost/pc ($)	4.80	2.70	1.50
Tool-change cost/pc ($)	0.08	0.08	0.08
Cutting-tool cost/pc ($)	0.02	0.02	0.03
Production rate/hr	11	18	29
Percent productivity improvement	0	64	164
Total cost/pc ($)	5.40	3.30	2.11
Percent cost savings	0	39	61

Table 1

*1. Schematic of worn coated tool shows principal areas of wear. Abrasive action causes flank wear **A**. High heat causes chemical wear to form crater **B** along cutting edge. Grooving and notching **C** and **D** occur at depth-of-cut line. Courtesy F A Dearnley and E M Trent, **Metals Technology**, Feb 1982, Vol 9.*

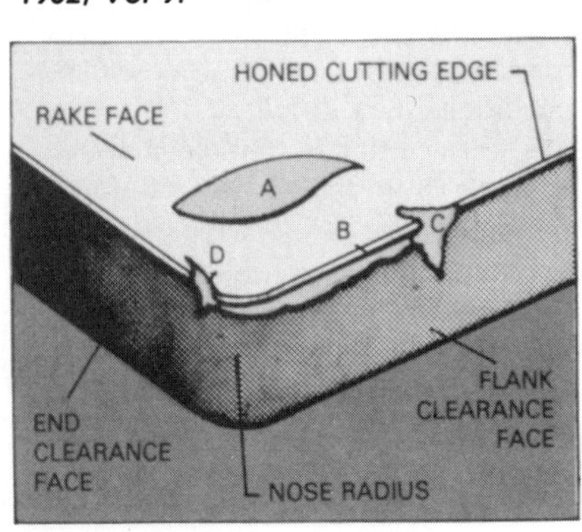

2. Cutaway view shows multilayer coating and special substrate of ProMax 550 inserts.

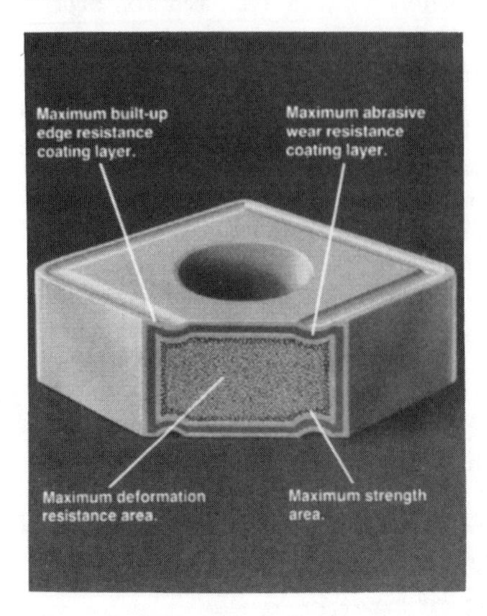

Reprinted from *Production*, October 1983

Multicoated Inserts Carve a Growing Niche

Valve manufacturer finds many reasons to fit new inserts into tooling inventory

A 50 percent increase in cutting speed resulting in a 25-35 percent reduction in machining time with no loss in tool life is being achieved at Anchor-Darling Valve Co., Williamsport, PA. There, a new grade of coated carbide indexable inserts manufactured by General Electric's Carboloy Systems Dept., Warren, MI, is being used. Furthermore, since switching to GE's ProMax 550 inserts for its stainless steel machining requirements, the valve manufacturer has eliminated at least one-fourth of its inventory of cutting tool inserts, resulting in a $15,000 cost savings.

The inserts have a layer of titanium carbide with an outer layer of titanium nitrate that perform well over the broad range of steels at this manufacturer's facility.

Anchor-Darling manufactures valves for nuclear and other critical service applications. Products include gate, globe and check valves which are made from steel, either carbon or special alloy depending on the service requirements. Five different types of carbon steel are used as well as 300 and 400 series stainless steels. Since most valves are specially engineered for the job, production runs are normally limited to three or four parts. Because of strict requirements for material integrity and the variety of materials machined, bar stock is often used instead of castings. This means there is normally a lot of metal to be removed from each part.

Until recently, Anchor-Darling's metalcutting productivity was restricted because of cutting speed limitations, especially on stainless steel parts. To compound the problem, Anchor-Darling had to purchase and maintain duplicate inventories of about 20 different insert sizes and styles in two cutting grades in order to meet its steel machining requirements.

Carbon steel jobs were run at a maximum 400 surface feet per minute (sfm), using C-5 carbide industry class indexable inserts. Tougher, C-2 class inserts were used for machining the more abrasive stainless steel parts. But cutting speed was limited to only 200 sfm. Despite these slow speeds there was downtime for tool changes caused by edge buildup on the cutting inserts.

Cycle Time Reduction. Constantly testing new products to upgrade production, the company found multicoated inserts could outperform conventional counterparts. After switching to the new insert, in one case on a 40-lb, 4.75-in.-dia. stainless steel drop-in seat valve body cutting speed increased 50 percent to 300 sfm. As a result, cycle time was reduced to 18 minutes, a 28 percent improvement. Machining was done on a 40-hp Warner & Swasey (Cleveland) SC-28 CNC lathe. When a conventional insert had been used the facing operation to machine the cover end of the

MULTICOATED INSERT ADVANTAGE CHECKLIST

- ☑ Longer tool life
- ☑ Faster cutting speeds
- ☑ Reduced inventory
- ☑ Reduced downtime

valve was run at .125-in. depth of cut, .015-in. feed per revolution (fpr), 134 rpm and 170 sfm. Finish boring the guide hole through the valve was done at .062-in. depth of cut, .010-in. fpr and 161 rpm for a cutting speed of 200 sfm. Including a trepanning operation, cycle time for machining this part averaged 25 minutes.

59 Percent Cutting Speed Increase. In another case, the ProMax inserts replacing a C-5 carbide cutting insert provided a cutting speed increase of 59 percent. Previously limited to 250 sfm at 48 rpm, the facing operation's cutting speed increased to 398 sfm at 76 rpm. According to David Bower, Anchor-Darling's NC programmer, satisfactory tool life is being achieved and the 250-microinch surface finish required for the seat face is actually a bit easier to hold at the accelerated cutting speed.

Previously, fracture of the conventional cutting insert was occurring when it was necessary to machine an angular face on a 20-in.-dia. carbon steel valve seat ring to prep the part for hard facing with stellite plasma spray. The part was chucked on a Warner & Swasey 5A manual lathe at a 3-degree angle which resulted in a difficult interrupted cutting condition and excessive tool failure. Using the machine's 25-hp low-range, the facing operation was performed at .20-in. depth of cut to remove up to 1.25 in. of stock.

According to Bower, cutting speeds on stainless steel parts have been increased by up to 50 percent to 300 sfm. "The increase in cutting speed for stainless steel parts results in a machining time savings of 25 to 35 percent, depending on the complexity of the part," he says. The faster cutting speeds in stainless steel allow the company to take advantage of the full productive capability of its machines. Machining speed for most carbon steel parts is still limited to 400 sfm because the asymmetrical shape of the valve parts and fixturing produces an out-of-balance condition which limits the rpm's the machine can handle.

Inventory Reduction. The broad application of the new inserts has allowed Anchor-Darling to consolidate its C-2 and C-5 cutting tool insert inventory. This in turn eliminates misapplications and allows smaller, simpler inventories. With only one grade required for general steel machining, inserts can be purchased in larger quantities with proportionately higher discounts, thus reducing inventory carrying costs. □

MJW

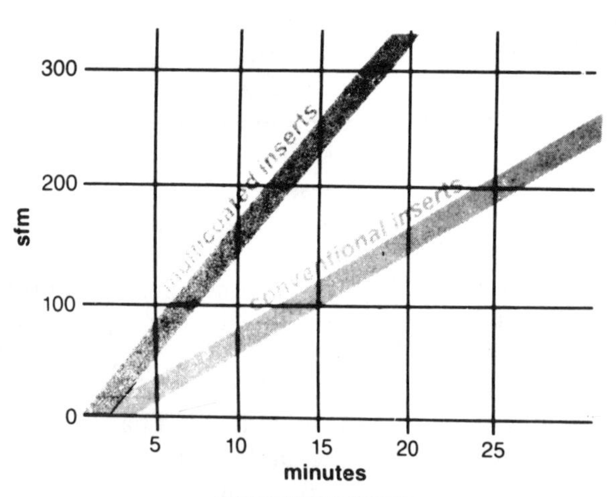

STAINLESS STEEL CUTTING SPEED COMPARISON

Machining an angular face on this 20-in.-dia. valve seat ring caused fracturing of conventional uncoated inserts. Cutting speed increased 59 percent with a new multicoated insert

Presented at the SME Cutting Tools Conference, May 1976

A General Look at Coated Carbide Tools

by John D. Christopher
Metcut Research Associates, Incorporated

Abstract

One of the most exciting new items to reach the field of metal cutting with any impact is the coated carbide insert. These tools have generally met with wide acceptance and are constantly being used in new operations. Along with new applications, additional research and development is continuing by the tool companies. Carbide inserts are now sold with single, double, and triple layers of coatings, depending on whose brand you select. This paper will take a general look at the performance of the coated carbide inserts in turning and milling steels.

Introduction

A cutting tool, to be successful needs two major properties:

1. high hot hardness to resist wear
2. sufficient transverse rupture strength to resist edge chipping

An additional feature would include a smooth, inert rake surface to retard the problem of cratering.

The coated carbide tool is an attempt, with good success, to incorporate these three benefits into an indexable insert. Generally, these improvements have been offered at a cost increase of less than 10% over the uncoated insert.

This paper will examine the performance of the first generation, titanium carbide tool as compared with a general purpose, C-6 grade, steel-cutting tool.

Figure 1 depicts the crater and flank wear regions of a coated insert and an uncoated insert in turning 1045 steel, 187 BHN. The failure mode of both tools was cratering through the end cutting edge. Machining at high feed rates as in this test, .030 in./rev., often results in crater growth through the end of the tool rather than through the side cutting edge. Notice that the flank wear on the uncoated tool was about three times as much as the coated tool. The large "gouge" at the depth of cut on the coated tool seems to be the result of secondary chip action, as it is somewhat detached from the main area of the crater.

Figure 2 shows the development of the tool wear for the two inserts shown in Figure 1. The slow rate of uniform wear on the coated tool compared to the uncoated tool is again very evident. The important consideration in this test is the tool life, 11 minutes on the C-6 carbide compared to 25 minutes on the coated carbide.

The tool wear curves of the uncoated C-6 carbide and the TiC coated carbide at 600 ft./min. and .010 in./rev. are shown in Figure 3. The slower rate of uniform wear and crater growth for the coated carbide again provides for three times longer tool life, 8 minutes compared to 25 minutes, in turning 1045 steel.

The coated and uncoated carbides are compared in Figure 4 with respect to cutting speed in turning 4340 steel, 331 BHN. The TiC coated tool provided an increase in cutting speed of 70% over the uncoated C-6 carbide. This increase gives evidence of the improved temperature resistance of the coated tool, as the cutting temperature increases proportionally with the cutting speed.

The general relationship between tool life and metal removal rate is shown in Figure 5 for a .062" depth of cut. The various points shown on the plot are not data scatter, but rather various combinations of speed and feed which produce different tool life values. This data indicates that, at this depth of cut, the maximum tool life attainable at 5 in.^{3}/min. would be approximately 20 minutes, or 5 minutes maximum tool life at 9 in.^{3}/min.

The performance of the TiC coated tool on 4340 steel at three levels of hardness is shown in Figure 6. Notice that an increase of only five points in hardness, from 35 to 40 R_c, results in a 30% reduction in cutting speed for a 25 minute tool life. Changing the hardness from 40 to 50 R_c, required a 50% decrease in speed to hold the tool life constant at 25 minutes. In addition, the feed was also reduced from .010 to .005 in./rev. resulting in an overall reduction in metal removal rate of 75%.

The performance of the coated and uncoated carbides in face milling 4340 steel, 331 BHN is shown in Figures 7 and 8. Figure 7 shows that in the range where the tool life per tooth is 60 minutes or longer, the coated tools easily withstand the interrupted cutting at a 30% increase in metal removal rate (table speed increase), resulting from faster cutter rotation.

The effect on performance of increasing the feed/tooth is shown in Figure 8. Notice how the tool life per tooth of the C-6 carbide remains essentially unchanged, while the life of the coated carbide varies from 60 to 100 minutes in the range of .010 to .007 in./tooth. Increasing the feed rate 43% resulted in a 40% drop in tool life for the coated tool.

In conclusion, the reader should remember that the depths of cut for the data shown range from .060" to .100" and that very light (.010") or very heavy (.250") cuts will likely alter the tool life and metal removal rate relationship significantly. Also, the data were from only one grade of TiC coated tool and that the performance of coated carbides will vary from one manufacturer to another. It is not within the scope of this paper to survey all the coated tools on the market, only to show general trends. It is not likely that coated carbides will replace uncoated carbides across-the-board. But in specific areas of metal cutting, the coated tools have proven to be of significant value in increasing tool life and metal removal rates.

Uncoated C-6 Carbide

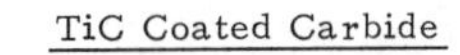

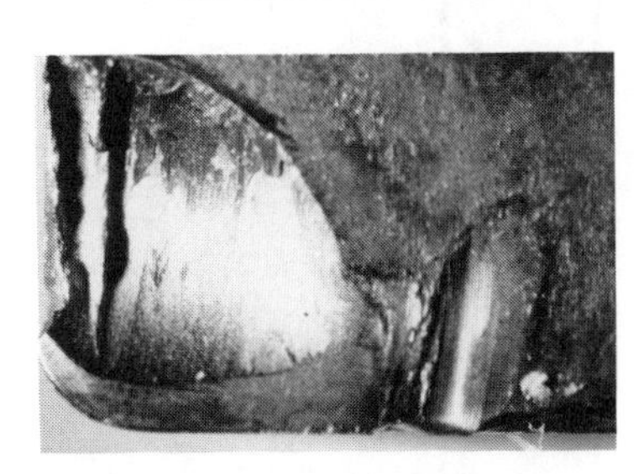

Crater

Flank Wear

Tool Life: 11.2 min.

Tool Life: 25.4 min.

Mag: 15X

Figure 1

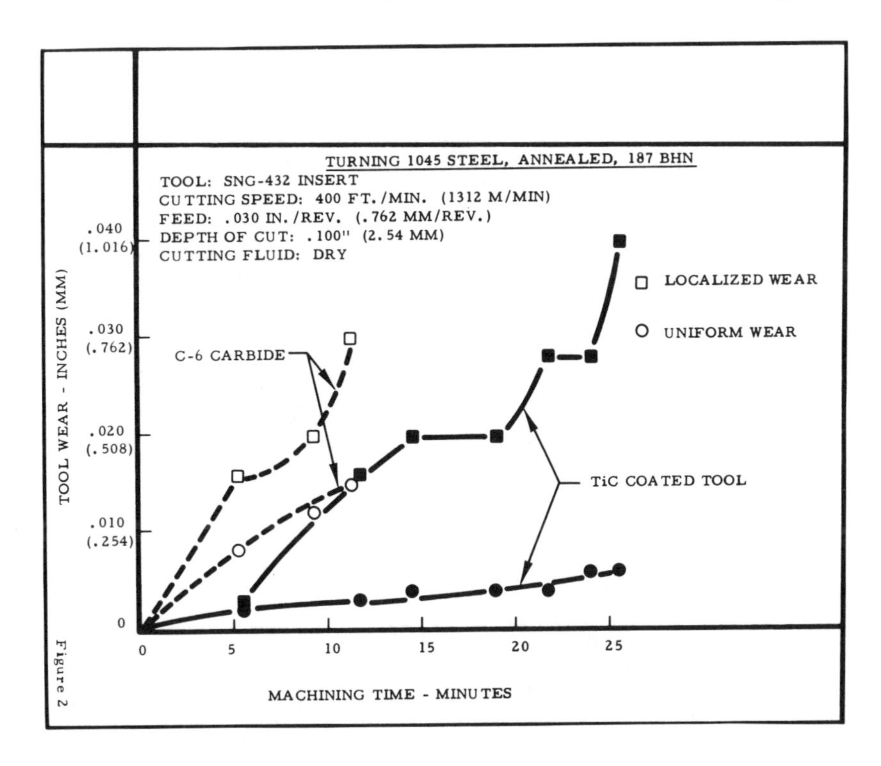

Figure 2

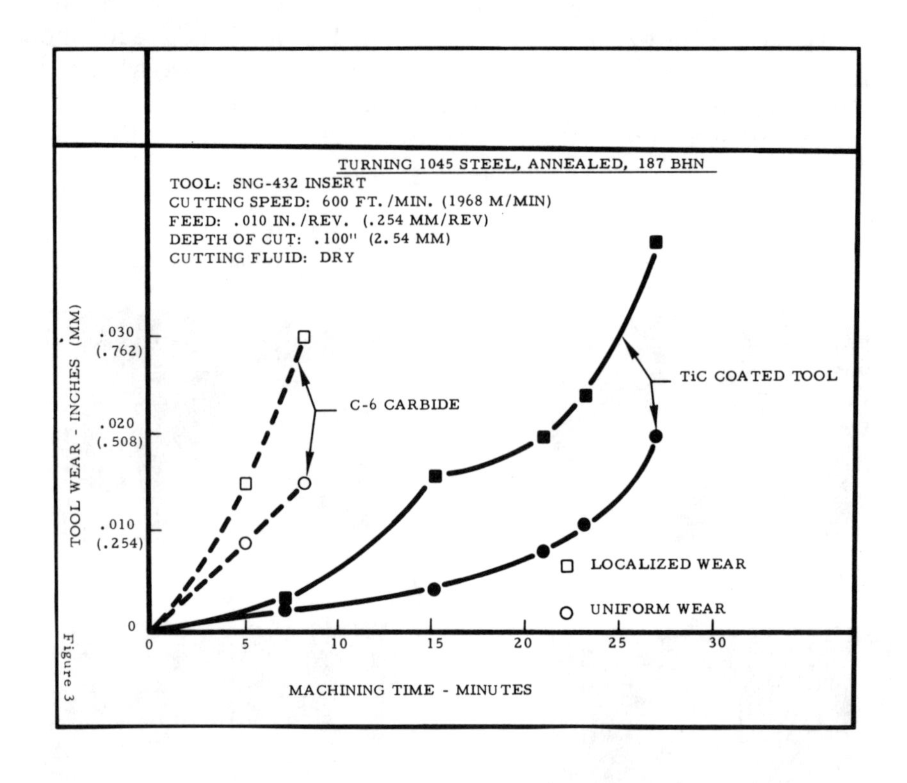

Figure 3

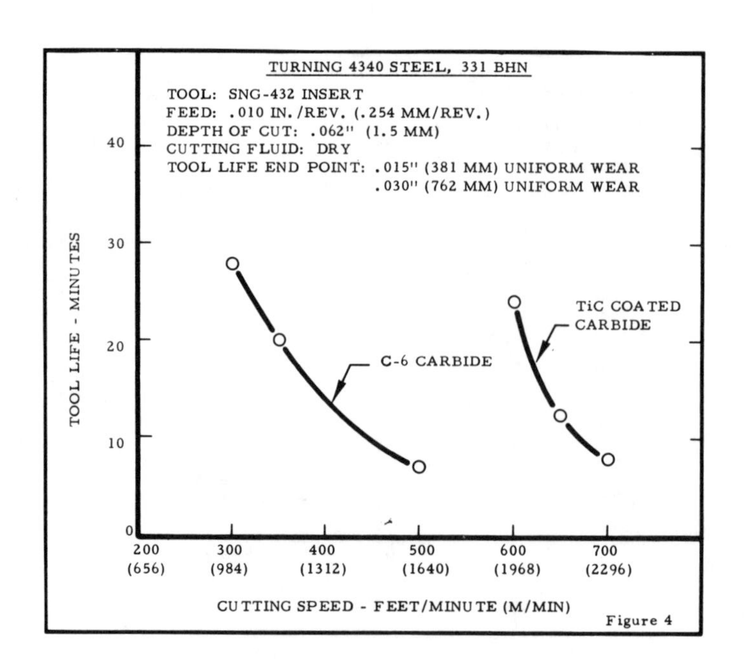

Figure 4

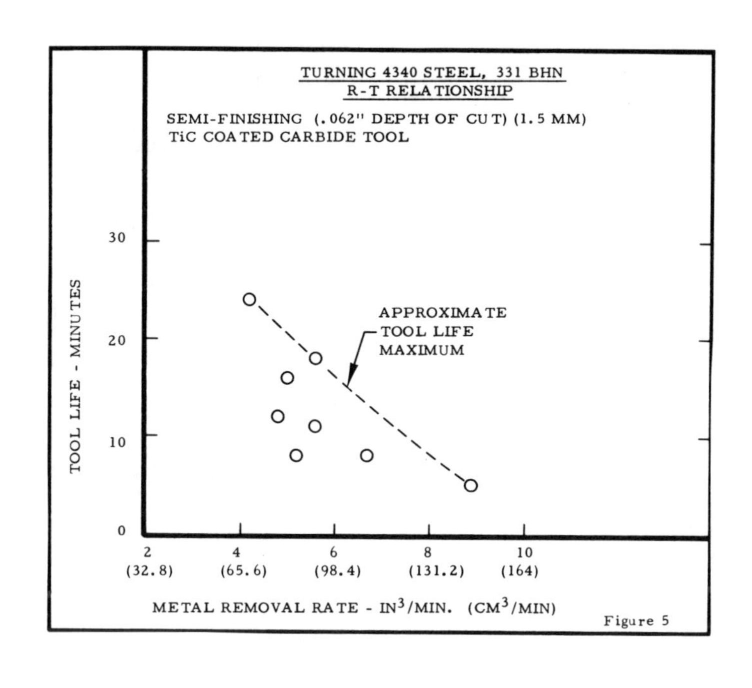

Figure 5

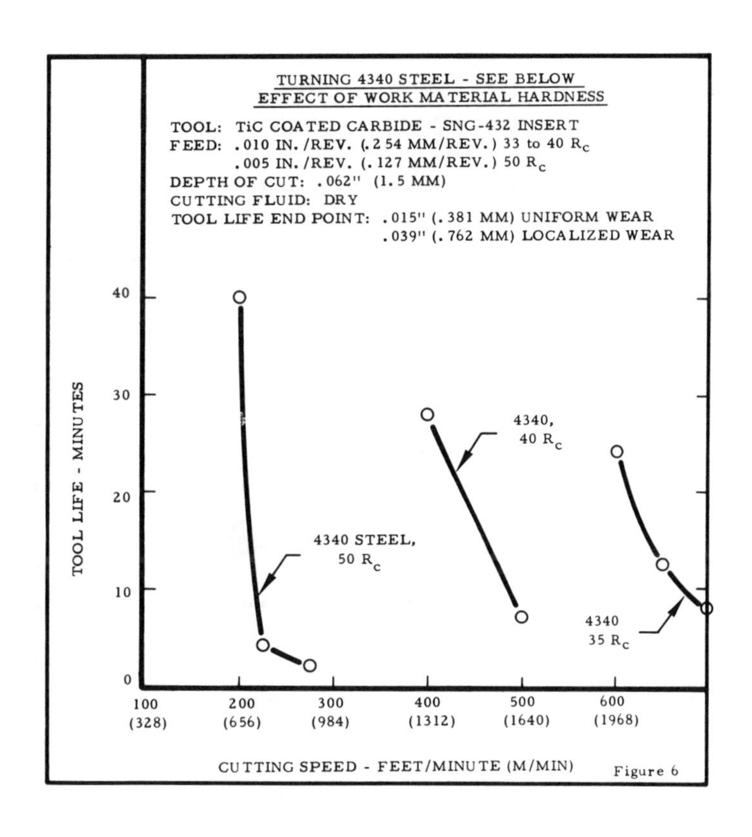

Figure 6

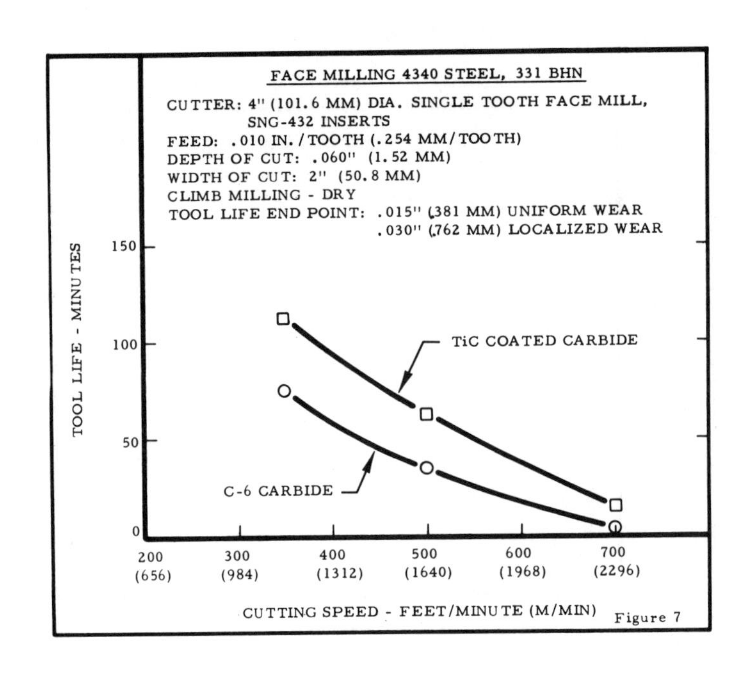

Figure 7

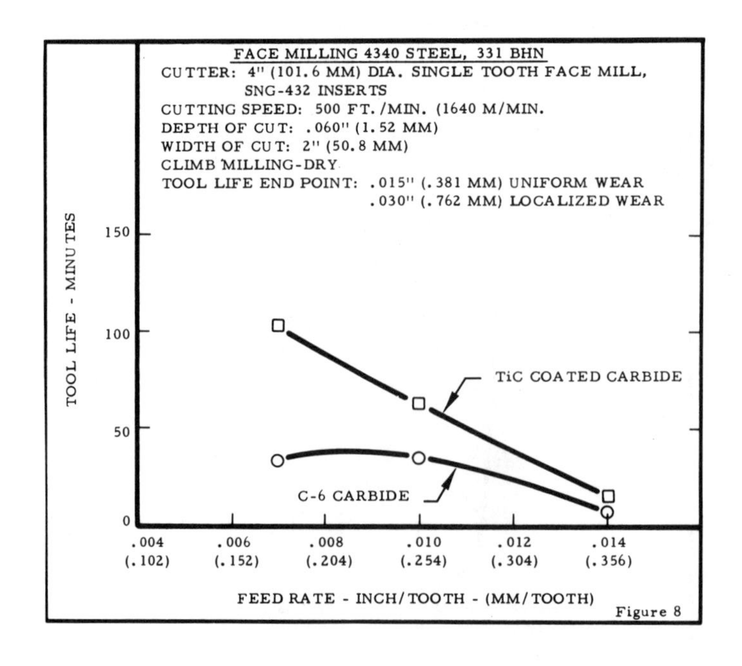

Figure 8

Presented at the SME Cutting Tools Clinic, March 1980

Coated Cutting Tools

by Dr. Bertil N. Colding
Sandvik, Incorporated

The last decade is characterized by the introduction of coated cutting tools, which presently take a large share of the cutting tool market. This involves, besides turning, milling and even drilling operations. The paper covers coated high speed steel and cemented carbides with nitrides and carbides of titanium, hafnium, zirconium and alumina and their application areas. The wear mechanism of TiC-coated carbides is believed to rely on the diffusion barrier of TiC, substantially reducing transfer of matrix particles to the chip. Coatings using still harder materials like cubic boron nitride and polycrystalline diamond are also covered.

In the 10 years since the introduction of coated carbides machining rates have increased between 2 and 10 times over cemented carbides. According to the SME Delphi forecast, this rate will have increased another 400% in 1985.

INTRODUCTION

Metal removal rates have continually increased over the years. Fig. 1 shows that, since 1900, the time required to rough turn a low alloy steel shaft of 4 inch (100 mm) diameter and 20 inch (500 mm) length has been reduced due to recently developed cutting tool material from approximately 105 minutes to one minute (1).

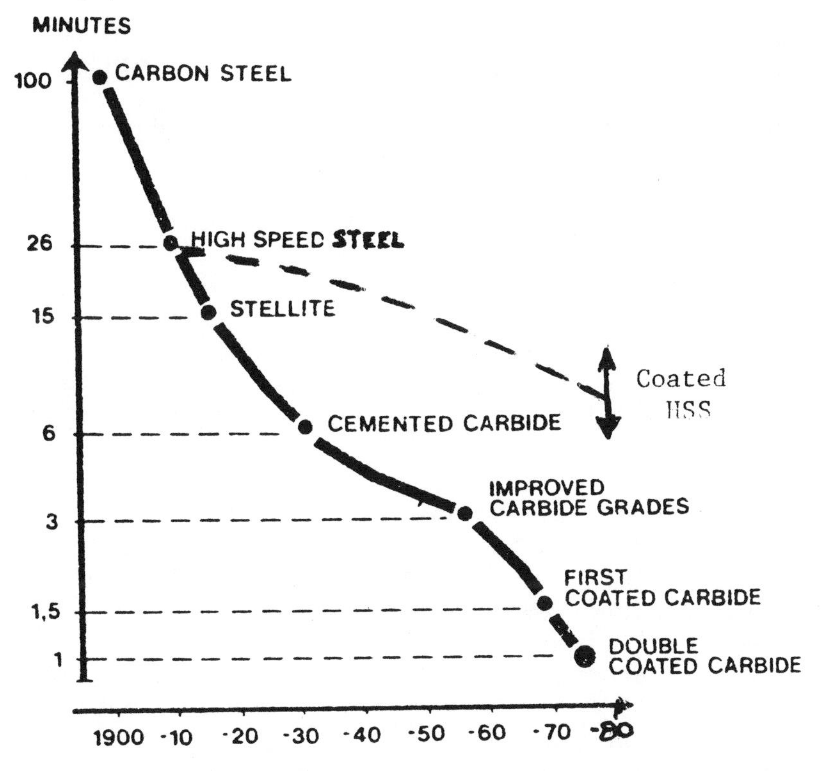

Fig. 1. Change in Cutting Time for Machining a 4-inch Diameter, 20-inch Length Steel Shaft with Different Tool Materials.

Today most chips are removed with tools having cemented carbide edges, either coated or uncoated. As shown by Fig. 2, the relative values of chip volume produced by different cutting materials vary considerably. From the cost point of view, HSS tools comprise about 65% of the world consumption, while cemented carbides amount to about 68% of the total chip volume removed.

	% of Total Purchased Value	% of Total Chip Volyme Produced
HSS	65	28
CEMENTED CARBIDE	33	68
CERAMICS	2	4
DIAMOND	<1	<1
OTHERS	≪1	≪1

Fig. 2. Cutting Tool Material 1977. Relative Purchasing value and relative chip volume.

Coated carbide tools were first introduced around 1969 and have been improved on since then. The development depicted in Fig. 1 should also include coated HSS which for a long time have been tried on the market. Coatings of cubic boron nitride (CBN) and polycrystalline diamond (PCD) were introduced in 1972 and begin nowadays to become economically justified in many special operations.

COATED HSS TOOLS

In the past many attempts were made to deposit thin layers of high wear resistance on high speed tools. Hard chromium plating has provided some good results, e.g., for deep-hole drilling, but has not been successful for lathe tools.

However, recently titanium carbide (TiC) and titanium nitrides (TiN) have been applied on to HSS with promising results in industry. Essentially, two methods yielding coating thicknesses of 2 to 6 µm were used: Physical Vaporation Deposit (PVD) and Chemical Vaporation Deposit (CVD) (2). The price of these new tools are some 2 to 6 times that of traditional HSS tools, but tool lives may be 5 to 10 times longer, or 50% to 100% greater metal removal rate for constant tool life. It is claimed that these tools perform better the higher the metal removal rate compared with HSS. They are also better the harder to machine the work materials are, see Fig. 3, which applies to the throw away inserts, end mills, form tools and gear hobs. Taps and drills are also coated by PVD or CVD techniques. The tools can be used with good performance when machining most work materials including, e.g., stainless steels and Cr-Mo-steels.

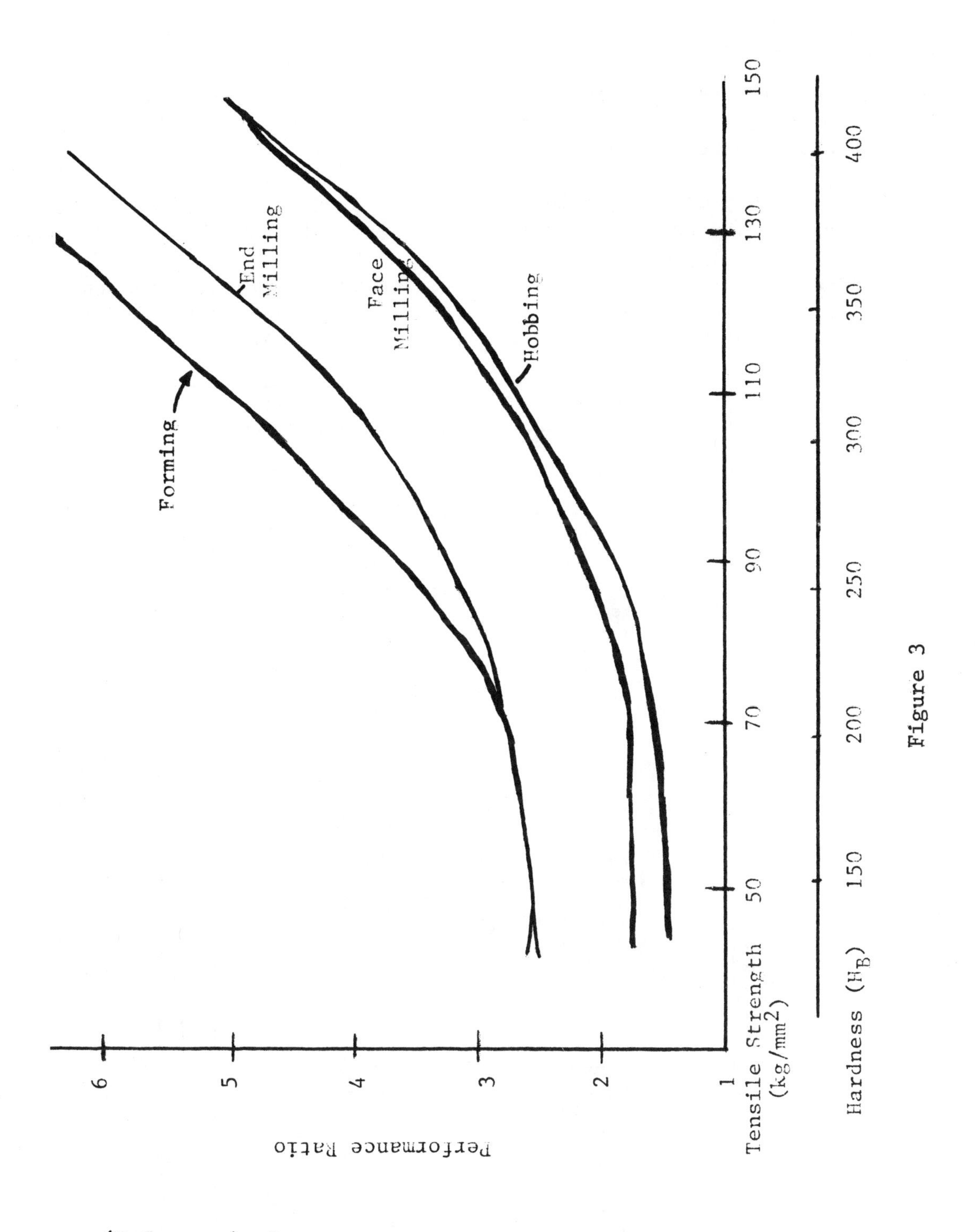

Figure 3

(Reference) Representative Workpiece Materials

S41	S45C	SCM	SKD	Titanium Alloy

Classification

The coated carbides ought to be classified by application areas according to ISO or to the US-C-system in conformity with the conventional carbide classification; see Fig. 4.

P-area: long-chipping materials such as steel and steel castings.
K-area: short-chipping materials such as cast iron, brass, aluminum, wood, plastic and stone.
M-area: positioned between P and K, includes difficult-to-machine materials such as stainless and heat resistant materials, austenitic steels and alloyed cast irons.

Main groups of chip removal				US-Designation	Direction of increase in characteristie	
Symbol	Broad categories of material to be machined	Distinguishing colours	Designation		of cut	of carbide
P	Ferrous metals with long chips	BLUE	P 01	C8	↑ Increasing speed / Increasing feed ↓	↑ Wear resistance / Toughness ↓
			P 10	C7		
			P 20			
			P 30	C6		
			P 40	C5		
			P 50			
M	Ferrous metals with long or short chips and non-ferrous metals	YELLOW	M 10		↑ Increasing speed / Increasing feed ↓	↑ Wear resistance / Toughness ↓
			M 20			
			M 30			
			M 40			
K	Ferrous metals with short chips, non-ferrous metals and non-metallic materials	RED	K 01	C4	↑ Increasing speed / Increasing feed ↓	↑ Wear resistance / Toughness ↓
			K 10	C3		
			K 20	C2		
			K 30	C1		
			K 40			

Figure 4 US C-System vs. ISO System Correlations

The grouping within the three material areas has been carried out according to the type of operation. Low demands for insert toughness start at 01 with extreme finishing and very stable conditions. At 50, unfavorable working conditions and intermittent tool engagement demand very tough cutting edges.

As the wear resistance generally decreases with increasing toughness, the ISO classification also indicates a scale showing how these two properties vary in relation to each other.

Grades for cutting steel are designed to be used at high cutting edge temperatures. They have high percentages of the solid-solution cubic carbide formed by tantalum, niobium and titanium. By decreasing the content of the cubic carbide phase (η phase) and increasing the content of the binder metal phase (γ phase), grades for more toughness demanding operations are designed. Cast iron grades have low or no additions of cubic monocarbides, which are almost entirely used for grain size control. The compositions of the grades for the demands of ISO main group M are between those for the cutting of steel and those for the cutting of cast iron.

Today, coatings provide a wide spectrum of applications within the ISO-system, with most of these in the area of turning inserts, but also for milling and drilling operations. Types of coating include:

Carbides and nitrides of mainly:

Titanium
Titanium carbon
Hafnium
Zirconium
Alumina (Al_2O_3)

The microhardness of carbide materials including the base substrate of ordinary carbides, WC, varies with temperature as seen in Fig. 5. While TiC is the hardest carbide at room temperature, tungsten carbide WC is the hardest at 1000°C, according to Westbrook and Stover (3).

The idea behind coated carbides was that a thin layer of a high temperature stable, hard constituent that was diffused into the surface of an ordinary cemented carbide would improve the wear resistance without lowering the important toughness property of a cutting insert. Fig. 6 illustrates this successful innovation: combining the wear resistance of a carbide grade A with the toughness of grade B into a new coated grade.

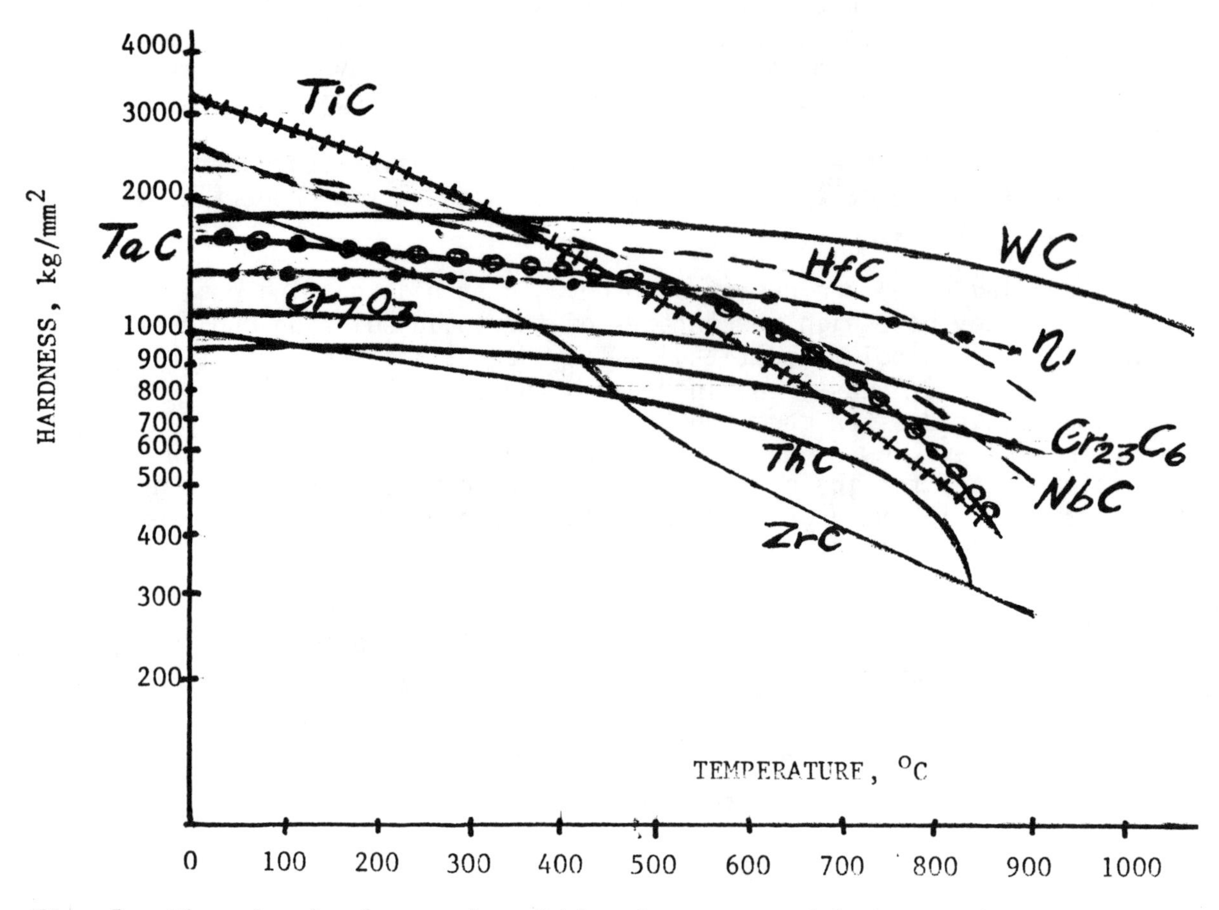

Fig. 5. The microhardness of carbides decreases with increasing temperature. At room temperature TiC is the hardest binary carbide but at 1000°C, WC is hardest binary carbide. (after Westbrook and Stover - from Toth)

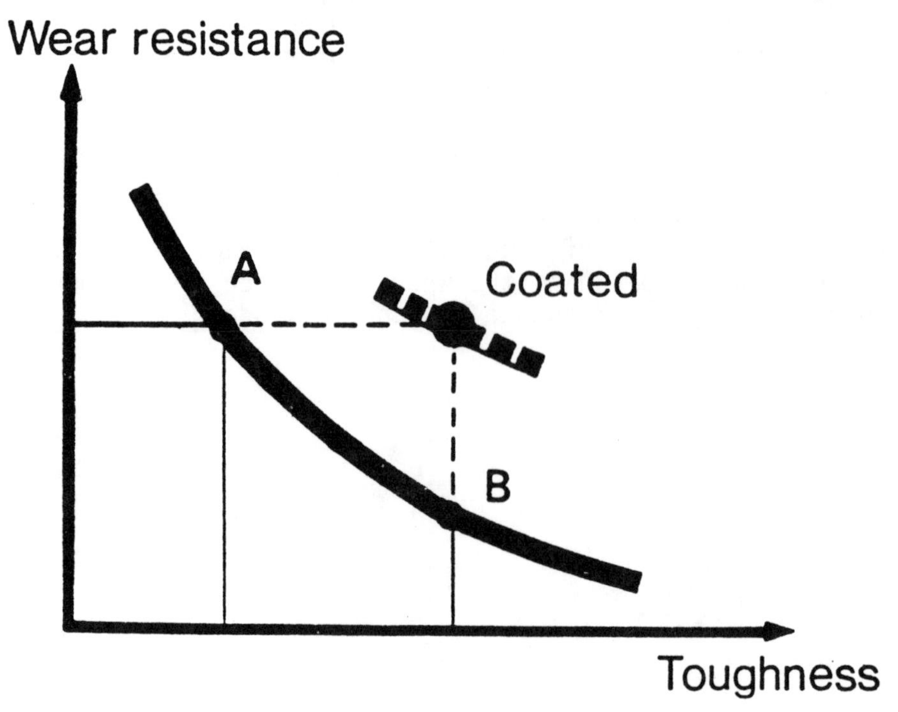

Fig. 6. Showing improved toughness. Wear Resistance Due to Coating.

The thin layer approach works because the cutting of steel and cast iron implies tool wear mainly caused by diffusion mechanisms. When dealing with essentially brittle materials, thin layers have superior elasticity and strength properties compared with solid bodies. Among different promising hard constituents, titanium carbide was first chosen because of its high chemical stability and its suitability as a constituent in conventional cemented carbide steel cutting grades.

In simple terms, this is what has happened. Using a newly developed technique, an extremely thin layer of hard-wearing titanium carbide was applied to an insert as the final operation in the production process. This means that good toughness and high wear resistance were combined in the insert. The wear resistance is in the surface layer; the toughness in the substrate.

The process gives an atomic and practically stress-free bond between the carbide coating and the tougher substrate. The thickness of the TiC layer is about 0.0002 in. (0.005 mm).

The first grades introduced in 1969 had substrates characteristic of conventional carbides, but later new substrates better suited for coating were developed. Also, thicker layers proved to yield a better performance. Improved manufacturing processes made it possible to apply double layers, such as aluminum oxide (Al_2O_3) on top of a TiC layer, see Fig. 7.

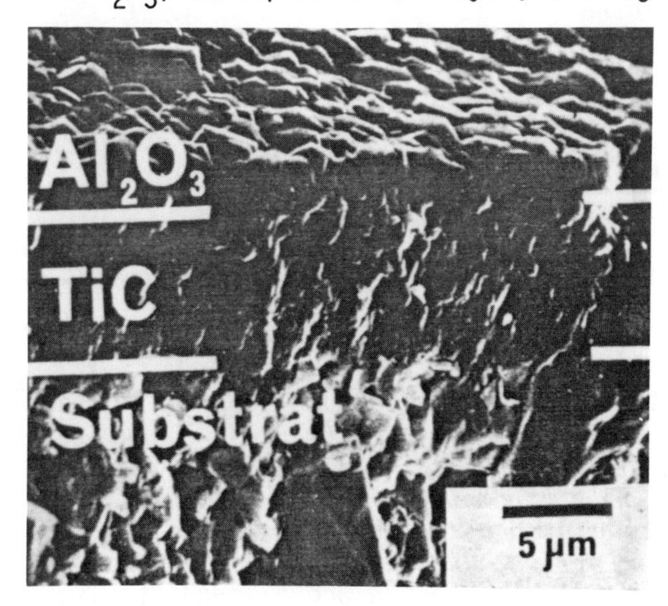

Fig. 7.

A cross section of a worn TiC-coated tool is shown by Fig. 8.

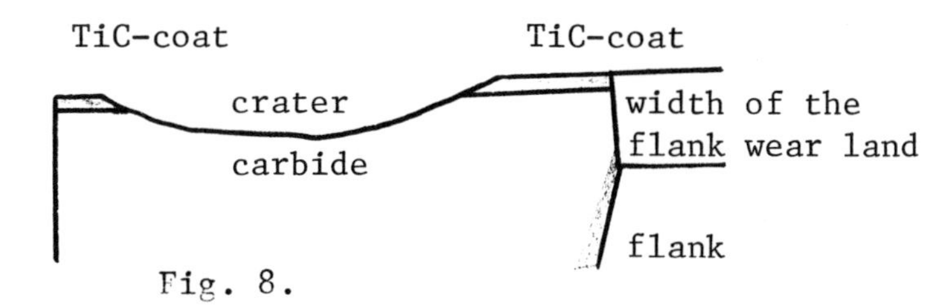

Fig. 8.

As seen, the wear pattern is similar to any carbide tool and the coating is worn away rapidly both on the rake and the flank face. Strangely enough, both kinds of wear occur at a slower rate than with an uncoated carbide under exactly the same cutting conditions as seen in Fig. 9.

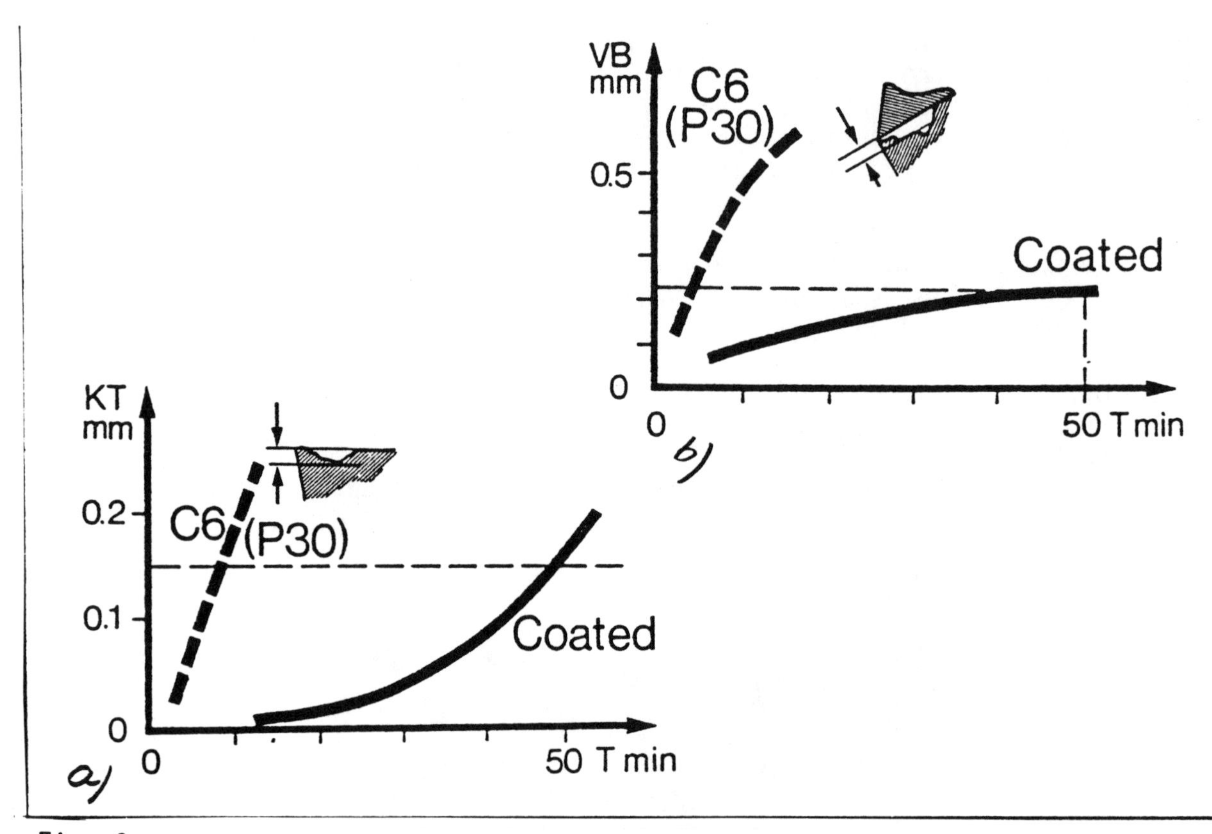

Fig. 9.

Here a P30 grade with a TiC coating is much superior to the standard carbide. This applies to the depth of the crater (KT) as well as to the flank wear (VB). Laboratory tests as well as field tests confirmed early substantial improvements in tool life of coated grades. Lab tests showed (4, 5) also that cutting edge temperature is considerably lower at all cutting speeds, see Fig. 10. The three cutting forces: tangential, axial and radial force are all lower according to Fig. 11, for carbon and alloyed steels, as well as for cast iron. This means on an average that the horsepower requirements are some 10 to 15% lower with TiC coated carbides. More recent tool life v-T-curves of German origin (6) in cast iron and steel turning tests show tool life improvements from 300 to 700%, see Figs. 12 and 13.

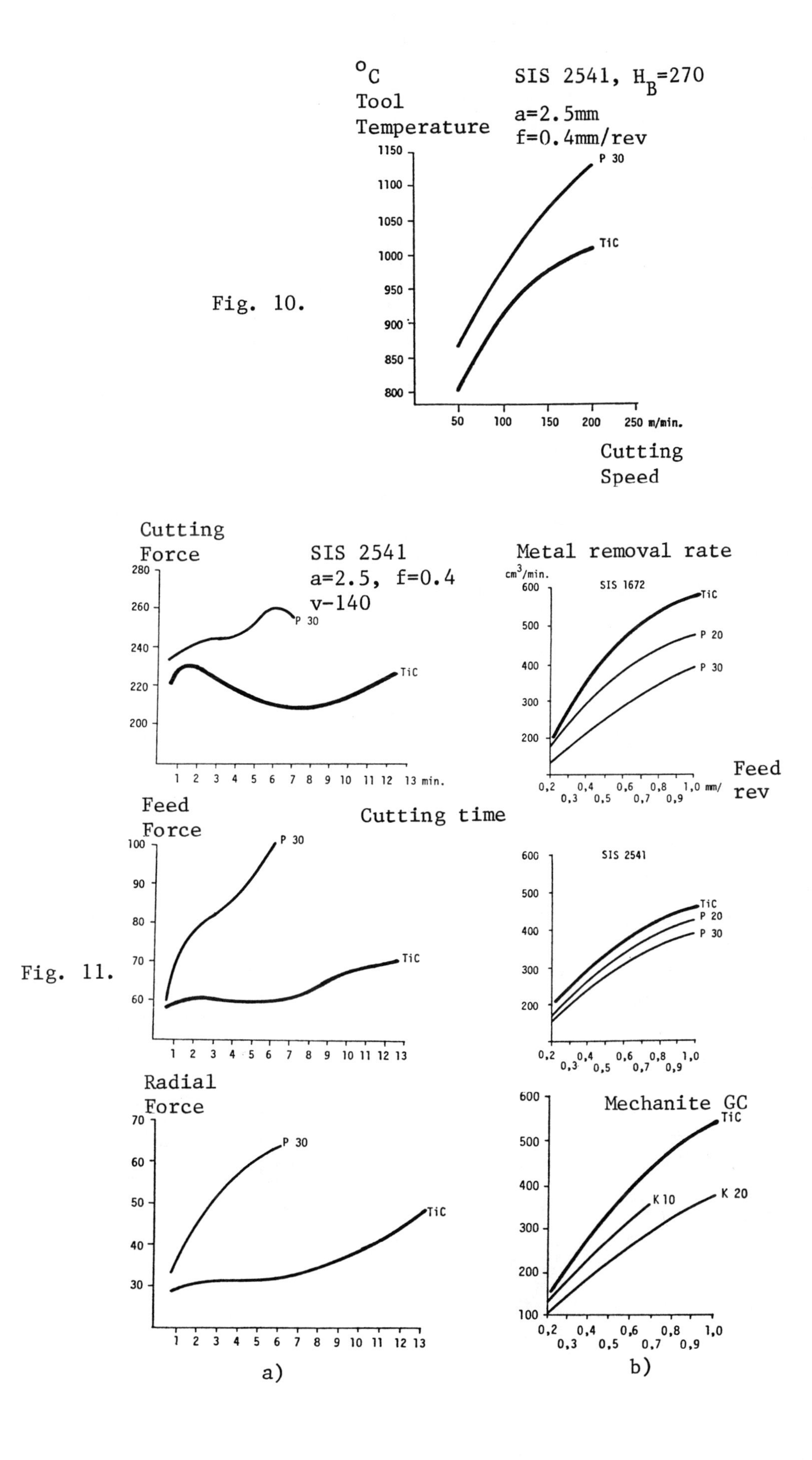

°C
Tool
Temperature
SIS 2541, H_B=270
a=2.5mm
f=0.4mm/rev
1150
1100
1050
1000
950
900
850
800
P 30
TiC
50
100
150
200
250 m/min.
Cutting
Speed
Fig. 10.
Cutting
Force
SIS 2541
a=2.5, f=0.4
v-140
280
260
240
220
200
P 30
TiC
1 2 3 4 5 6 7 8 9 10 11 12 13 min.
Cutting time
Feed
Force
100
90
80
70
60
P 30
TiC
1 2 3 4 5 6 7 8 9 10 11 12 13
Fig. 11.
Radial
Force
70
60
50
40
30
P 30
TiC
1 2 3 4 5 6 7 8 9 10 11 12 13
a)
Metal removal rate
cm³/min.
SIS 1672
600
500
400
300
200
TiC
P 20
P 30
0,2 0,4 0,6 0,8 1,0 mm/
0,3 0,5 0,7 0,9
Feed
rev
SIS 2541
600
500
400
300
200
TiC
P 20
P 30
0,2 0,4 0,6 0,8 1,0
0,3 0,5 0,7 0,9
Mechanite GC
600
500
400
300
200
100
TiC
K 10
K 20
0,2 0,4 0,6 0,8 1,0
0,3 0,5 0,7 0,9
b)

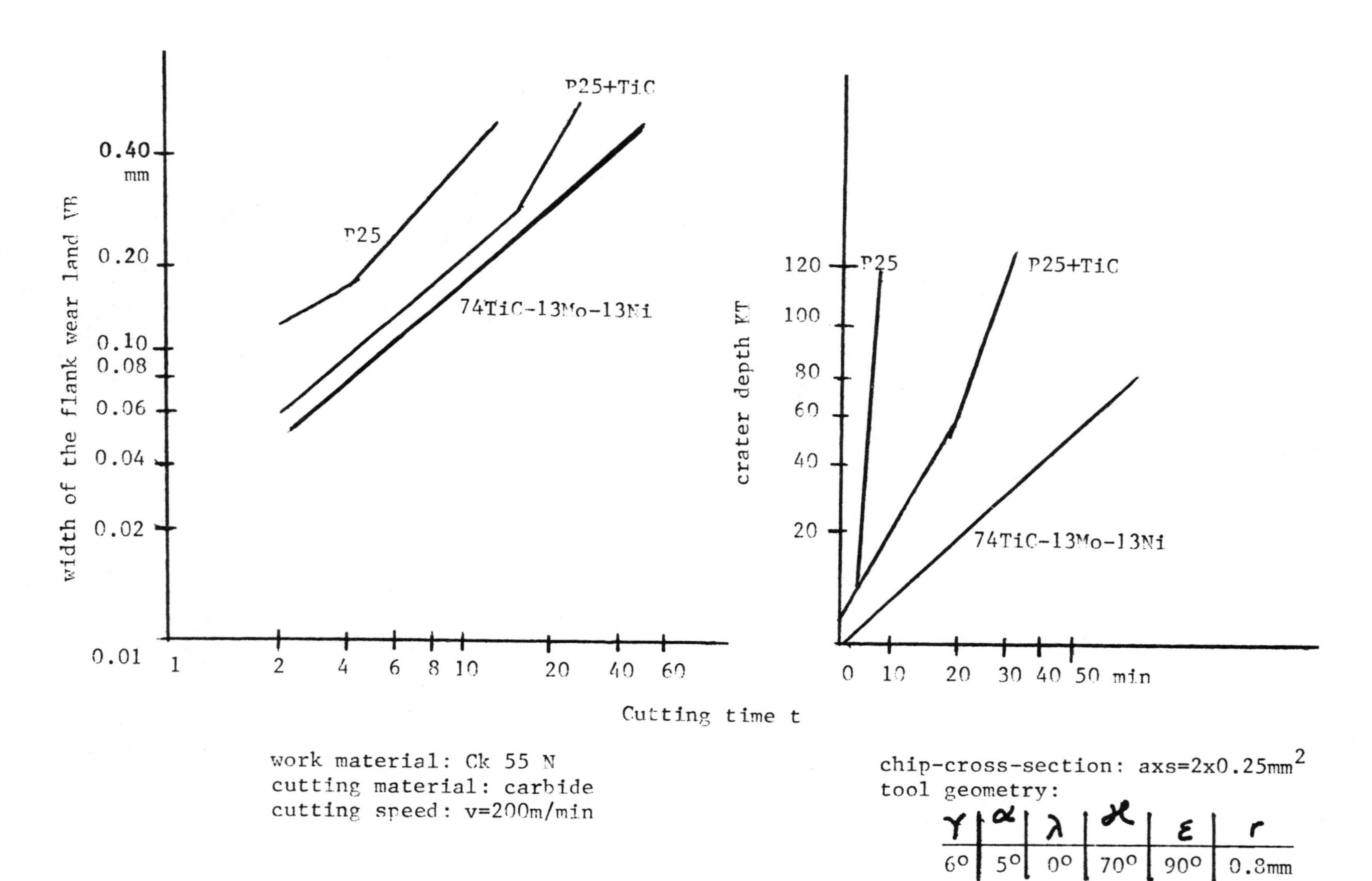
width of the flank wear land VB
0.40 mm
0.20
0.10
0.08
0.06
0.04
0.02
0.01
1
2
4
6
8
10
20
40
60
P25
P25+TiC
74TiC-13Mo-13Ni
crater depth KT
120
100
80
60
40
20
0
10
20
30
40
50 min
P25
P25+TiC
74TiC-13Mo-13Ni
Cutting time t
work material: Ck 55 N
cutting material: carbide
cutting speed: v=200m/min
chip-cross-section: axs=2x0.25mm2
tool geometry:
γ | α | λ | ϰ | ε | r
6° | 5° | 0° | 70° | 90° | 0.8mm

Fig. 12

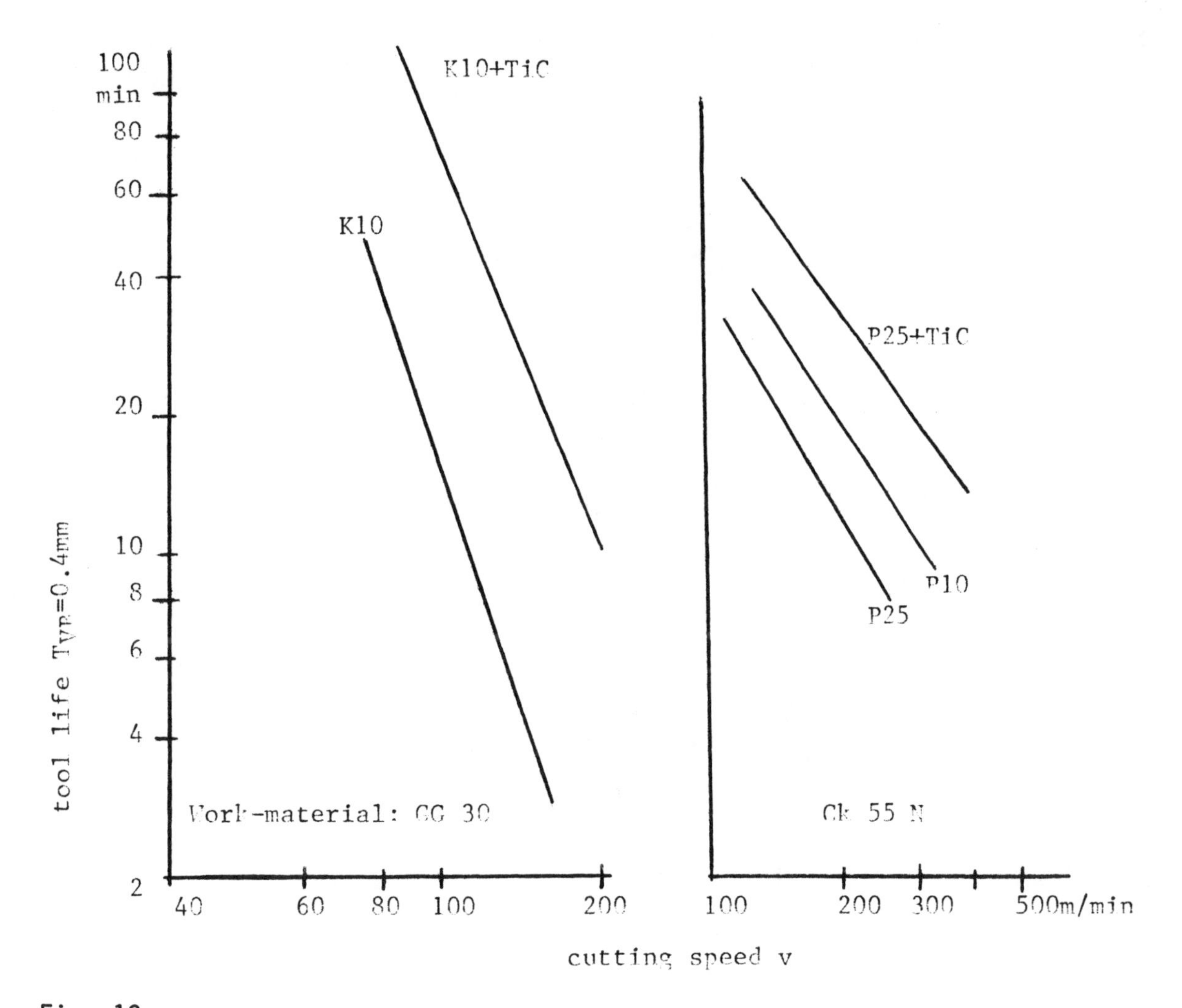

Fig. 13

WHY ARE COATINGS BETTER?

There is rather conclusive evidence that the TiC coating acts as a diffusion barrier between the chemical elements of the material cut and that of the carbide substrate. In 1969, the Author (4) investigated the wear characteristics of TiC coated carbides using several techniques:

1. Tool tip temperature
2. Apparent friction angle
3. Cutting forces
4. Crater wear by a Talysurf
5. Crater wear by interferometry using a new laser technique
6. Absolute volumetric and total wear by a radioactive method
7. Relative volumetric wear of flank and rake face by an autoradiographic technique

The radioactive and autoradiographic methods, where the tungsten is otope W187 was the active element, showed conclusively that no tungsten carbide penetrated the TiC coating during the first 10 to 20 seconds of wear. The Talysurf method and the laser interferometry showed that this time corresponds to crater depths of the same magnitude as the thickness of the coating (4 to 5 μm), see Fig. 14.

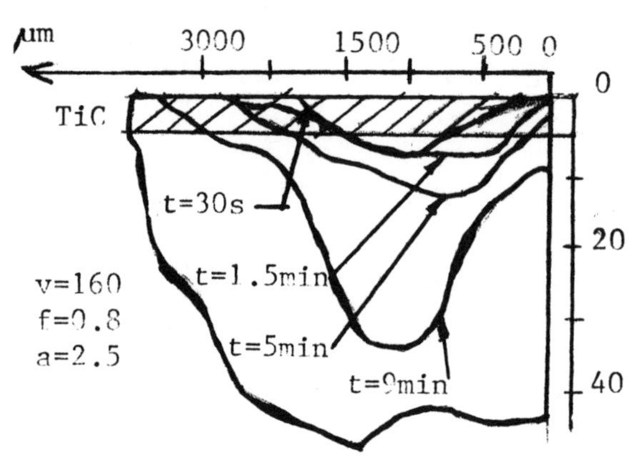

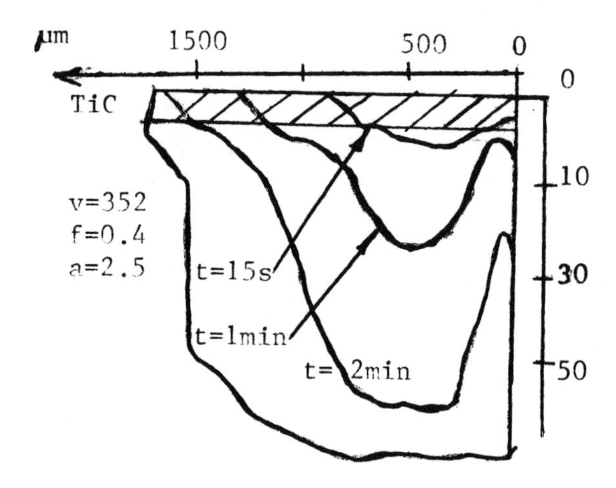

Fig. 14.

The reason for the effectiveness of the TiC coating during the whole life of the tool may have several causes:

1. The chip is supported on the periphery of the crater which contains the hard and thin TiC coating. The mechanical loading on the crater surface may therefore be relatively smaller than is the case for pure cemented carbide.

2. The rear end of the crater acts as a very good chip breaker due to the hard TiC edge.

3. The TiC coating is successively pressed down into the matrix during the formation of the crater[x], the latter thus containing TiC in spite of the fact that the thin TiC film is broken through only after 10 to 20

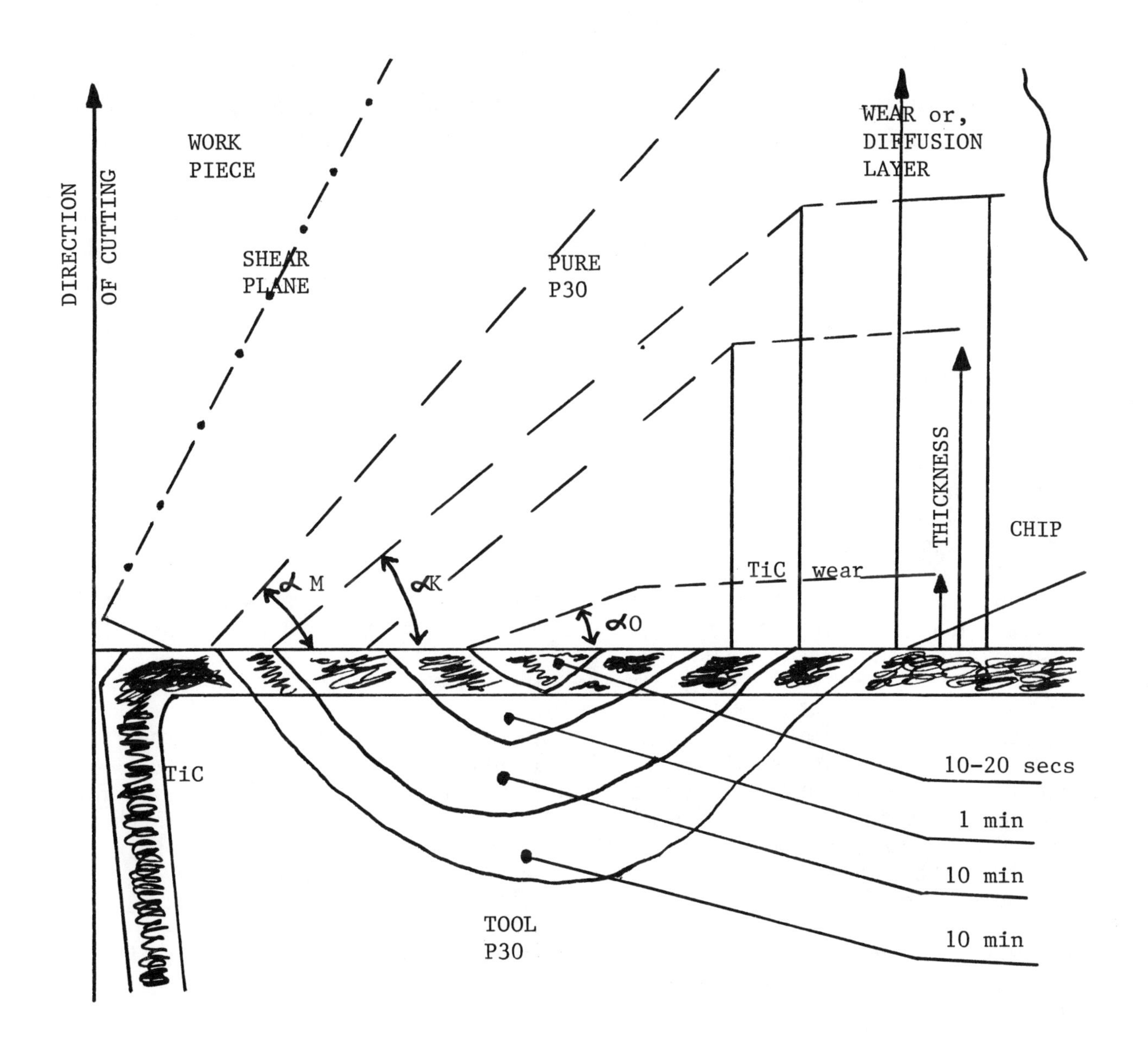

Fig. 15. Stages of wear rate (α), or thickness of diffusion layer of TiC coated carbide.

sec. of cutting. This prevents e.g. W to penetrate into the chip at the same rate as occurs with conventional carbides.

x (Some of the matrix elements must also change place with the penetrating TiC.)

4. The TiC coated carbide forms oxide layers with the work material. During the crater formation, these oxides are formed close to the tip of the tool and are transported by the chip into the crater zone.

Although all of the above-mentioned causes to a larger or less extent may contribute to the superiority of TiC coated carbides, the following picture seems to be the most reasonable in explaining this question.

In the case of standard carbides, Gappisch (7) found that the flow layer under the chip (in which a ferrite-austenite transformation occurred) became thicker the higher the carbon content of the chip. At the same time, more carbon diffusion was observed from the carbide matrix than with steels of a lower carbon content. Although here the diffusion process was initiating tool wear, the strength of the formed layer was the most important factor contributing to tool wear.

In the case of a work piece of relatively high carbon content, the shear primarily took place within the carbide portion of the layer. For a slow carbon steel, the shear process took place in the relatively weak zone close to the chip. The wear became thus less for the low carbon steel.

It is postulated here that the TiC coating close to the tip of the edge is deposited onto the chip at a lower rate than would be the case for e.g. a P30 carbide. This thin transformation zone will eventually pass across the crater in the way shown in Fig. 15. Now this zone will come into contact with the matrix carbide (approximately the same composition as P30) and a reaction will take place meaning that this zone becomes progressively thicker due to pick up of pf W. After leaving the crater, the same point of the chip will continue to rub against the rear TiC coating. In Fig. 15, three stages of this proposed wear process are shown. In the first stage, at 10-20 sec. of cutting, the first W atoms are transferred, as disclosed from the radioactive and from the autoradiographic tests. The W wear rate is here α_0. In the second stage, at 1 minute, the wear rate has attained its almost constant value ($\alpha_K > \alpha_0$), valued also after e.g. 10 minutes of cutting. The reason for the wear rate being less than for P30 is thus due to this TiC barrier in the flow layer of the chip, which prevents the other constituents in the matrix to deposit at the same rate as P30. However, after a considerably longer time, when there is no TiC left at the tip of the tool, the wear rate K will increase to that of M, corresponding to P30 matrix. This transition corresponds to the time where VP and KT increase at a steeper slope, see Fig. 12. Also, the reasons postulated under points 3 and 4 may contribute to the process described in connection with Fig. 15.

APPLICATION AREAS

At present, almost 50% of all turning inserts are coated extending over almost the entire range of the ISO system.

Coated grades today encompass milling applications in both the P-groups and the K-group. The situation of cast iron milling in the metalworking industry has been and will be one of change. The push for higher machining rates and the development of tool materials bring about mass-production machinery improved in way of spindle design and speeds, stability and rigidity, as well as workpiece clamping. Fig. 16 reflects the past trend and the point where industry stands at today, in milling tool materials and machine capability as far as the mentioned points are concerned.

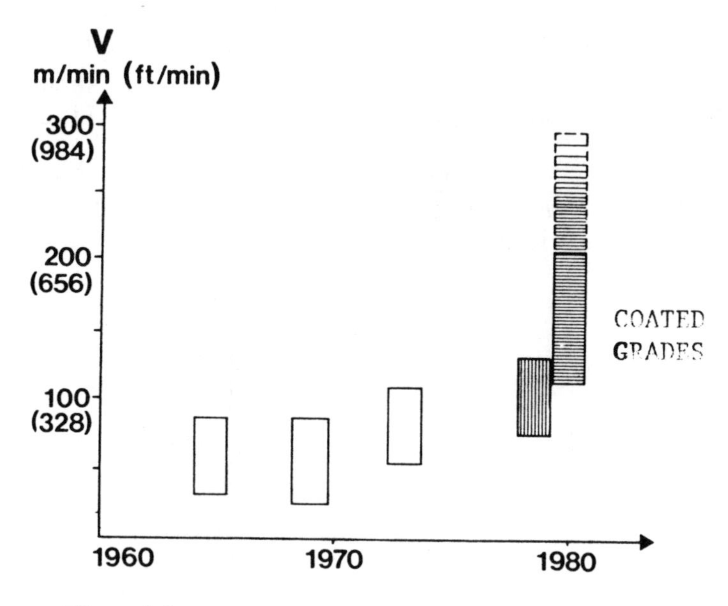

Fig. 16.

Coated carbides compete favorably with ceramic tools as exemplified by a machining case from the Japanese automotive industry. The company in question was trying out ceramic inserts for milling cylinder blocks. They found, however, that milling in their transfer-line required more toughness than that available in ceramic inserts, see Fig. 17.

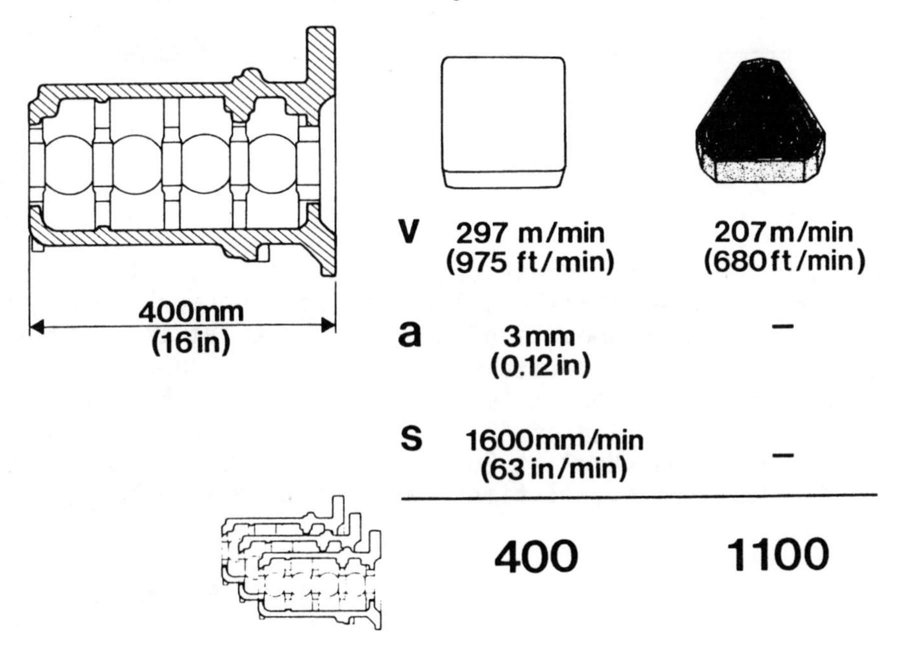

Fig. 17.

The cutting data used in the transfer line was for ceramic inserts: v: 207 m/min (975 ft/min), a: 3 mm (0.12 in) and s: 1 600 mm/min (63 in/min). The result for ceramic inserts were 400 components per cutting edge set. Coating cutting data was: v: 207 m/min (680 ft/min) with the same feed and cutting depth. Coated results were 1 100 components per cutting edge set. The limiting criterion in both cases was edge frittering on the cylinder block.

Even in tough operations such as drilling, coated carbides can be used to great advantage. A new type of carbide drill is here used with 2 trigon-shaped inserts, as shown by the right-hand picture in Fig. 18.

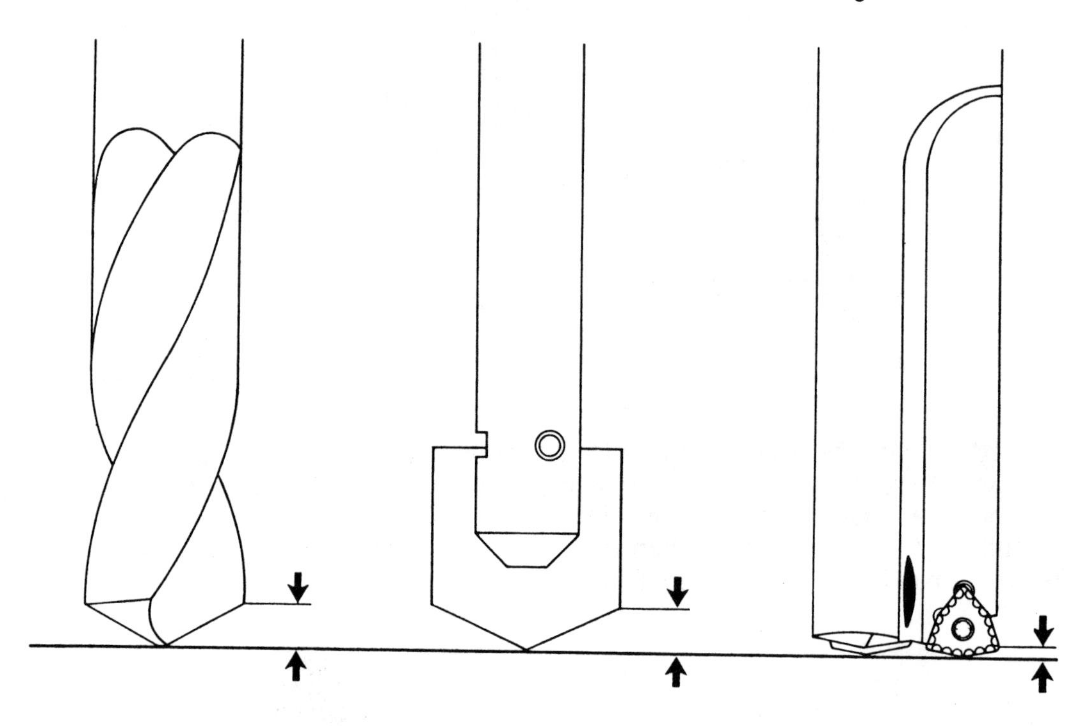

Fig. 18.

Compared to the long point on a twist drill or spade drill, this design means also that the feed distance before drill entry into the work piece is relatively short. The large amounts of chips produced at the high metal removal rates are here evacuated with cutting fluid at high pressure. A comparison of the speeds and feed rates used with that of HSS twist drills is depicted in Fig. 19. Four holes of 42 mm (1.63 in) diameter and 6 mm depth were drilled in high carbon steel at 264 m/min (860 sfm) and a feed rate of 500 mm/min (19.7 in/min), i.e., 11 times faster than with HSS.

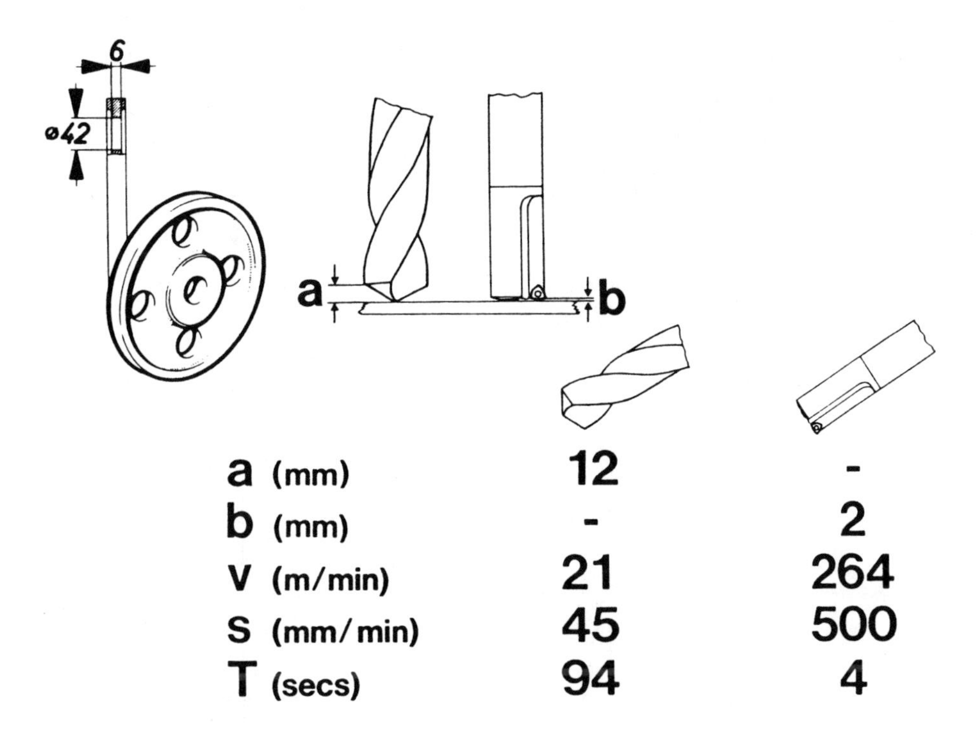

Fig. 19.

CUBIC BORON NITRIDE (CBN)

Inserts of this hard material may be entirely made up by boron nitride, or consist of a 0.5-1.0 mm (0.02-0.04 in.) thick layer on a substrate of cemented carbide. CBN is produced at ultrahigh pressures and high temperatures yielding a product almost as hard as diamond, but more temperature stable than the latter. CBN tools seem to be more prone to brittle behavior than diamond tools, so the cutting edge has to be manufactured with great care. The price of an insert is extremely high: about 200-300 times that of a coated carbide, or about the same as polycrystalline diamond inserts. However, in many special cases, CBN is economically feasible. CBN is used to turn hardened steels (45PC or harder), chilled cast irons and nickel and cobalt-based super alloys. A Japanese investigation (8) shows that the performance of CBN tools differs largely depending on the sintering method. Cutting temperatures are lower than that of carbide tools. This temperature even decreased with work piece hardness when exceeding a certain hardness limit. CBN is chemically stable for iron and diffusion wear seems to be relatively low. Two tool grades: CBN 1 (Japanese, ceramic bound) and CBN 2 (American, metal bound) were tested using identically chamfered edges. In hardened steel, the crater wear occurs right after the cutting edge, see Fig. 20, and at least CBN 2 experiences plastic deformation at the tip.

The actual wear land value is rather constant, even though its value measured from the initial edge increases with cutting time. This helps to keep the cutting edge sharp and the tool lasts for a long time. Fig. 21 pertains to turning of three differently heat treated high speed steels for 5 minutes

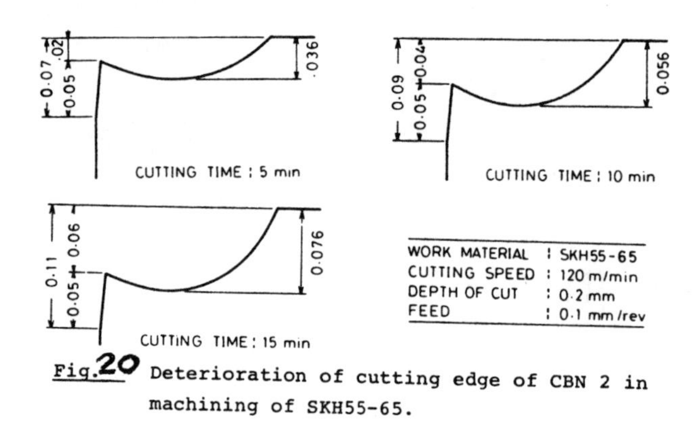

Fig.20 Deterioration of cutting edge of CBN 2 in machining of SKH55-65.

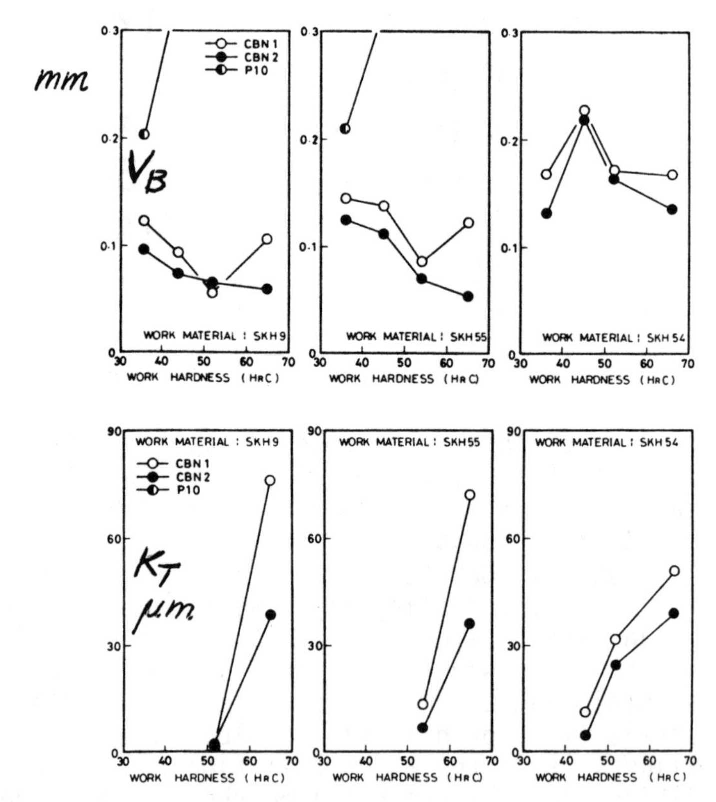

Fig. 21. Tool wear after machining variously heat treated high speed steels for 5 minutes.

using CBN 1 and CBN 2 comparing with P10 carbide. The cutting speed was 120 m/min, feed = 0.1 mm/rev and depth = 0.2 mm. As seen, P10 wears out very rapidly or immediately, whereas CBN 2 performs somewhat better than CBN 1. Above HRC = 45 to 50 the flank wear VB becomes smaller with increasing hardness, while the opposite occurs for the crater depth KT.

Two industrial examples conclude this section: one involving machining of a centrifugal cast roll, Fig. 22 and one a flame sprayed crankshaft, Fig. 23. The cast roll hardened surface has to be turned in one cut rather than grinding it. However, it was not possible to obtain the required surface finish with either cutting data A or B. In spite of this, the extra grinding needed took 45 min instead of 6 hours. This means that the total machining time was reduced by 3 hours using alternative A. The tool cost became only a fraction of the machining cost.

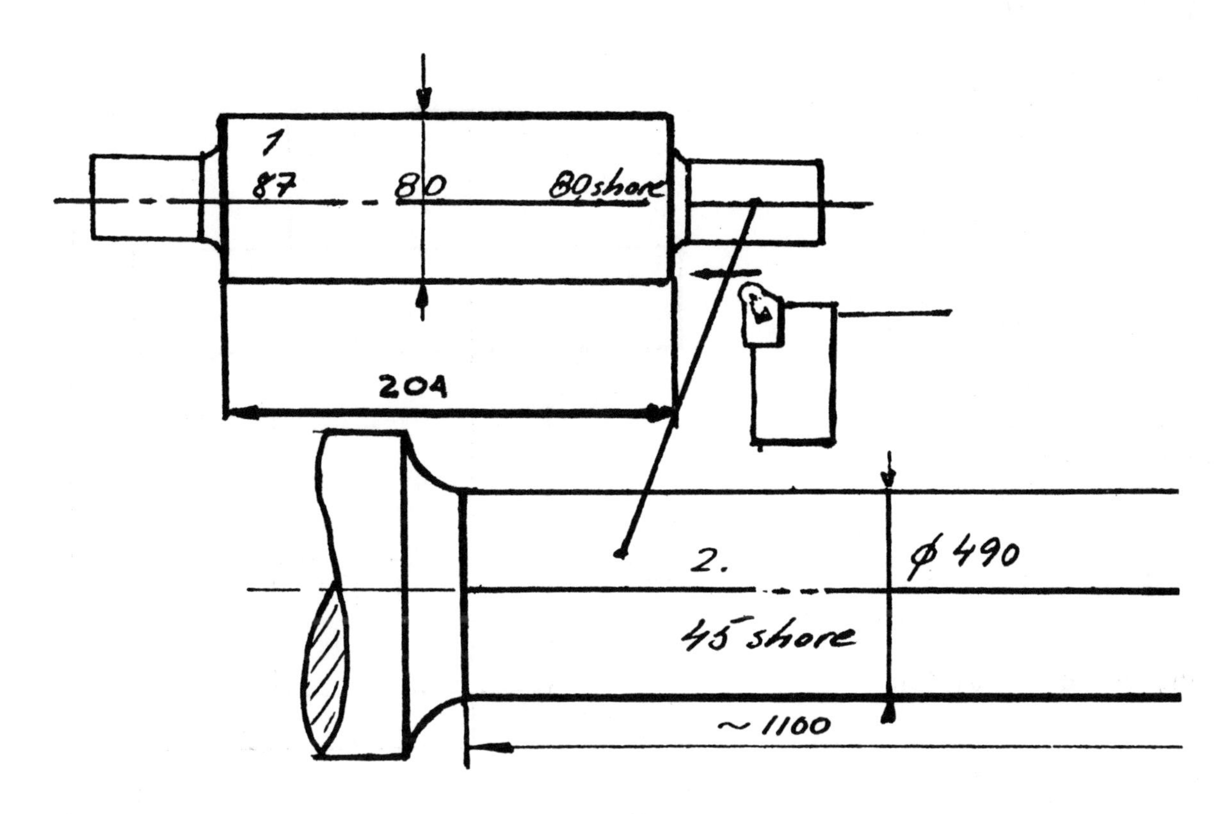

Fig. 22. Machining a Centrifugal Cast Roll.

	1. Turning of hardened surface		2. Turning shaft	
Hardness:	30 Shore		45 Shore	
Toolholder:	Special		Special	
Insert:	RNGN 12 04 00		RNGN 12 04 00	
Cutting data:	A	B	C	D
Cutting speed:	66 m/min	238 m/min	260/m/min	563 m/min
Depth of cut:	0.5 mm	0.5 mm	1.0 mm	1.3 mm
Feed per rev:	0.37 mm/rev	0.46 mm/rev	0.5 mm/rev	0.4 mm/rev
Feed per min:	15 mm	45 mm	85 mm	120 mm
Tool life:	135 min	28 min	12	15
Surface finish:	H=3 µm	H=5 µm	H=5 µm	H=3.3 µm
Tool wear:	VB=0.3 mm		VB not visible	VB=0.3 mm

The turning of the sprayed crankshaft in Fig. 23 was done in one setup, still at a comparatively small tool cost. The operation took 40 min compared to 168 min grinding time.

Material: Metco 160, Mi-based, flame sprayed

Hardness: 60 HPC (650 HB)

Earlier operation: grinding

Cutting conditions:
Speed v = 116 m/min
Feed s = 0.12 mm/rev

Depth of cut: a = 0.5 mm

Cutting fluid: Emulsion

Insert: SNGN 12 04 12
$K = 75^0$

Required surface finish:
$R_a = 1.0\ \mu m$

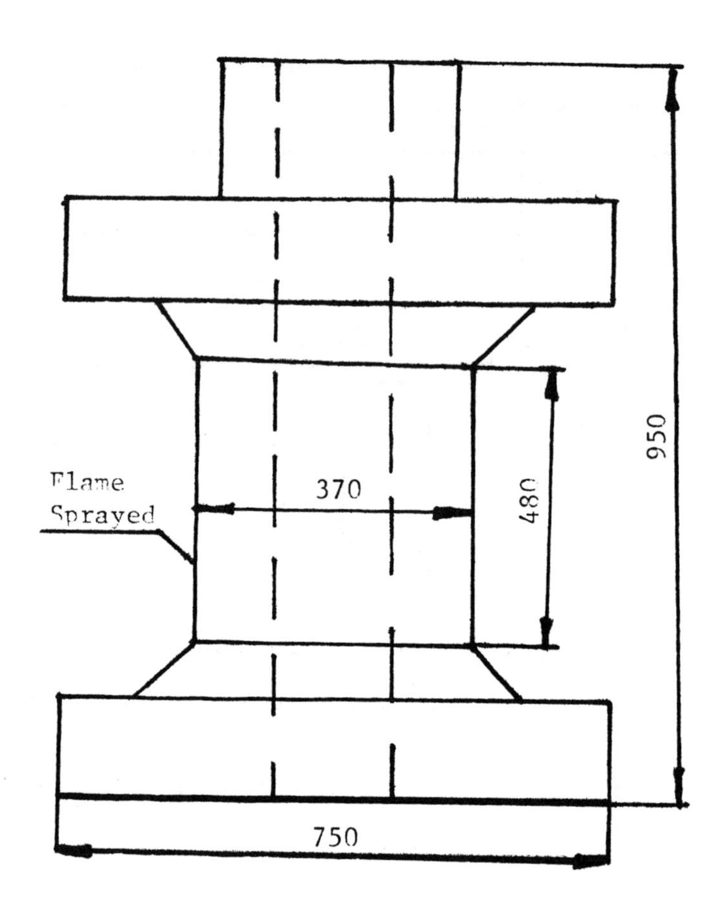

Fig. 23 Machining a Flame Sprayed Crankshaft

POLYCRYSTALLINE DIAMOND (PCD)

Like CBN, polycrystalline diamond is available as a high temperature synthesized constituent as a 0.5-1 mm thick layer on a substrate of cemented carbide. Remembering that diamond is one kind of carbon, it is unsuited for machining materials with which carbon reacts, such as iron and steel. The application field is light metals and aluminum alloys, plastics ceramics and high silicon aluminum alloys at cutting rates causing relatively low tool temperatures up to about 650°C. In Table 1 are summarized practical cutting conditions using PCD in different materials (9).

TABLE 1.

STARTING CONDITIONS FOR TURNING WITH DIAMOND POLYCRYSTALLINE TOOLS

Workpiece Material	Speed ft./min. (m/min.)	Depth of Cut in. (mm)	Feed Rate in./rev. (mm/rev.)
Aluminum alloys	3000-5000 (915-1524)	0.005-0.020 (0.13-0.50)	0.002-0.008 (0.05-0.20)
Copper alloys (including bronze)	1500-3500 (457-1067)	0.005-0.020 (0.13-0.50)	0.001-0.006 (0.03-0.15)
Sintered cemented tungsten carbide	500-1500 (152-457)	0.0005-0.005 (0.01-0.13)	0.001-0.003 (0.03-0.08)
Glass fiber/plastic composites	400-3600 (122-1098)	0.001-0.003 (0.03-0.08)	0.001-0.010 (0.03-0.25)
Carbon/plastic composites	500-2000 (152-610)	0.010-0.100 (0.25-2.50)	0.005-0.015 (0.13-0.38)
High-alumina ceramics	1500-3000 (457-915)	0.0005-0.005 (0.01-0.13)	0.001-0.004 (0.03-0.10)

FUTURE TRENDS

In the 10 years since the introduction of coated carbides, a lot has happened to machining rates. Improved carbides, better geometries and more efficient toolholders have combined to increase performance and widen application ranges of cutting tools. The chucking component represents machining as it stands today and how a modern machine should be utilized. The tooling represents a pick of common turning operations being performed (Fig. 24) in machine shops everywhere. With it, we have indicated cutting times typical of 10 years ago, just before the introduction of coated carbides, and cutting times as performed with today's coated carbides and cutting edge geometries.

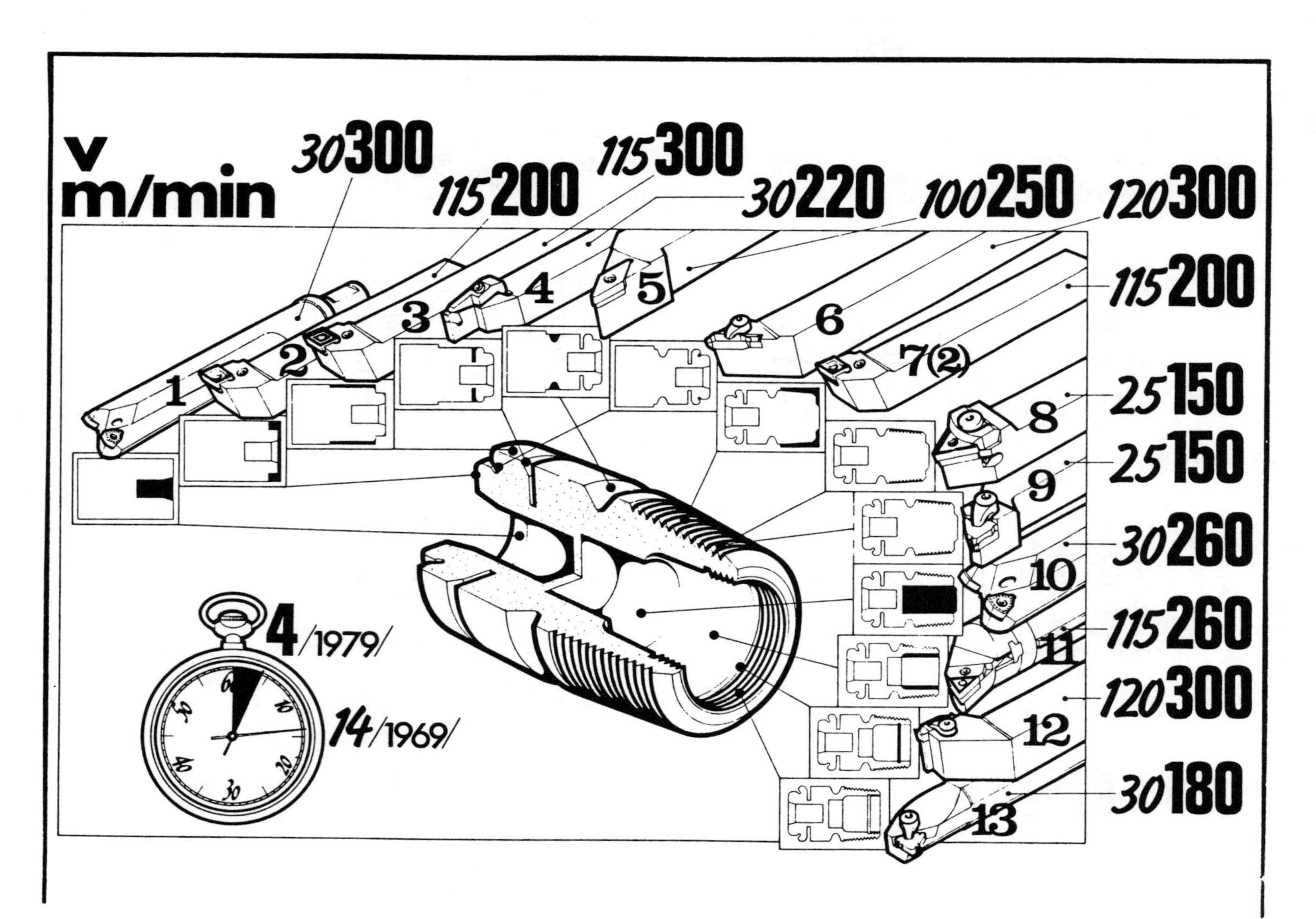
v
m/min
30 300
115 200
115 300
30 220
100 250
120 300
115 200
25 150
25 150
30 260
115 260
120 300
30 180
1
2
3
4
5
6
7(2)
8
9
10
11
12
13
4/1979/
14/1969/

Fig. 24

According to the SME-University of Michigan conducted Delphi Forecast of Manufacturing Technology (10):

1. Machining cutting speeds used in production will be increased to between 1200 m/min to 1800 m/min in 1984.

2. New abrasive materials other than Al_2O_3, SiC, BN, ZrO_2 and diamond will be introduced in production in 1987.

According to an international survey in 1977-1978 (9), there existed only two cases of production applications where the 1984 figure, according to the Delphi study, was matched or exceeded: one for cast iron at 1200 m/min. (3935 ft/min) and one for steel at 1480 m/min (4855 ft/min) using ceramics or coated carbides. Table 2 summarizes high speed machining data presently used by different organizations for coated carbides. Considerable research is carried out at still higher cutting speeds in order to investigate future possibilities. It is evident that non-machining times, such as loading, unloading and rapid traverse times, have to be reduced in order to make considerably shorter cutting times economically feasible. As the progress in this field of automation towards in-process gaging and adaptive control is rather intense there is no question about the need for new tool developments with increased wear resistance and for higher metal removal rates.

BIBLIOGRAPHY

1. Metal Cutting, Sandvik Coromant, A. Andreasson, Section 2, Tool Materials and Cutting Fluids, to be published.

2. Coated High Speed Steel Tools Are Now Put to Practical Use on a Business Basis, Nikkan Kogyo Shimbun, July 4, 1978.

3. Transition of Metal Carbides and Nitrides, L. E. Toth, Academic Press, New York, 1971.

4. Wear Characteristics of Coated Carbides, B. N. Colding, Int. Cutting Tool Day, Sandviken, Sweden, Oct. 1969, or Fertigung 1/70, Hallwag, Switzerland.

5. Wear Resistant Tool Material by TiC-Coating, N. Hallberg, Verkstaderna, Stockholm, June, 1970.

6. New Tool Materials--Wear Mechanism and Applications, W. Konig and K. Essel, Annals of the CIRP. Vol. 24/1, 1975.

7. Wear of Carbide Tools, K. Cappisch, JKA, Stockholm, 1963:1.

8. Tool Wear and Cutting Temperature of CBN Tools in Machining of Hardened Steels, N. Narutaki and Y. Yamane, CIRP Annals, Vol. 28/1, 1979.

9. High Speed Machining Possibilities and Needs, J. F. Kahles, M. Field, S. Harvey, CIRP Annals, Vol. 27/2/1978.

10. Delphi Forecast of Manufacturing Technology: Manufacturing Systems and Material Removal, B. N. Colding, L. V. Colwell, D. Smith, Society of Manufacturing Engineers, May, 1977.

TABLE 2. PRODUCTION DATA FOR HIGH SPEED MACHINING

Operation	Work Material	Speed ft./min. (m/min.)	Feed in(mm)per rev or per tooth	Depth of Cut in(mm)	Tool Life min.
Turning	Carbon Steel 220 HB	985-1150 (300-350) 1150-1475 (350-450)	.020-.039 (.6-1.0) .031-.039 (.8-1.0)	Up to .393(10) .020-.039 (.5-1.0)	5-15
Turning	4620 Steel, 200 HB	800(244) 1200-1500 (366-457)	.025(.64) .012(.3)	.250(6.4) .030(.76)	12-15 Rough 30 Finish
Turning	Cast Iron	1000-3200 (305-976)	.015-.020 (.38-.5)	.375-.3125 (9.5-7.9)	Up to 30 per edge
Turning	Steel, 250 HB	700-1200 (213-366)	.015-.030 (.37-.76)	.375-.500 (9.52-12.7)	30+ per edge
Turning and Milling	Aluminum	2000-5000 (610-1524)			10,000-60,000 in^2
Face Milling	Aluminum 380 diecast housing	12,645 (3855)		.065-.015 (1.65-.38)	
Boring	Aluminum 380 diecast housing	4967 (1514)		.080-.015 (.2-.38)	
Chamfering	Aluminum 380 diecast housing	4967 (1514)		.080-.015 (2-.38)	
Face milling	Aluminum 330 diecast outboard engine blocks	10,000 (3050)	100 ipm		18,000 castings

Reprinted from the *Carbide and Tool Journal*, September-October 1981

Ceramic Cutting Tools: Application Guidelines

by

Ronald D. Baker
Product Manager-New Materials
Metalworking Products Group
Kennametal Inc.
Raleigh, North Carolina

Productivity will be the keynote of the 80's. With industry confronted with higher manufacturing costs, the trend is toward more coated carbides and ceramic cutting tools.

Ceramic cutting tools possess high hot hardness and good chemical resistance; this allows them to be operated at much higher cutting speeds than conventional cemented or coated carbides. Ceramics are limited by their lower resistance to thermal/mechanical shock; another tool life criterion - fracture - assumes a prominent role.

Proper grade selection is achieved when the appropriate crater resistance, edge-wear resistance, and shock resistance required for the cut is matched with the physical/chemical properties of the tool material. Failure mechanisms are discussed and ceramic application guidelines are presented for machining cast iron, steel, and nickel base alloys.

INTRODUCTION

With industry confronted with higher manufacturing costs, machine tool users are becoming more concerned with increasing productivity and old equipment is being replaced with modern machine tools. Today, for example, numerically controlled (NC) products represent more than one-third of all annual dollar value sales in metal cutting machine tools (1). At the cutting edge, the trend is toward more coated carbides and ceramic cutting tools.

Ceramic cutting tools possess high hot hardness and good chemical resistance; this allows them to be operated at much higher cutting speeds than conventional cemented or coated carbides.

When operating conditions are changed to increase the metal removal rate, tool life can be expected to decrease. The key to profit in the metalworking industry is the optimum metal removal rate relative to the tool cost, machining cost, tool changing cost, and the handling cost (2, 3, 4).

It is common practice to use normal abrasive (flank) wear (Figure 1) as the tool life criterion for cemented carbide, coated carbide and ceramic cutting tools. Several authors have plotted ceramic flank wear data and constructed Taylor's Tool Life Equation (3, 5, 6, 7, 8):

$$VT^n = C \qquad \text{(Eq. 1)}$$

where V = Cutting speed (ft/min)
T = Tool life (min)
n = Exponent varies with tool/work material system (slope of tool life line)
C = Constant (represents the cutting speed that yields a one minute tool life)

In machining economics, the parameters of Taylor's Tool Life Equation (n,C) are often used to calculate:

1. The cutting speed for minimum cost (2, 3, 4).

$$V_c = \frac{C}{[\frac{1}{n} - 1]\, t_c + \frac{C_t}{C_o}} \qquad \text{(Eq. 2)}$$

2. The tool life that corresponds to the cutting speed for minimum cost.

$$T_c = [\tfrac{1}{n} - 1]\, t_c + \frac{C_t}{C_o}$$

where: C = Constant (Taylor's equation)
n = Exponent (Taylor's equation)
t_c = Tool changing time (min.)
C_t = Cost per cutting edge ($/edge)
C_o = Operating cost... direct labor rate plus overhead rate ($/min.)

Since the preceeding equations are based upon a normal abrasive (flank) wear mechanism, they are particularly useful when the economical parameters for high speed steel and cemented carbide cutting tools are being investigated...these tool materials are very resistant to chipping and fracture caused by thermal/mechanical shock. Ceramics, however, are limited by their lower resistance to thermal/mechanical shock; another tool life criterion - fracture - assumes a prominent role. Thus, Taylor's Tool Life Equation and the equations for economical production cannot be efficiently applied to ceramic cutting tools unless the selected parameters minimize thermal/mechanical stress.

The following discussion will be limited to selected wear/failure mechanisms; guidelines will be recommended for applying ceramic cutting tools:

1. Cast iron
2. Plain carbon and alloy steel
3. Stainless steel and nickel base alloys

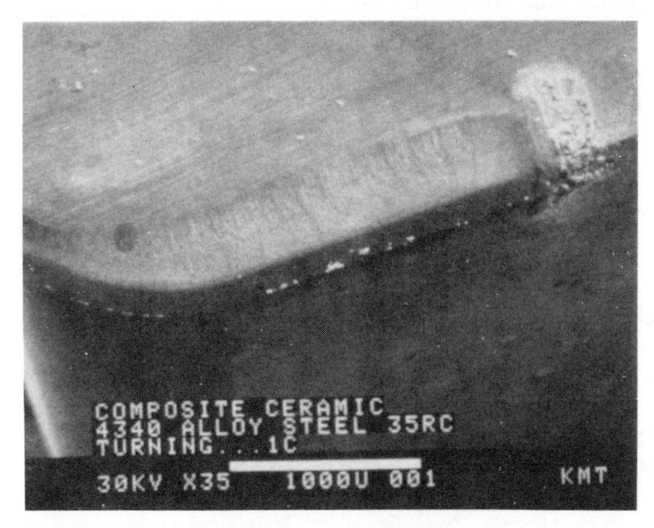

Figure 1. Cutting edge of SNG433 composite ceramic used to turn 4340 steel 45 RC with KSBR-854 toolholder. 750 sfm, 0.0145 ipr, 0.100-inch depth, cutting time = 5.0 minutes, flank wear = 0.009-inch, crater wear = 0.0008-inch. (Kennametal 35X)

TRANSVERSE RUPTURE STRENGTH — — —
COMPRESSIVE STRENGTH ———
TRANSVERSE RUPTURE STRENGTH (KSI): 0, 80, 160, 240, 320, 400
COMPRESSIVE STRENGTH (KSI): 200, 400, 600, 800, 1000
CARBIDE CERAMIC

Figure 2

DISCUSSION

Cast Iron

The wear advantage of ceramic is more evident for cast iron than for machining steel. Since cast iron has a lower shear strength than steel, chipping/fracturing is less troublesome (under similar conditions...feed, speed, depth of cut).

When compared to carbides, ceramics have lower transverse rupture strength (Figure 2) and higher hot hardness. For this reason, lower feed rates and higher cutting speeds are used. Cutting speeds can often be used to the limit of the machine and the work holding device. There are definite limitations, however, on the maximum feed rate that can be applied. Under normal conditions, feed rates exceeding 0.020 ipr (0.508 mm/rev) should not be used. The following test illustrates this point.

Test Conditions - Feed Force (Cast Iron)
Test Lathe: Lodge and Shipley (60 hp motor)
Model 2516 Superturn
Test Bar Material: Preturned Gray Cast Iron 147BHN
Cutting Speed: 1500 ft/min (450 m/min)
Depth of Cut: 0.100 inch (2.54 mm)
Feed Rate: Variable
Coolant: Dry
Cutting Tool Geometry:

a. Insert - SNG 433
  1. Hot pressed composite ceramic ($70A1_20_3/30TiC$)
  2. Ceramic coated carbide
b. Toolholder - Kennametal KSBR-854
  1. Lead angle +15°
  2. Back rake angle -5°
  3. Side rake angle -5°

TABLE 1
A Comparison of Feed Sensitivity
Hot Pressed Composite Ceramic VS Ceramic Coated Carbide

GRADE	SPEED ft/min (m/min)	DEPTH OF CUT inch (mm)	FEED in/rev (mm/rev)	VOLUME REMOVED in^3 (cm^3)	COMMENTS
Hot Pressed Composite Ceramic ($70A1_20_330TiC$)	1500 (455)	0.100 (2.540)	0.024 (0.610)	20.0 (327.7)	Fracture (Fig. 3)
Ceramic Coated Carbide	1500 (455)	0.100 (2.540)	0.024 (0.610)	21.0 (344.1)	0.0059 inch Flank wear
			0.0029 (0.737)	20.4 (334.3)	
			0.036 (0.914)	20.4 (334.3)	
			0.042 (1.067)	19.6 (321.2)	0.0087 inch Flank wear
			0.049 (1.245)	19.6 (321.2)	
			0.055 (1.397)	18.9 (309.7)	
			0.064 (1.626)	18.9 (309/7)	0.0135 inch flank wear (Fig. 4)

TABLE II
A Comparison of Wear Resistance
Hot Pressed Composite Ceramics VS Ceramic Coated Carbide

GRADE	CUTTING TIME min	FLANK WEAR inch (mm)	NOTCH WEAR inch (mm)	VOLUME REMOVED in^3 (cm^3)
Hot Pressed Composite Ceramic ($70Al_2/0_3/30TIC$) Refer to Figure 5	1.0	90.004 (0.102)	0.004 (0.102)	14.0 (229.0)
	10.0	0.006 (0.152)	0.006 (0.152)	144.0 (2360.0)
	22.0	0.008 (0.203)	0.014 (0.356)	317.0 (5195.0)
	30.0	0.009 (0.229)	0.014 (0.356)	432.0 (7079.0)
	42.0	0.010 (0.254)	0.023 (0.584)	605.0 (9914.0)
	50.0	0.012 (0.305)	0.026 (0.660)	720.0 (11799.0)
	62.0	0.013 (0.330)	0.031 (0.787)	893.0 (14634.0)
	76.0	0.020 (0.508)	0.036 (0.914)	1094.0 (17927.0)
Ceramic Coated Carbide Refer to Figure 6	1.0	0.005 (0.127)	0.005 (0.127)	14.0 (229.0)
	2.0	0.011 (0.279)	0.016 (0.406)	29.0 (475.0)
	3.3	0.072 (1.829)	Blend to 0.072 (1.829)	48.0 (786.0)

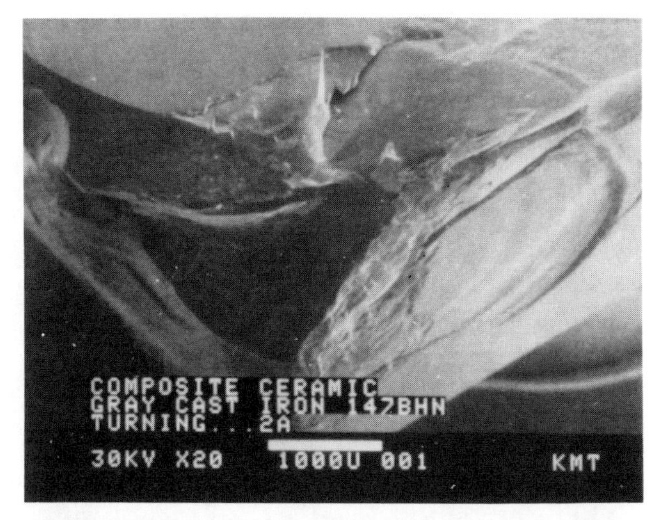

Figure 3. High feed rate (0.024 ipr) fractured the cutting edge of SNG433 composite ceramic used to turn gray cast iron 147BHN with KSBR-854 toolholder. 1500 sfm, 0.100-inch depth, coolant = dry. (Kennametal 20X)

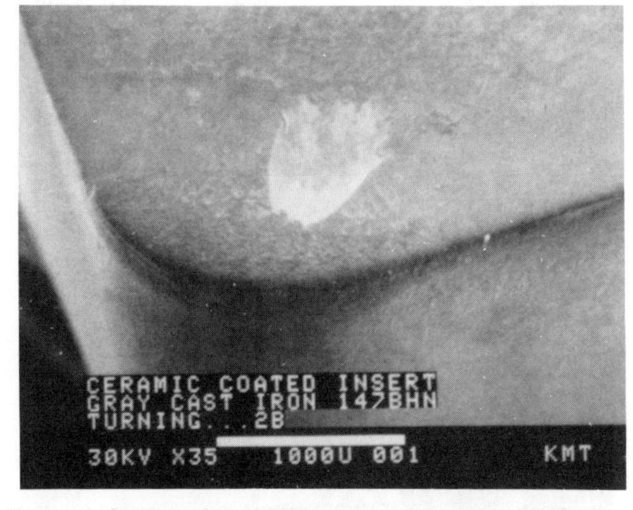

Figure 4. Cutting edge of SNG433 ceramic coated carbide that withstood progressive feed rates up to 0.064 ipr. Insert was used to turn gray cast iron 147 BHN with KSBR-854 toolholder 150 sfm, 0.100-inch depth, coolant = dry. (Kennametal 35X)

The data represented in Table I clearly illustrates the tivity of hot pressed composite ceramic ($70Al_2O_3/30TiC$) to feed. At the higher feed rates (above 0.024 ipr, 0.610 mm/rev) ceramic coated carbides offer a definite performance advantage. The following test compares the wear resistance of these grades; Table II illustrates the performance advantage of hot pressed composite ceramic when lower feed rates are used on continuous cuts (Figures 5, 6).

Test Conditions - Wear Resistance (Cast Iron)

Test Lathe: Lodge and Shipley (60 hp motor)
Model 2516 Superturn

Test Bar Material: Preturned Gray Cast Iron 190BHN
Cutting Speed: 1200 ft/min (365m/min)
Depth of Cut: 0.100 inch (2.540 mm)
Feed Rated: 0.010 in/rev (0.265 mm/rev)
Coolant: Dry
Cutting Tool Geometry:

a. Insert - SNG433
  1. Hot pressed composite ceramic ($70Al_2O_3/30TiC$)
  2. Ceramic coated carbide

b. Toolholder - Kennametal KSBR-854
  1. Lead angle +15°
  2. Back rake angle -5°
  3. Side rake angle -5°

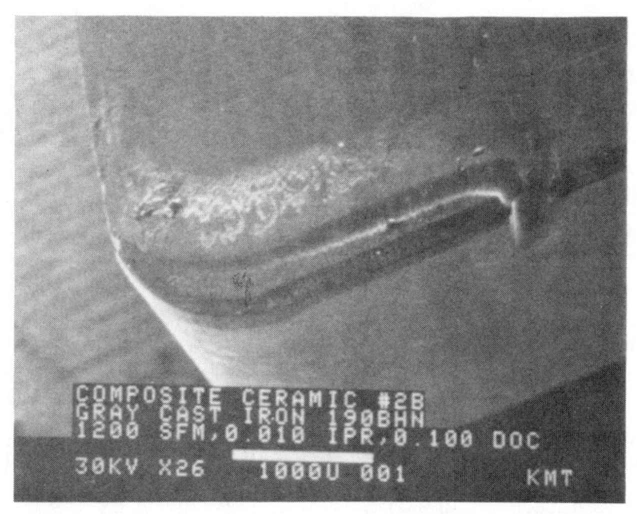

Figure 5. Cutting edge of SNG433 composite ceramic used to turn gray cast iron 190 BHN with KSBR-854 toolholder. Coolant = dry, cutting time = 76.0 minutes, flank wear = 0.020 inch, notch wear = 0.036 inch. (Kennametal 26X)

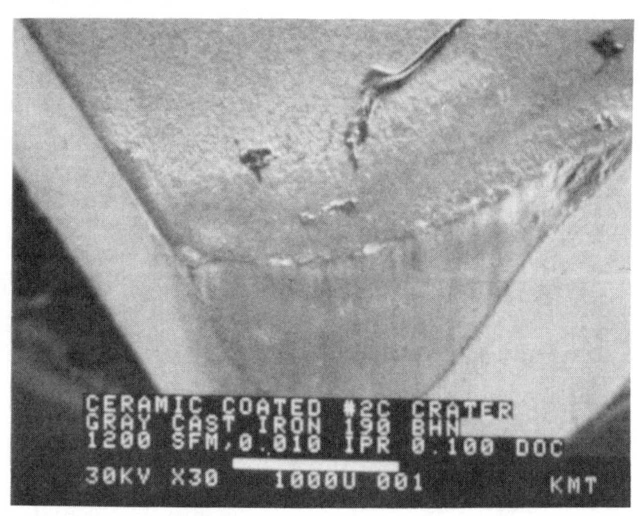

Figure 6. Cutting edge of SNG433 ceramic coated carbide used to cut gray cast iron 190 BHN. Coolant = dry, cutting time = 3.3 minutes, flank wear = 0.072 inch. (Kennametal 30X)

TABLE III
A Comparison of Crater Resistance by Grade

GRADE	SPEED ft/min (m/min)	FEED in/rev (mm/rev)	DEPTH OF CUT inch (mm)	CUTTING TIME min	VOLUME REMOVED in³ (cm³)	FLANK WEAR in (mm)	CRATER WEAR in (mm)	REFERENCE PHOTOGRAPH
C6 Uncoated Carbide	750 (230)	0.008 (0.203)	0.100 (2.540)	1.0	7.0 (118.0)	0.024 (0.610)	0.0031	Figure 7
Ceramic Coated Carbide	750 (230)	0.008 (0.203)	0.100 (2.540)	3.0	22.0 (360.0)	0.039 (0.991)	0.0006	Figure 8
Hot Pressed Composite Ceramic	750 (230)	0.008 (0.203)	0.100 (2.540)	20.0	144.0 (2360.0)	0.021 (0.533)	0.0019	Figure 9

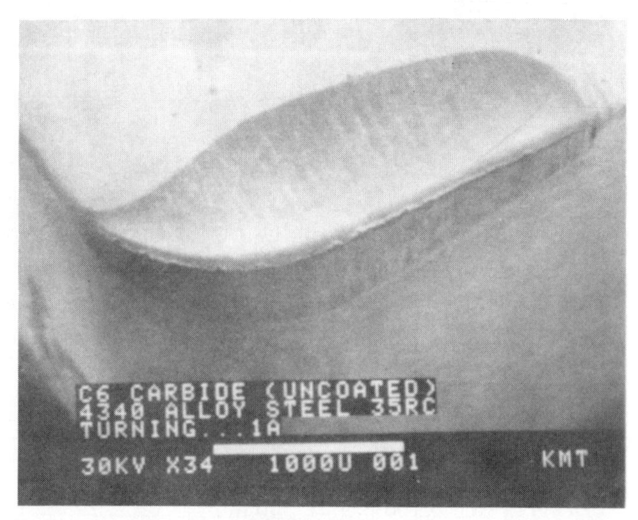

Figure 7. Cutting edge of SNG433 uncoated carbide (C6) used to turn 4340 steel 35 RC with KSBR-854 toolholder. 750 sfm, 0.008 ipr, 0.100-inch depth, coolant = dry, cutting time = 1.0 minute, flank wear = 0.024-inch, crater wear = 0.0031-inch. (Kennametal 34X)

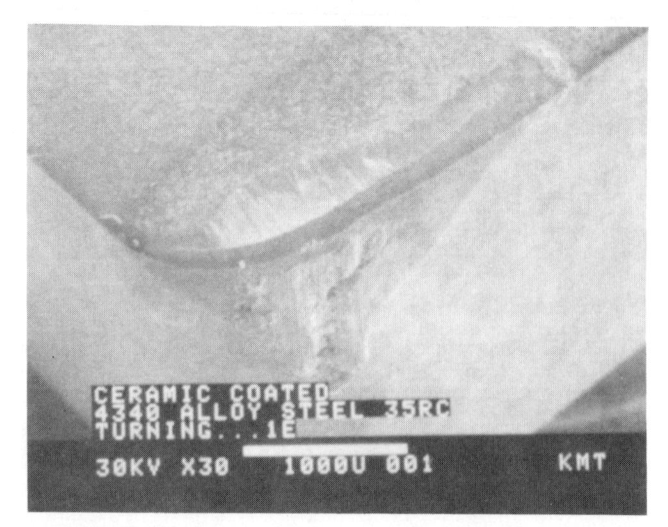

Figure 8. Cutting edge of SNG433 ceramic coated carbide used to turn 4340 steel 35 RC with KSBR-854 toolholder, 750 sfm, 0.008 ipr, 0.100-inch depth, coolant = dry, cutting time = 3.0 minutes, flank wear = 0.039-inch, crater wear = 0.0006-inch. (Kennametal 30X)

Steel

Reliable/ predictable performance is a major criterion for the selection of any tool material. Ceramic cutting tools have experienced significant technological advances, however, severe interrupted cuts should still be avoided.

Crater wear is a common wear mechanism that often leads to failure when uncoated carbide is used to machine the plain carbon and alloy steels.

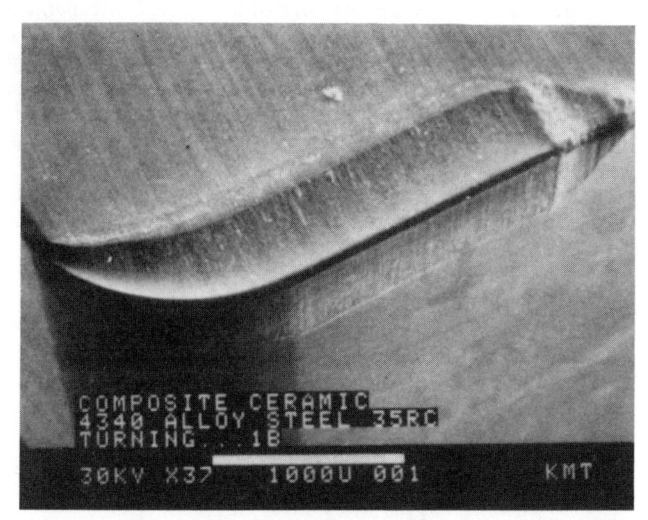

Figure 9. Cutting edge of SNG433 composite ceramic used to turn 4340 steel with KSBR-854 toolholder. 750 sfm, 0.008 ipr, 0.100 inch depth, coolant = dry, cutting time = 20.0 minutes, flank wear = 0.021-inch, crater wear = 0.0019-inch. (Kennametal 37X)

Figure 10a. Thermal/mechanical cracks are invisible under low magnification. (Kennametal 30X)

Test Conditions - Crater Resistance (Alloy Steel)

Test Lathe: Lodge and Shipley (60 hp motor)
Model 2516 Superturn
Test Bar Material: Preturned 4340 Alloy Steel 35 Rc
Cutting Speed: 750 ft/min (230m/min)
Depth of Cut: 0.100 inch (2.540 mm)
Feed Rate: 0.008 in/rev (0.203 mm/rev)
Coolant: Dry
Cutting Tool Geometry:
- a. Insert - SNG433
- b. Toolholder - Kennametal KSBR-854
  1. Lead angle +15°
  2. Back rake angle -5°
  3. Side rake angle -5°

A review of the data in TABLE III reveals that hot pressed composite ceramics (70A1$_2$0$_3$30TiC) offer significant improvements in both crater and flank wear resistance.

The premature failure of ceramic cutting tools is often chipping/fracturing caused by a combination of thermal/mechanical shock. When ceramic cutting tools are used to turn steel, small parallel cracks often appear in the crater region (Figures 10, 11) perpendicular to the cutting edge. Pekelharing observed that this cracking occurs when various combinations of speed and feed are exceeded (9). In a few cases, fracture occurred "...in such a way that the break made use of one or more of these cracks". Similar tests have been made in Kennametal's laboratory.

Test Conditions - Thermal/Mechanical Cracks (Alloy Steel)

Test Lathe: LeBlond (125 hp motor)
Model 2516 NF Test Lathe
Test Bar Material: Preturned 4140 Alloy Steel (220BHN)
Cutting Speed: 150 ft/min
Depth of Cut: 0.100, 0.200 inch
Feed Rate: 0.010 ipr, 0.020 ipr
Cutting Time: 1.0 minute
Coolant: Dry
Cutting Tool Geometry:
- a. Insert - SNG454
  Hot pressed composite ceramic (70A1$_2$0$_3$/30TiC)
- b. Toolholder - Kennametal KSKNR-206D
  1. Lead angle +15°
  2. Back rake angle -5°
  3. Side rake angle -5°

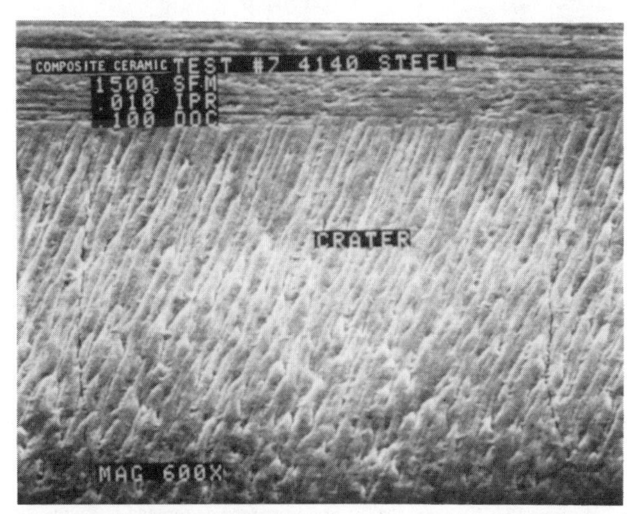

Figure 10b. Thermal/mechanical cracks generated by light load are visible under high magnification. For comparison refer to Figure 11b. (Kennametal 600X)

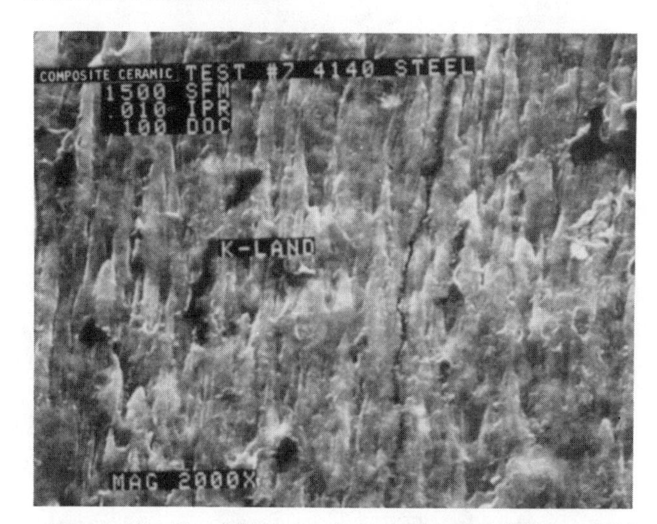

Figure 10c. In addition to the crater region (Figure 10b), thermal/mechanical cracks are also generated in the K-land area. For test conditions refer to Figure 10. (Kennametal 2000X)

Figure 10. Thermal/Mechanical Cracks (Alloy Steel). Cutting edge of SNG454 composite ceramic used to turn 4140 steel 220BHN with KSKNR-2060 toolholder. 1500 sfm, 0.010 ipr, 0.100-inch depth, cutting time = 1.0 minute, coolant = dry.

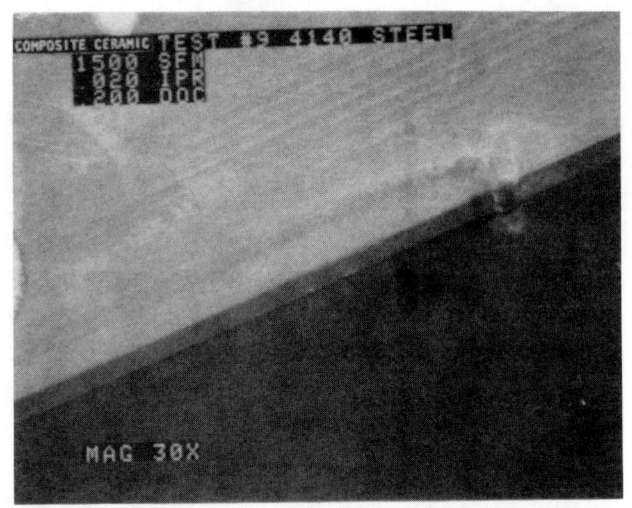

Figure 11a. Thermal/mechanical cracks are invisible under magnification. (Kennametal 30X)

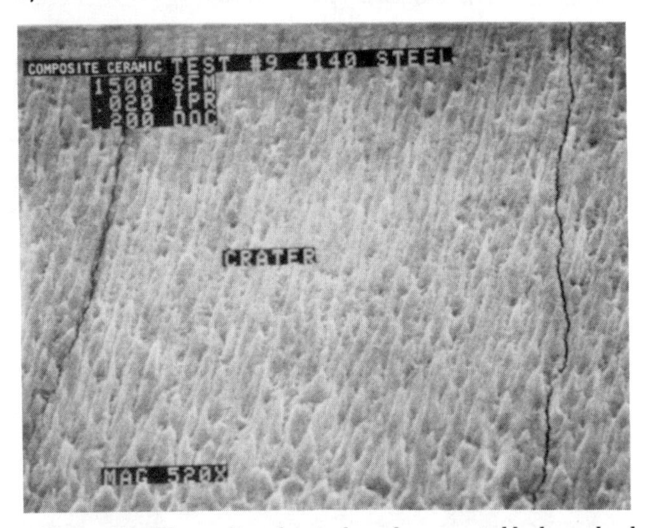

Figure 11b. Thermal/mechanical cracks generated by heavy load are visible under high magnification. For comparison refer to Figure 10b. (Kennametal 520X)

Figure 11c. In addition to the crater region (Figure 11b), thermal/mechanical cracks are also generated in the K-land area. For test conditions refer to Figure 11. (Kennametal 1000X)

Figure 11. Thermal/Mechanical Cracks (Alloy Steel). Cutting edge of SNG454 composite ceramic used to turn 4140 steel 220BHN with KSKNR-206D toolholder. 1500 stm, 0.020 ipr, 0.200-inch depth, cutting time = 1.0 minute, coolant = dry.

In general, the number and size of the thermal/mechanical cracks increase with respect to time and load. Eventually the cracks will break into each other and macroscopic particles of tool material will become dislodged. This results in an increased rate of chipping that may lead to fracture. Optimum performance is achieved by reducing the load (feed, depth of cut) at the cutting edge.

This failure mechanism is illustrated in the following scanning electron photomicrographs of the ceramic inserts used for the Test Conditions - Thermal/Mechanical Cracks (Alloy Steel):

1. Figures 10a, 11a
   Magnification 30X
   Thermal/mechanical cracks are invisible under low magnification.
2. Figures 10b, 11b
   The cracks generated by the heavier load (520X) are larger than those generated by the lighter load (600X). At the same magnification, the difference in crack width would be more pronounced.
3. Figures 10c, 11c
   Cracks are also generated in the K-land area.

Nickel Base Alloys

Ceramic cutting tools, primarily the hot pressed composite ceramics ($70Al_2O_3/30TiC$), are gaining acceptance for the machining of stainless steels and nickel base alloys (Inconel 718, Incoloy 901, etc.). Typical conditions range from 400 ft/min (120 m/min) to 800 ft/min (245 m/min), 0.006 ipr (0.152 mm/rev) to 0.014 ipr (0.356 mm/rev, and 0.030 inch (0.762 mm) to 0.150 inch (3.810 mm) depth of cut. In order to improve surface finish and reduce notching at the depth of cut line, steady flood coolant (water miscible) is normally used.

When ceramic cutting tools are used to machine this group of work materials, the edge preparation should be a small K-land (typically 0.008 inch X 20 degrees) or a small radius hone (typically 0.002 to 0.004 inch). This smaller edge preparation provides a cutting action that more closely resembles the shearing effect of positive rake tooling.

Since castings/forgings of nickel base alloy are very expensive, every effort should be made to avoid fracturing the cutting tool. For this reason, a tool change criterion of 0.040 inch (1.016mm) depth of cut notch is recommended. As the width of the depth of cut notch increases, the probability of fracture also increases. In Figure 12 a severe crack developed after a 0.059 inch (1.499 mm) depth of cut notch was generated:

1. Test Lathe - LeBlond (125 hp motor)
   Model 2516 NF
2. Test Bar Material - Preturned Incoloy 901 (42 Rc)
3. Cutting Speed - 750 ft/min (230 m/min)
4. Depth of Cut - 0.080 inch (2.032 mm)
5. Feed Rate - o.008 inch (0.203 mm)
6. Coolant - Trim Regular 30:1
7. Cutting Tool Geometry -
   a. Insert - SNG433
      Hot pressed composite ceramic ($70Al_2O_3/30TiC$)
   b. Toolholder - Kennametal KSBR-854
      1. Lead angle -15°
      2. Back rake angle -5°
      3. Side rake angle -5°

CONCLUSIONS

Productivity will be the Keynote in the 80's. Efficient metal cutting consists of three elements acting as a single chip-making system. The workpiece, machine tool, and cutting tool are interrelated; these elements make up the productivity triangle.

Proper grade selection is achieved when the appropriate crater resistance, edge wear resistance, and shock resistance

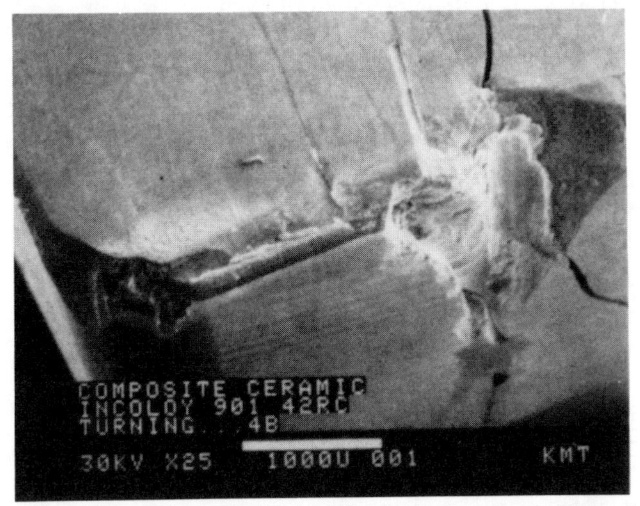

Figure 12a. Severe crack caused by depth of cut notch. View of flank face. (Kennametal 25X)

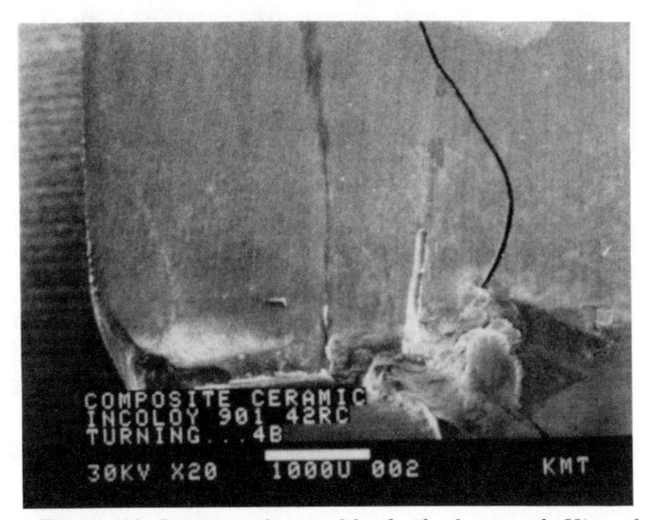

Figure 12b. Severe crack caused by depth of cut notch. View of rake face. (Kennametal 20X)

Figure 12. Severe crack developed after a 0.059-inch depth of cut notch was generated. Cutting edge of SNG433 composite ceramic used to turn Incoloy 901 42 RC with KSBR-854 toolholder. 750 stm, 0.008 ipr, 0.080-inch depth, coolant = Trim Regular 30:1.

TABLE IV
APPLICATION GUIDELINES
STEEL AND CAST IRON

	FINISHING (.006-.010 ipr)	SEMIFINISHING (.010-.014 ipr)	ROUGHING (.014-.020 ipr)
STEEL — 3,000 - 1,000 sfm Below 35 RC	.006-.008 x 30°/20° (1st - K060) (2nd - K090)	.008-.010 x 30°/20° (1st - K060) (2nd - K090)	.012-.014 x 30° (1st - K090) (2nd - K060)
STEEL — 1,500 - 500 sfm Above 35 RC	.006-.008 x 30°/20° (1st - K090) "A" hone optional (2nd - K060)	.008-.010 x 30°/20° (1st - K090) "A" hone optional (2nd - K060)	.012-.014 x 30° plus "A' hone (1st - K090) (2nd - K060)
CAST IRON — 4,000 - 1,000 sfm Below 250 BHN	"A" hone or .006-.008 x 30°/20° (1st - K060) (2nd - K090)	.008-.010 x 30°/20° (1st - K060) "A" hone optional (2nd - K090)	.012-.014 x 30° plus "A" hone (1st - K060) (2nd - K090)
CAST IRON — 2,000 - 700 sfm Above 250 BHN	"A" hone or .006-.008 x 30°/20° (1st - K060) (2nd - K090)	.008-.10 x 30°/20° K060 or K090 "A" hone optional	.012-.014 x 30° plus "B" hone (1st - K090) (2nd - K060)

*K060 - Cold Pressed Pure Al_2O_3
*K090 - Hot Pressed $70Al_2O_3/30TiC$

*Trademarks owned by Kennametal Inc.

TABLE V
APPLICATION GUIDELINES
NICKEL BASE ALLOYS AND STAINLESS STEELS

	FINISHING (.003-.006 ipr)	SEMIFINISHING (.005-.008 ipr)	ROUGHING (.008-.014 ipr)
NICKEL BASE ALLOYS 40-45 Rc 1,000 - 500 sfm	.002-.004 x 20° or "A" hone (K090)	.006-.008 x 30°/20° or"B" hone (K090)	.010-.012 x 20° or"C" hone (K090)
STAINLESS STEELS 300-400 Series 1,500 - 700 sfm	.002-.004 x 20° or "A" hone (K090)	.006-.008 x 30°/20° or "B" hone (K090)	.010-.012 x 20° or "C" hone (K090)

*K060 - Cold Pressed Pure Al_2O_3
*K090 - Hot Pressed $70Al_2O_3/30TiC$

*Trademarks owned by Kennametal Inc.

required for the cut is matched with the physical/chemical properties of the tool material. Guidelines for applying ceramic cutting tools are presented in Table IV (steel, cast iron) and Table V (stainless steel, nickel base alloys).

REFERENCES

1. Green, R.V., "New Era in Manufacturing," Tooling and Production (January 1, 1981), pp. 66-71
2. Gilbert, W.W., "Economics of Machining," Machining - Theory and Practice, (Published lectures), American Society for Metals, 1950,, pp. 465 - 485
3. Brierley, R.G. and Siekmann, H.J., **Machining Principles and Cost Control,** New York: McGraw-Hill Book Company, 1964
4. Bhattacharyya, A. and Ham, I., Design of Cutting Tools, American Society of Tool and Manufacturing Engineers, 1969
5. Ogawa, K., Furukawa, M. and Hora, Y., "Cutting Performances and Practical Merits of Carbide Ceramics", Nippon Tungsten Review, pp. 96 - 104
6. Whitney, E.D., "Ceramic Cutting Tools", Powder Metallurgy International, Volume 6, Number 2, 1974, pp. 73 - 76
7. King, A.G. and Wheildon, W.M., **Ceramics in Machining Processes,** New York, Academic Press, 1966
8. Doi, Dr. A., "New Cutting Tool Materials and Designs", Metalworking Engineering and Marketing, (March 1980), pp. 50 - 53
9. Pekelharing, A.J., "A Story About the Cracking of Ceramic Tools When Cutting Steels", CIRP Ann. XI, pp. 25 - 36

Presented at the SME 1975 International Tool and Engineering Conference, April 1975

Productivity Performance of Ceramic Cutting Tools

by Charles H. Kahng
and W.C. Koegler
Michigan Technological University

Productivity performance of ceramic cutting tool materials was reviewed for turning gray cast iron, malleable iron, and low carbon steel, as well as for face milling cast iron.

The superiority of ceramics over cemented carbide was confirmed for limited applications.

INTRODUCTION

The major goal of the development in metal cutting tool materials is to achieve a reduction in production cost. Increased wear resistance, higher hot hardness and greater impact toughness are essential to this reduction. Ever since the alumina ceramics were developed in the 1940's as a new cutting tool material, great interest has been shown in this particular material. The combination of very high hardness, chemical inertness, and low toughness limits its application. However, remarkable improvements have been made in the past decade through research, and the position of this tool material in the metal cutting field has become better established.

It was felt that a review of the status of the ceramic cutting tool from the productivity performance point of view would be significant to the metal cutting industry. In this paper, a description is made of the recent status of the ceramic cutting tool material, based on published data, both American and foreign, and our own investigation.

1. Composition and Properties

Ceramics have a unique composition compared with other cutting tool materials. Opitz [1], as reported in Table 1 indicates a brief comparison in the chemical composition of various cutting tool materials. As shown in the table, ceramics have aluminum oxide as the major composition instead of metallic carbides, and the metallic bonding agent has totally disappeared. The significant advantages of sintered oxide are very low heat conductivity and extremely high compressive strength. However, the disadvantage is relatively low transverse rupture strength. Generally, however, where shock and vibration are not factors, ceramic tools can outperform carbide cutting tools.

In recent years, the ceramic tool has been remarkably developed in order to give a wide range of physical properties. By combining metallic carbide with aluminum oxide the tool gains higher toughness without changing the wear resistance characteristics. The relative physical properties of ceramic tools, as compared with other tool materials, offer a good guide to their applications.

Table 2 indicates important physical properties of different cutting tool materials. The specific weight of ceramics and the newly developed types of ceramic tools are lower than high speed steel.

Microhardness, which characterizes the solidity of total particle, has similar values for cemented carbide tools at room temperature. However, ceramic cutting tools retain their high hardness value up to 1400°F, while the hardness of other tool materials is seriously decreased at this temperature. The refractory property of ceramics is used to advantage at higher cutting speeds where temperatures as high as 1500°F can be encountered. The bending strength, which can be used as a parameter for impact toughness, is relatively low in ceramic tools compared with other materials. This means ceramic tools are very sensitive to impact operations, especially interrupted cutting.

The compressive strength is also lower in ceramics than cemented carbide. When both bending and compression stresses are applied together on the tool tip, a relatively small amount of force will cause the failure of ceramics.

The modulus of elasticity of ceramics is generally lower than cemented carbide but higher than high speed steel. Since the elastic limit of ceramic tools is low at normal room temperature, the failure is caused by exceeding the elastic limit. Ceramics have a large thermal expansion ratio, with this value lying between those of carbide and high speed steel.

Specific heat capacity in ceramics is between that of high speed steel and carbide at room temperature, however, when the temperature is raised very high, this capacity dropped by half of that of the other tool materials. This capacity of dropping by only half of other tools is an advantage for ceramic cutting tools because a large portion of generated temperature in metal cutting is carried away by the chip so that the tool keeps a relatively low temperature.

To summarize, the ceramic tool has high hardness, chemical inertness and low toughness limits, and is always an economical cutting tool material when properly applied.

2. Trend of Development in Ceramic Cutting Tools

According to Dworak [2], the development of ceramics indicated a unique trend as years have progressed. The grain size was decreased from 5μm to 2μm during the past 20 years and bending strength has been doubled. At the beginning of the 1970's, the impact toughness at higher temperatures was remarkably improved by applying several additives such as TiO, MgO, TiC, Cr_3C_2, WC + Co, NiO, etc.

As shown in Figure 1, the cutting speed of a turning operation has been steadily increased as the new developments in cutting tool materials have been made. At the present, ceramic cutting tools can generally obtain 500 m/min (1640 fpm) of cutting speed. In special cases, for example the turning of brake drums with ceramic tools, can reach up to 740 m/min (2427 fpm) using a multispindle automatic lathe, and 1050 m/min (3445 fpm) with N/C machine tools [3].

3. Causes of Failure of Ceramic Tools

Wear characteristics and causes of failure of ceramic cutting tools are unique. Thijssen [4] investigated the causes of several failures of ceramic tools.

The first area of consideration is the feed groove on the end cutting edge face. When a feed rate exceeded more than 0.5 mm/rev (0.019 ipr) after a short cutting time, a groove could be observed at a point as is shown in Figure 2. This point is on the face at the junction between the previously machined surface and the portion of the workpiece being cut. This groove is caused by concentrated stress due to a hardness increase on the previously machined surface. The groove causes, in many cases, a complete failure in combination with thermal stress.

There are several methods to eliminate this feed groove; one is controlling feed rate during cutting using a variable feed drive. Another is using a larger nose radius by two or three times the fixed feed rates. Through this arrangement concentrated and very highly heated points and crater zones will be separated and thermal stresses can be eliminated.

A narrow negative land, honed onto the cutting edge of the tool, substantially decreases the tendency for the tool tip to fracture. This negative land effect also distributes thermal stresses and increases the strength of the edge. Another advantage is optimal influence on the chip curl formation.

Often axial failure through the crater zone on the tool tip is observed. This failure has a very random figure and has very little repeatability. The cause could be an increased feed rate, resulting in a twisting moment, e.g., a concentrated stress on the point where crater wear began. The other cause would be the error in flatness on the surface of the ceramic tip. A lack of a proper supporting seat for the cutting tool tip could also be a cause of the failure.

When a large feed rate is applied in turning, a failure similar in appearance to axial seprating failure can be seen on the rake face. This appearance has a different origin and is called failure on the rake face, because it begins from a point on the rake face. Too narrow a clamp plate for the tool tip could be the cause of the failure.

4. Performance of Ceramic Tools

a) Surface finish and dimensional accuracy

Besides superiority of wear resistance, ceramic cutting tool materials have unique advantages such as maintainability of dimensional tolerance and confined variation of surface finish through long cutting times. According to Mori and Watanabe [5], a comparison of two kinds of tool materials with regard to progress of: dimensional tolerance and surface roughness are shown in Figure 3 and Figure 4, respectively.

After a cutting length of 20 km (65,620 ft.), the dimensional variation shows more than 0.1 mm (.0039 in.) in diameter. When turning AISI 4140 steel with a cemented carbide tool for the same length of cut, a ceramic cutting tool could reduce this variation by half. From this result, it can be seen that very small radial wear of the tool tip assures a sufficient dimensional accuracy, which is a most important advantage for long run and especially precision machining operations, due to elimination of tool point set up.

Generally, the feed groove on the side cutting edge increases its depth as cutting time is increased. The generated surface roughness will be influenced by the depth of the feed groove throughout the cutting operation. Ceramic cutting tools demonstrated a steady increase in their generated surface roughness, while cemented carbide showed a large variation. There are many research reports stating that a ceramic tool has numerous advantages in terms of the produced surface roughness. The reason for this can be explained by the fact that higher wear resistance and the honed land edge may cause reduction in the depth of the feed groove.

b) Turning gray cast iron

The German automobile industry, in 1973, utilized ceramic cutting tools for 8% of their total hard cutting tool materials.

Orelio [6] reported that an automobile factory had great success with the following operation:

Work: Gray cast iron (BHN = 195)
Brake drum length 43 mm (1.67 in.)
and diameter 250 mm (9.75 in.)

Cutting Conditions:

Cutting speeds - between 1000 m/min ~ 1400 m/min (3300 ~ 4593 fpm)
Feed rate - 0.4 ~ 0.5 mm/rev (.0157 ~ .0187 ipr)
Tool life - VB (average) 0.16 mm (.006 in)
VB (max.) 0.27 mm (.010 in)

Production rate- 35 drums/hour

Anschütz [7] reported that a ceramic cutting tool can reach a cutting speed when turning cast iron, of up to 2000 m/min (6562 fpm) and feed rates up to 0.4 mm/rev (.0157 ipr). However, in order to minimize impact stress at this condition, tool position against the workpiece should be so arranged that the entry and exit cuts keep their slope.

c) Turning malleable iron

A systematic investigation for turning malleable cast iron (HV600) with a ceramic cutting tool was reported by Klicpera [8]. He used a tool of hot pressed aluminum (oxide) mixed with titanium carbide for the investigation into the cutting of very hard cast iron.

The effect of rake angle on the tool life under varied cutting speeds was investigated and the optimum angle was found to be -13°. The back rake angle was determined to be best at -6° with the side cutting edge being maintained as small as possible and allowing the total length of cutting edge to apply cutting action.

A tool life comparison test was made with cemented carbide grade of K01 (C-6) under varied feed rates, and it was found that cutting speed is four times faster than for a carbide tool to reach the same tool life. Because a ceramic tool has less bending strength than a carbide tool, the feed rate influenced the tool life for a ceramic tool more than for a carbide tool.

Figure 8 presents a summary of optimum values in connection with hardness of cast iron. Based on the result from Figure 6, the calculation was made under the following assumptions:

Labor and overhead cost = 32 DM/hr ($12.80/hr)
Tool changing time = 3 min.
Tool costs, ceramics = 8 DM ($3.20)

The optimum tool life is independent of the hardness of the work material. Since optimal tool life as a function of the machining cost is almost steady in a large range of hardnesses, fixed cutting conditions for a certain type of cast iron can be applied to machining of other types of materials. The cutting speed and feed rate have a decreasing trend as hardness increases. The machining cost based on the cutting conditions showed a rapid increasing trend as the hardness is increased. For example, the machining cost of 0.80 DM/dm^3 ($$.0053/in^3$) gray cast iron having HV200 rose to 5.00 DM/dm^3 ($$.033/in^3$), when machining malleable iron having HV600.

d) Performance comparison with other cutting tool materials

Michigan Technological University conducted a comparison test turning AISI 1019 (R_B = 75) using three different tool materials [9]. The results indicated that, after a cutting time of 4 minutes, the cemented carbide tool had already reached its tool life criterion, while ceramic tools lasted more than 15 minutes before having the same amount of tool wear. Titanium carbide tool had superior performance for this material. The tests were repeated twice and each test showed very close results. However, when finish turning cast iron, under a cutting speed of 1300 fpm, the ceramic tool demonstrated the most superior performance; i.e., the obtained tool life was 5 times that of cemented carbide and 3 times that of titanium carbide. This result is limited to finish turning conditions only and a comparison of the three cutting tool materials in rough cutting cast iron can be evaluated only after extensive investigation.

Heavy cutting of gray cast iron (BHN = 220) also indicated a superior performance of ceramic tools compared with cemented carbide. Approximately 3 times faster cutting speed is obtained by a ceramic tool against a carbide tool under the same tool life, as shown in Figure 11 [10].

e) Face Milling Performance

The recent development of mixing ceramics with titanium carbide performed very well as a throwaway insert for face milling. Hot pressed ceramics had longer tool life, especially when compared to sintered ceramics at room temperature with a cutting speed below 300 m/min (984 fpm) [11].

The general trend for ceramic cutting tools is that they are becoming capable of operating in interrupted cutting.

CONCLUSIONS

1. Ceramic cutting tools have unique chemical and physical properties which give them their superior performance as well as disadvantages as cutting tools.

2. Several causes of failure of ceramic tools were reviewed based upon publications.

3. Ceramic tools have favorable performance from the point of view of dimensional accuracy and surface roughness variations.

4. Research results from foreign literature for turning malleable iron were introduced and it was concluded that ceramic cutting tools are able to run at cutting speeds of four times faster than cemented carbide tools under a fixed tool life.

5. Turning low carbon steel and cast iron with ceramic, cemented carbide, and titanium carbide tools confirmed the superiority of ceramics over other tool materials.

6. The ability of ceramic tools to take interrupted cuts has shown remarkable improvement.

REFERENCES

1. H. Opitz, P. Brammertz, K. F. Meyer, "Untersuchungen an keramishen Schneidstoffen," 1963, Westdutscher Verlag, Köln and Opladen, Germany.

2. V. Dworak, "Herstellung und Eigenschaften der Schneidkeramik," Werkzeugmaschine International 2 (1972) 4, pp. 24-26.

3. H. Tully and R. Schütze, "Stand und Tendenzen der Spanenden Formgebung durch Drehen," Industrie-Anzeiger, Vol. 95 (1973), No. 42, pp. 872-875.

4. H. Thijssen, "Schruppbearbeitung von Stahl mit oxidkeramischen Schneidwerkzeugen," Werkstatt und Betrieb, Vol. 102 (1969), No. 2, pp. 79-82.

5. J. Mori and T. Watanabe, Oyokikaikogakul (in Japanese), Sept. 1973, pp. 114-118.

6. J. M. B. Orelio, "Research for Industry/Machining Cast Iron with Ceramics," Annals of the C.I.R.P., Vol. 21 (1972), pp. 35-36.

7. E. Anschütz, "Erweiterte Anwendungsbereiche für Schneidkeramik," Werkzeugmaschine International, 1974 (4), pp. 27-29.

8. U. Klicpera, "Die Anwendung von Schneidkeramik beim Zerspanen von Hartguss," Industrie Anzeiger, Vol. 96 (1974), pp. 357-358.

9. R. L. Larson, "Investigation of the Performance of Three Cutting Tool Materials," M.S. Thesis, Michigan Technological University, 1974.

10. M. Tanaka, Science of Machine (in Japanese), Vol. 23 (1971), No. 7, pp. 35-38.

11. M. Tanaka, Science of Machine (in Japanese), Vol. 29 (1971), No. 8, pp. 42-44.

TABLE 1

IMPORTANT COMPOSITIONS OF CUTTING TOOL MATERIALS
(% in weight)

Composition / Tool Material	C	Cr	Mo	V	W	Ti	Ta	Fe	Co	Al_2O_3
High speed Steel	0,75 ~1,5	3,5 ~4,5	0 ~9	1 ~5	2 ~20	-	-	60 ~85	0 ~15	--
Carbide	4 ~10	-	-	-	30 ~90	0 ~34	0 ~10	-	5 ~30	-
Ceramic	0,1 ~3,0	-	0~ ~15	-	0 ~50	0 ~10	-	-	-	45~ ~99,7
Diamond	100	-	-	-	-	-	-	-	-	-

TABLE 2

PHYSICAL PROPERTIES OF CUTTING TOOL MATERIALS

	Spec. Weight	Vickers Hardns.	Bending Strength	Comp. Strength	Thermal Expans. Ratio	Specific Heat Capacity
	$\frac{p}{cm^3}$	$\frac{kp}{mm^2}$	$\frac{kp}{mm^2}$	$\frac{kp}{mm^2}$	$\frac{10^{-6}}{°C}$	$\frac{cal}{cm \cdot sec \cdot °C}$
HSS	8,0-8,8	1300 ~ 1800	200 ~400	250 ~400	9 ~12	0,04 ~0,06
WC	8-15	1200 ~ 3000	75 ~260	350 ~590	5 ~ 7,5	0,04 ~0,21
$A\ell_2O_3$	3,6-6,9.	1200 ~ 2900	20 ~ 60	280 ~320	6,3 ~ 9,0	0,01 ~0,05
Diam.	3,52	3000 ~10000	30 -	200 ~600	0,9 ~ 1,9	0,33 ~0,38

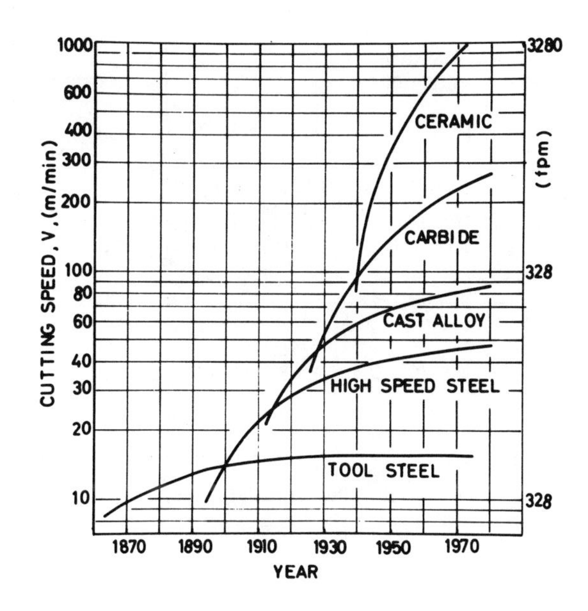

Fig. 1. Trend of Cutting Tool Material Development

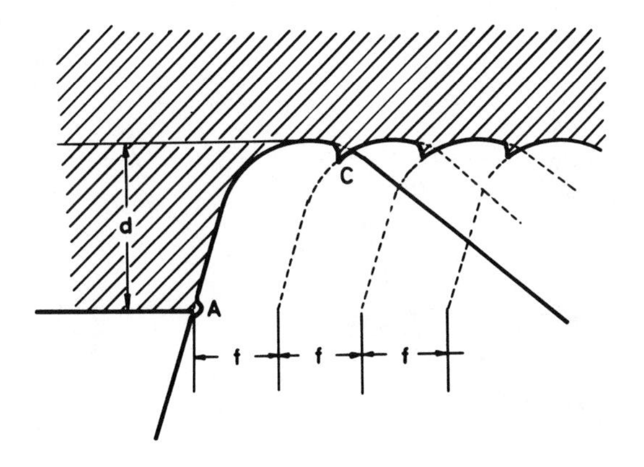

Fig. 2. Concentrated Wear on the Turning Cutting Tool

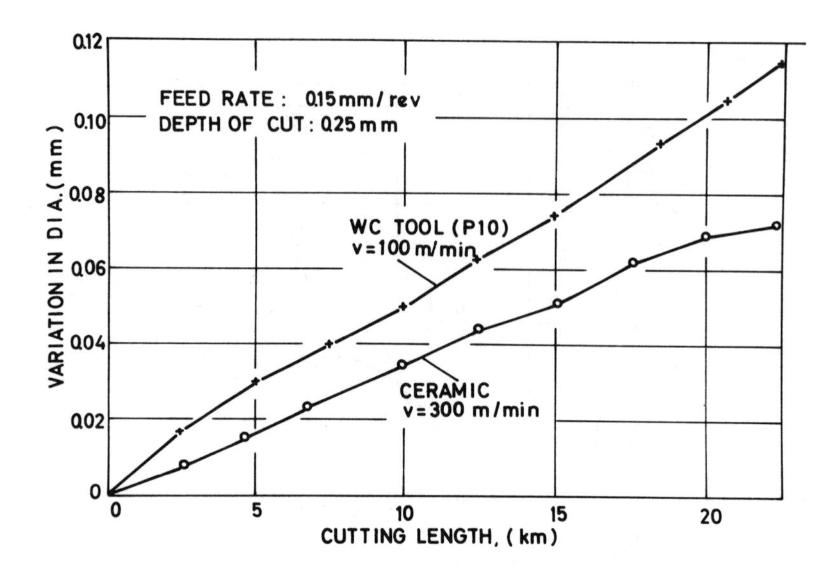

Fig. 3. Comparison of the Tolerance as a Function of Cutting Time for Two Cutting Tools
Work: AISI4140 (BHN 212)
Rake Angle: $\alpha = 0°$ Ceramic
$\alpha = -5°$ WC P10

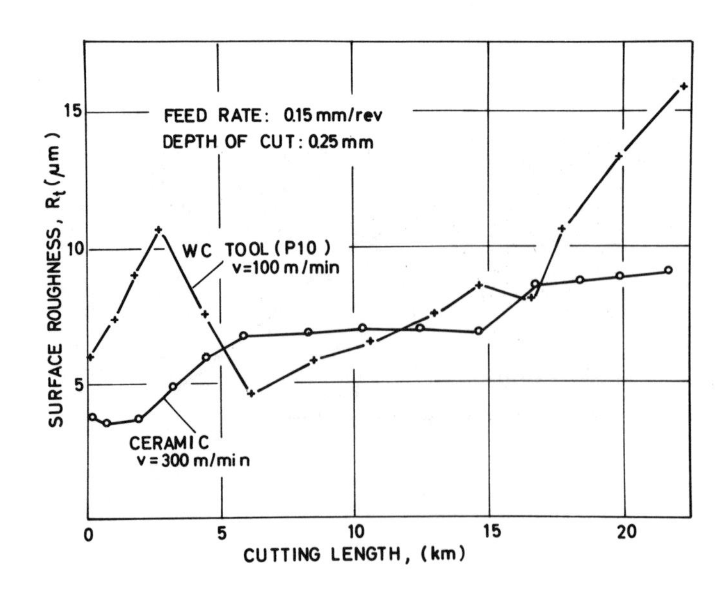

Fig. 4. Comparison of Two Cutting Tools for Surface Finish as a Function of Cutting Time.
Conditions: Same as Fig. 3

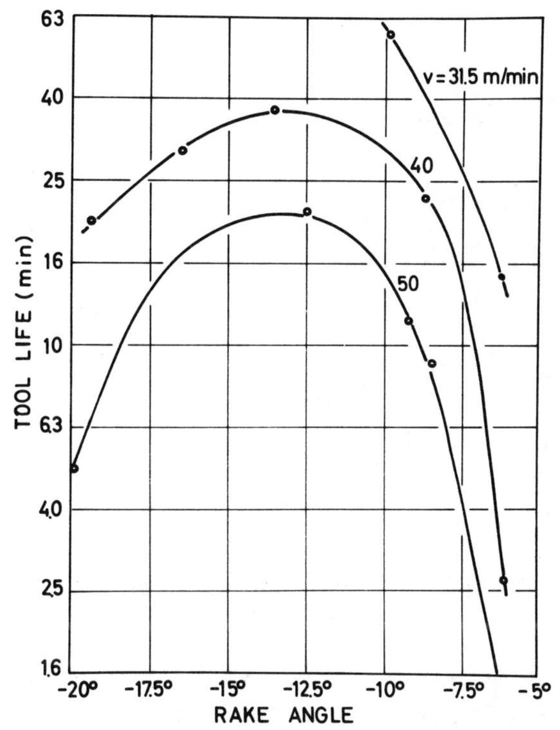

Fig. 5. Effect of Rake Angle on Tool Life
Work: Malleable Iron, HV 520
Tool: Ceramic, SHT1 $\alpha = -6°$
Feed: $h_1 = 0.33$ mm

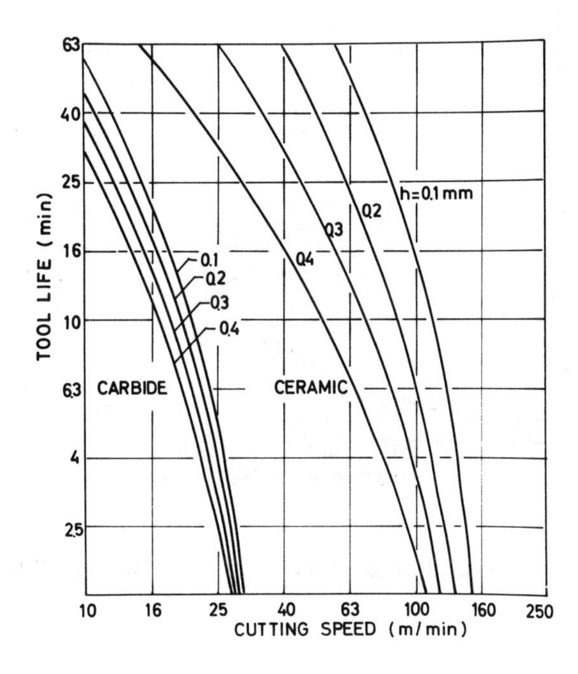

Fig. 6. Comparison of Tool Life - Cutting Speed Relationship for Two Cutting Tool Materials
Tool: Same as Fig. 5
Work: Malleable Iron, HV = 580

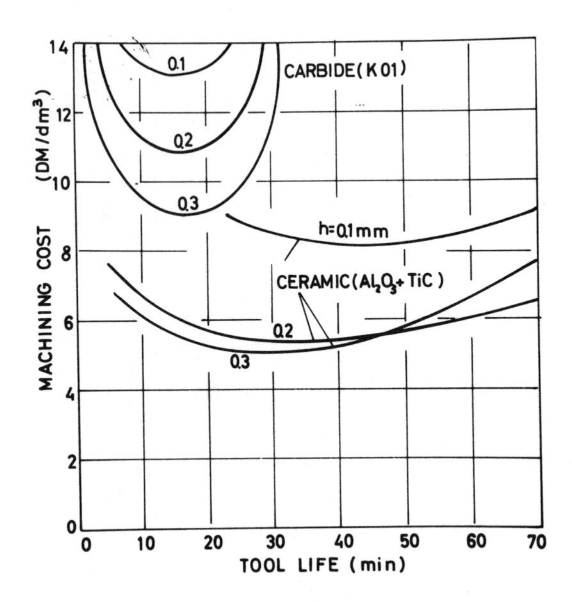

Fig. 7. Relationship between Machining Cost and Tool Life for Two Different Cutting Tool Materials

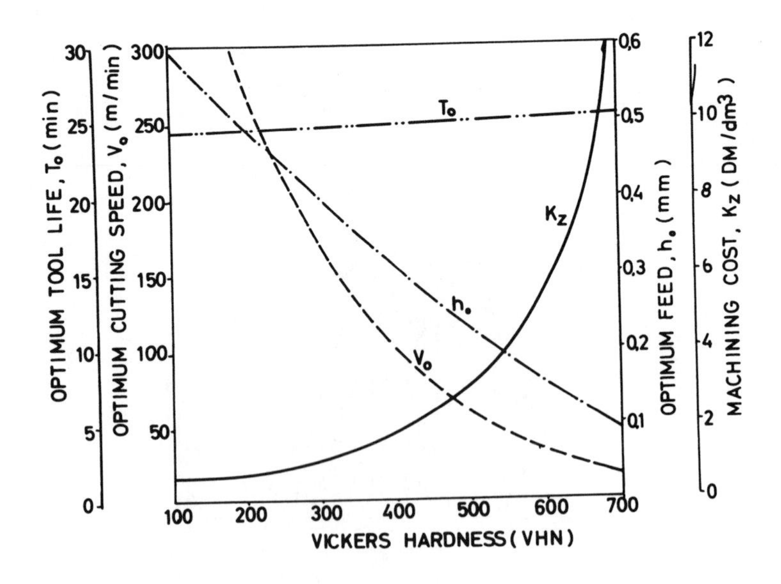

Fig. 8. Optimum Conditions versus Hardness of Cast Iron

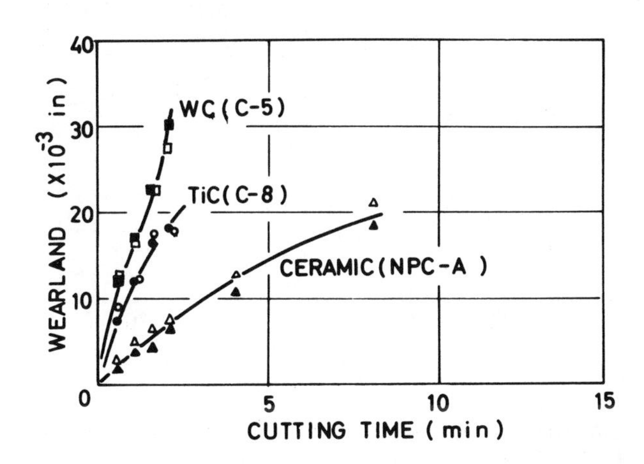

Fig. 9. Performance Comparison of Three Different Cutting Tool Materials when Turning AISI 1018 (R_B = 75), V = 1300 fpm, f = 0.0104 ipr, d = 0.030 in.

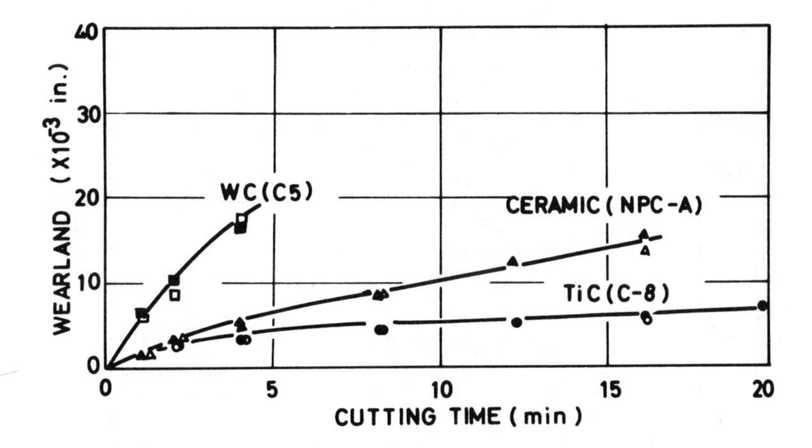

Fig. 10. Performance Comparison of Three Different Cutting Tool Materials when Finish Turning Gray Cast Iron (R_B 83), V = 1300 fpm, f = 0.0104 ipr, d = 0.005 in.

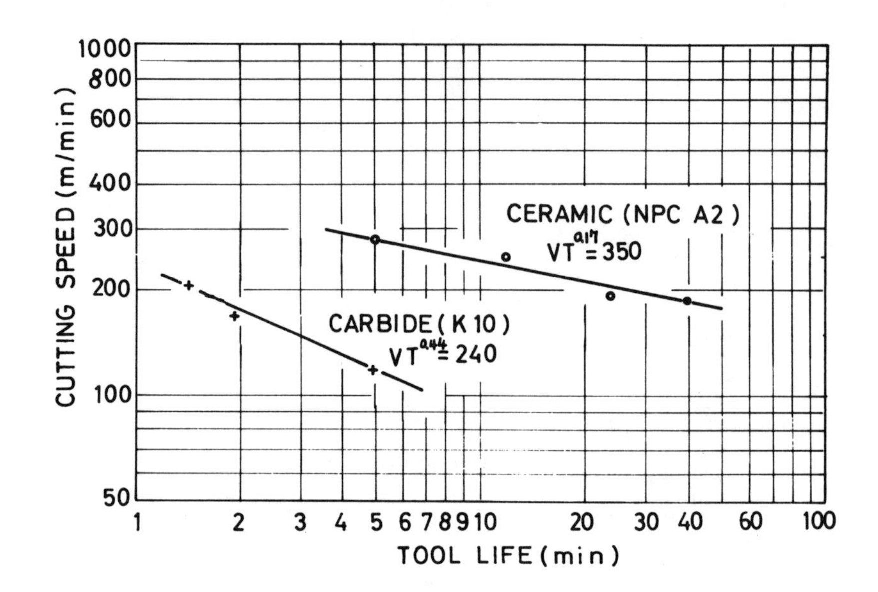

Fig. 11. Performance Comparison of Two Cutting Tool Materials when Turning Gray Cast Iron (BHN 220) [10]
f = 0.97 mm/rev, d = 4 mm
Tool life: Wearland VB = 0.4 mm

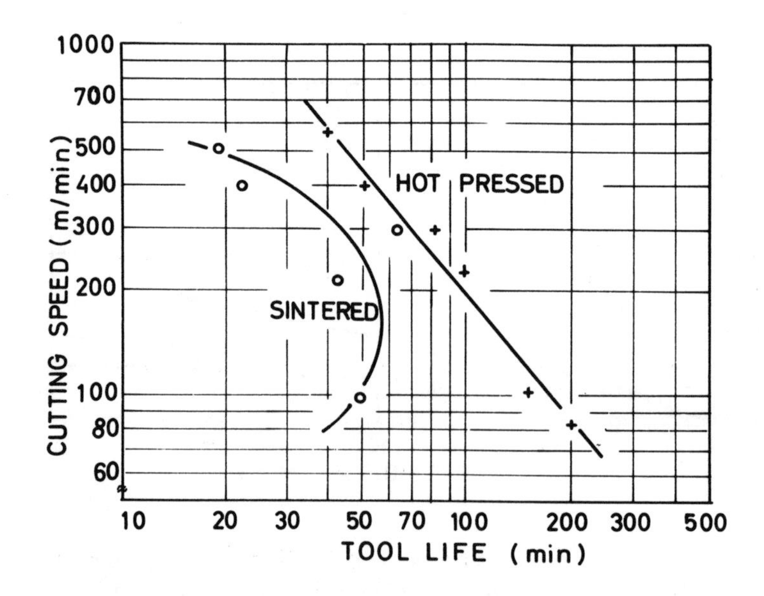

Fig. 12. Performance Comparison of Two Different Face Milling Cutter insert materials [11].
Work: Gray cast iron (BHN 220)
f_t = 0.11 mm/tooth, d = 0.1 mm
Tool life: VB = 0.2 mm

Reprinted from the *Carbide and Tool Journal*, September-October 1979

New Metal Removal Rates for Turning Machines

by

Roger O. Schultz
Western District Manager
Warner & Swasey Company
Cleveland, Ohio

SECTION I

INTRODUCTION

The effect of cutting speed, feed rate, and depth of cut in NC Turning applications will be discussed individually and collectively in this first section.

On conventional machines, as opposed to NC, it is generally left to the operator's discretion to select speed, feed, and depth of cut and in the main, they are determined by a seat-of-the-pants method.

The higher initial investment of NC equipment precludes this method of determining machining parameters. A more scientific approach must be used to obtain the required return on investment.

The machine user has one objective, to produce a part within the size and finish tolerance specified for the least possible cost. It is not a simple matter of increasing speed, feed, and depth of cut to a point just shy of cutting tool failure.

The Research Center of The Warner & Swasey Company has developed a computer program called "Comad", which is Computerized Machineability Data. This program permits 16 or more variables to be introduced into the program which subsequently gives an output of speed, feed, and depth of cut. There are 8 additional variables which are economic parameters that produce output information regarding optimum tool life.

Recognizing that we want maximum productivity, along with optimized tool life, let us now look at speed, feed, and depth of cut in relation to these desired results.

SPEED

Since cutting speed is the rate at which the surface of the work moves past the cutter, it is measured in surface feet per minute (SFM) and can be determined from the formula SFM = .262 x Diameter x RPM.

Affect of SFM on Tool Life

An investigation conducted by GE's Metallurgical Products Department has shown that a 50% increase in cutting speed results in approximately 80% loss in tool life as measured on the flank or clearance face of the tool. It follows that increased cutting speed beyond a certain point, while possible, is uneconomical in terms of tool and setup costs.

Since increasing cutting speed results in dramatic loss in tool life, let us look at the various cost elements that go into a turning operation.

1. Setup cost (non-productive time)
2. Machining cost
3. Tool-changing cost

The total machining cost is the sum of these three elements. If these three elements of cost are plotted against cutting speed, the total machining cost passes through a minimum value because:

1. Setup cost is constant for any cutting speed
2. Machining cost decreases with increasing cutting speed
3. Tool-changing cost increases with increasing cutting speed

HORSEPOWER

Since the amount of horsepower required for a cut is proportional to the cutting torque times the RPM of the spindle, and since speed has no effect on the cutting torque, then horsepower varies directly with spindle speed. Friction loss in transmitting horsepower from the motor to the work part can result in up to 30% reduction of horsepower at the cutting tool. This must be considered in determining actual horsepower available since all cutting calculations are dealing with the horsepower at the cutting area.

Efficiency of the power train can be measured by the cubic inches of metal removed per horsepower consumed. One horsepower per each cubic inch of metal removed per minute is a good rule-of-thumb for comparing horsepower available with metal removal capability.

FEED RATE

In turning applications, rate of feed is generally measured in thousandths of an inch for every revolution of the spindle (IPR). The investigation by GE's

Metallurgical Products Department, mentioned previously, found that a 50% increase in feed rate results in approximately 60% loss in tool life. Therefore, like SFM, increasing feed rates beyond a certain point is uneconomical in terms of tool and setup costs.

As the rate of feed increases, so does the horsepower required at the cutting tip. For an approximation of horsepower required, the formula HP = D x F x S x C can be used where

D = Depth of cut
F = Feed rate, IPR
S = Speed, SFM
C = H.P. factor for material to be cut

Feed rate has the most significant affect on the finished surface of the work piece, however, other factors, such as material to be cut and tool geometry, cannot be ignored, but for this discussion we will look at the affect of feed rate only.

Since a turning cut creates peaks and valleys, much like a fine thread, and surface finish is generally measured by how close together the peaks and valleys of the turning cut are, it follows that surface is directly proportional to feed rate. A good rule-of-thumb to establish feed rate for a required surface finish is to use the numerical surface finish value as the feed rate in thousandths of an inch per revolution. For example, 63 RMS required uses a feed rate of .0063 IPR.

DEPTH OF CUT

In metal turning applications, depth of cut is the amount of tool penetration into the work piece. Unlike SFM and IPR, it has been demonstrated that the depth of cut has very little affect on tool life except when it is less than 10 times the feed rate, likewise surface finish, since depth of cut is relatively constant for finish cuts.

THE COMBINED AFFECT OF SPEED, FEED, AND DEPTH OF CUT

Obviously all three machining parameters are closely related and must be accurately arrived at when preparing the machining process. If we assume the machine tool has adequate horsepower and rigidity, then the tool-life factor becomes of most concern. There are two basic types of cut in turning applications, rough and finish. Let us consider each separately and the effect of the three machining parameters.

FINISH CUTS

Since tolerance and finish dictate feed rate and depth of cut for finish cuts, we are left with speed as the variable within the range dictated by material hardness and so forth. Because it is a close tolerance cut, it is expensive, in terms of time, to change and reset a tool, despite the use of NC offsets and indexable inserts. Therefore, it is better to cut time in favor of tool life. Since we are dealing with low to medium volume production runs, it would not be unusual to complete a batch of parts with no finish tool changes.

ROUGH CUTS

In this case, the tool life factor becomes a question of cutting edge cost because, unlike finish cuts, we are not concerned with finish or tolerance, thus, tip changing time is negligable.

We have previously shown that depth of cut has the least affect on tool life followed by feed rate and then cutting speed. Therefore, to increase the rate of metal removal, with the least affect on tool life, the three machining parameters should be increased in the order mentioned above, and if the cutting speed is increased, the gain in production must be enough to pay for the higher cost of replacement cutting edges.

SUMMARY

Speed has the most serious affect on tool life. Feed and depth of cut are the best opportunity for increased productivity on roughing cuts. Since todays cutting tools permit higher cutting speeds, we see a diminishing proportional time of finish cuts to rough cuts. This encourages our machine designers to concentrate on higher horsepower and increased rigidity in order to offer NC Turning Machines that offer maximum metal removal capabilities on roughing cuts.

SECTION II

MEASURING METAL REMOVAL RATES

The interplay of speed, feed, and depth of cut will be continued with an emphasis on ways to easily calculate metal removal rates by measuring cubic inches of metal removed per minute.

American Machinist (January, 1979) published an article entitled "A Push for Practicality", which discussed how the emphasis in cutting is shifting to higher feeds and better chip control in search for faster metal removal. Machine tool builders consider longer tool life, reliability, improved surface finish and versatility important factors in design, however, "the biggest payoff is where the tool meets the work in making chips faster".

Making metal removal rate the key to evaluating machine productivity is truly the best measure of machine performance. A practical method of determining metal removal amounts can be accomplished by the use of a simple formula and the understanding of the inter-relationships between the three variables which make up the formula. With increasing costs in machines, labor, and tooling, a seat-of-the-pants selection in machining parameters is too costly a gamble. Machinists, manufacturing engineers, programmers, and supervisors should have a good working knowledge of what is needed to obtain the optimum level of performance out of the equipment in their shop. It is our intention to provide a short and easy method of determining whether the cuts taken on your equipment are operating at maximum potential.

Metal removal rate is the product of cutting speed (SFM), feed rate (IPR), and depth of cut in inches. Any change in one of these parameters produces a proportional effect in chip volume per unit time. There are a number of considerations other than what is contained in the formula in the search for maximum productivity. Subjects such as tool life, work holding devices, work piece configurations, and machine horsepower are some examples of forces which affect the final productivity level. With a number of these related subjects being discussed in detail later in this paper, this section will concentrate on metal removal.

There is a simple formula to determine metal removal rate.

$$M = 12VFD$$

Example: M = Metal Removal Rate (cubic inches per minute)
V = Speed, SFM
F = Feed Rate, IPR
D = Depth of Cut in Inches

V = 350
F = .015
D = .300

M = 18.9 cu. in./min.

Now that we know how to determine metal removal rate, let us look at the effect changing the variables has on over-all productivity. These three variables are again cutting speed (SFM), feed rate (IPR), and depth of cut in inches. (It has also been stated that a change in any one will have a proportional change on metal removal). With this in mind, let's review the advantages and disadvantages encountered by changing any one of the three individually.

CUTTING SPEED

For quite some time now, cutting speed has been receiving most of the press as newly introduced cutting tools continue to demand higher RPM's of existing machines. We are now learning that higher speeds tend to create new problems in the areas of vibration, spindle bearings, work holding devices, general operator safety and most importantly, tool life. Specifically in the area of tool life, cutting speed tends to have the most pronounced detrimental affect. A G.E. Metallurgical Products Department study showed that a 50% increase in speed equals approximately a 80% reduction in tool life, whereas a 50% increase in feed decreased tool life about 60% and a 100% increase in depth of cut showed only a 25% reduction in tool life. For a number of reasons it seems that increasing speed to increase productivity may be the least desirable route to follow.

FEED RATE

Now that we are aware of the numerous inherent problems of increasing productivity with speed, let's now investigate feed as another alternate source. The increasing of the feed rate seems to be a better approach to improving metal removal for both the machine tool manufacturer as well as the tooling manufacturer. As long as the machine tool has adequate horsepower, slide thrust, and rigidity to prevent deflection, increased feed offers fewer problems than speed. Increasing the feed rate seems to be a good compromise between higher speeds and deeper cuts. Increased feeds seem to affect to a lesser degree, work piece rigidity and workholding device stability.

DEPTH OF CUT

The effect of varying the depth of cut is directly tied to the feed rate. When the depth of cut is doubled and the feed reduced by one-half, the cutting speed can usually be increased by approximately 30 to 40%. This increase is due to the fact that there is a narrower chip and twice as much cutting edge in contact with the work piece, consequently, the cooling area for conducting away heat is larger. Therefore, if the cutting speed were increased 30% by doubling the depth of cut and reducing the feed by half, the cubic inches of metal removal will also increase by 30%. Let's take this example a step further and apply it to time. A common question faced by many shops is "Should we rough turn with one deep cut, or take two shallower cuts?". If the part, holding device, and machine will allow, it is usually preferable to take one deep cut. For example, if the diameter of a part is to be reduced 1/2″ by roughing, we could take two 1/8″ cuts, or one 1/4″ cut. Let us see the difference in time between the .250 deep cut at .015 feed and 100 RPM, vs two .125 deep cuts at .030 feed and 70 RPM.

To compute the time in minutes, divide the length of cut by the RPM, times the feed rate.

Example: Assuming a 12″ length of cut.

One cut $\dfrac{12}{100 \times .015} = 8$ minutes.

Two cuts $\dfrac{12}{70 \times .030} \times 2 = 11.4$ minutes.

Even though with one cut we were running at a lower feed rate, the increased depth of cut allows a 30% increase in RPM. Assuming the equipment has the horsepower and rigidity to take heavier cuts, it is important to determine whether present machining methods take full advantage of the equipment.

SUMMARY

Recognizing how best to manipulate speed, feed, and depth of cut by measuring the resultant cubic inches of metal removed per minute, it becomes apparent that the next variable is the NC Turning Machine itself. We have determined that a tendency toward increased depth of cut followed by feed rate and lastly spindle RPM or speed, to be the best sequence to follow in obtaining maximum productivity and tool life. This is possible only if the machine tool is designed with the concept of high horsepower, adequate slide thrust, overall machine rigidity, plus sufficient spindle RPM, to take advantage of the requirements to todays cutting tools.

SECTION III

DESIGN

There are over a dozen domestic manufacturers of NC Turning Machines, and over two dozen more designs available from Foreign builders. Selecting the proper NC Turning Machine for specific requirements becomes a very large task. Although no one design is correct for all applications, the evaluation of the capabilities of an NC Turning Machine can best be evaluated by dissecting the design in the following areas.

HORSEPOWER

Conventional Turret and Engine Lathes are usually equipped with AC motor drives. NC Turning Machines, for the most part, are equipped with DC motors. A DC motor offers the advantage of variable speed, but does not develop its full horsepower capabilities until it is running at its upper RPM range. Another peculiarity of the DC motor is that it will attempt to perform any task it is called upon to do, to the point of actually destroying itself. By monitoring the heat generated by the DC motor, overload protection is provided. By providing blowers, the DC motor can be cooled to operate at higher than normal horsepower loads. The rating of the DC motor is best evaluated in three areas. One being the maximum continuous horsepower available on a full uninterrupted load basis. This is usually somewhat less than the actual horsepower rating of the motor. The second rating is the machine tool duty cycle, which assumes some high horsepower cuts, and some below maximum horsepower loads. The third rating is actually the maximum overloaded condition that the motor can sustain for five minutes out of any given hour of running. An example of this in the Warner & Swasey rating system of spindle drive motors is a 20 H.P. motor rated at 30 H.P. maximum for five minutes, 20 H.P. machine tool duty cycle, and 15 H.P. continuous.

POWER TRAIN

In addition to the motor itself, the power train transmitting the motor horsepower to the spindle must be considered. To transmit the power directly through belts or direct gearing from motor to spindle, would result in an inexpensive spindle drive train, however, the horsepower available would only be at the top spindle RPM ranges and, therefore, inefficient in the areas of lower RPM which can also require a high horsepower in cutting examples such as drilling. By using the upper one-third range of the DC motor RPM, run through a high-low clutch, offers the ability to have a wide range of speeds along with an almost straight line curve of horsepower available throughout the spindle RPM range. This design offers minimum transmission components and offers the maximum benefits of transmitting the power to the spindle with greatest efficiency and minimum loss through friction.

SPINDLE SPEEDS

There is virtually no limit to the RPM at which a spindle can be driven from a design standpoint. However, from a practical standpoint, when considering machine life and operator safety, the practical limit comes down to a relatively moderate top spindle RPM. When comparing an NC Turning Machine to an NC Machining Center, the major element that is different is that the Machining Center utilizes a rotating tool, whereas the Turning Machine has a rotating work part. The tool can be a constant, whereas the work part is always a variable. Later in this paper, a comparison will be made with the various types of work parts and the effects of work part configuration on the cutting abilities. New work part holding devices, such as countercentrifugal chucks and imposed bar feeds, have given Turning Machine designers the ability to increase spindle RPM substantially without loss of quality of work part, or safety hazards. New bearing designs also have permitted high horsepower to be transmitted through the spindle and still affording the long life to the machine tool. Prior to the new developments in holding equipment, the rule-of-thumb for maximum RPM was approximately 60 miles per hour at the outside diameter of the chuck, that is, 5,280 surface feet per minute at the chuck O.D. Today our machines are operating at over 7,000 surface feet at the outside diameter of the chuck.

Another point that should be considered is that cutting force is independent of cutting speed, therefore, an increase in horsepower and cutting speed can offer increased cutting capabilities relative to speed, however, it has already been proven that increasing feed rate and depth of cut can offer greater machine productivity and longer tool life, which then requires greatly increased machine rigidity.

SLIDE RIGIDITY

A common fallacy in evaluating a machine's rigidity is to look at the overall weight of the machine. More important than how much a machine weighs, is where is the weight utilized. Some Turning Machines have clip pans that are made of cast iron, rather than sheet metal which add greatly to the weight, but have no affect on actual rigidity.

Other things to look for in slide rigidity are the distance between the ways. Wide spanned ways offer greater rigidity and less chance for skewing or deflection during heavy cuts. Another important things to note is a machine that has close coupled members. Short ballscrews and minimum travel beyond that which is required for the work part envelope, offer greater rigidity and less opportunity for deflection. Slide protection is also very important in that longer life is obtained if the slides can be protected from bombardment of chips, and coolant which would otherwise wash the lubricant from the ways.

The easiest way to evaluate a machines rigidity is to follow a theoretical forceline from the cutting tool through the tool holder through the turret to the slide to the machine base, and evaluate how great a distance this forceline must travel before it is actually on the solid platform of the machine base. The shorter this forceline is, and the more direct it is, the more rigid the machine will be. Also the mass of the machine should represent a pyramid whereby the greatest strength is at the base and diminishes as it gets to the cutting tool. Some machines have massive units that offer rigid strong appearances, but have a much less rigid section between the massive area and the actual machine base.

SLIDE THRUST CAPABILITY

In order to take heavy cuts at high feed rates and heavy depth of cuts, the Turning Machine must have adequate slide thrust capability to drive the tool. Since most NC Turning Machines today are operated by electric DC servo motors, we have the same situation in driving the slides as we do with the main spindle drive motor, in that power requirements of the motor are converted to heat which can restrict the operation of the servo drive motor. Utilizing blowers on the servo motors permits increased capability of the motor. High response servo motors, which operate at high RPM through a gear train, offer high thrust capability and high rapid traverse rates, yet sensitivity to very fine feed rates.

TOOLING

The elements that have been described above offer a machine design that will provide optimum capabilities regarding speed, feed, and depth of cut, however, the application requirements of NC Turning Machines today call for extremely high versatility in that many different types of jobs are to be run on the machine with very fast setup changeover. This requirement brings us into the tooling area of the machine, which actually is probably the most unique part of an individual machine's design. Some NC Turning Machines have two turrets on one slide with one turret being

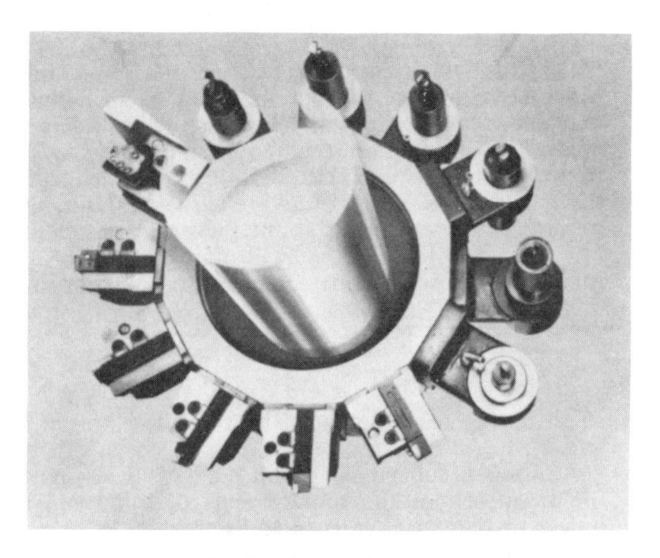

Figure 1. Twelve interference free tooling positions.

Figure 2. Single plane tooling.

utilized for external machining, and the other for internal machining. This design offers limitations in that, first of all, there is a great deal of additional travel that the slide must make to move from internal to external tooling. The programmer must always be aware of which tool is in operating position of the other turret, so that that tool will not interface with a portion of the machine or the work part itself. Another limitation is the fixed number of stations available for internal or external machining. Recognizing these limitations, Warner & Swasey attempted to overcome these problems by developing the Column Type NC Turning Machine, known as the 1-SC, 2-SC, and 3-SC Machines. The tooling concept on this line of machines has twelve tool positions on a single turret, which can hold any combination of internal and external tool holders based on the application requirements. By having a large turret, the tooling is presented to the work part on a very large platform, which is interference free in terms of any potential adjacent station interferences. With this design, all tools can be set in line with each other, which minimizes the machine idle time in moving from one tool point to another. (See Figure 1). Another benefit of this design is that the operator and programmer both know automatically where the tools are located, since they are always at a common point with the Single Plane Tooling. (See Figure 2)

VERSATILITY

Versatility is also a primary requirement of a highly productive NC Turning Machine, because of the demands on the machines ability to produce a wide range of work parts. Bar, shaft, and chucking parts that can be produced with minimum setup changeover time is

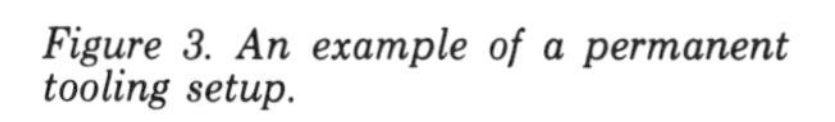

Figure 3. An example of a permanent tooling setup.

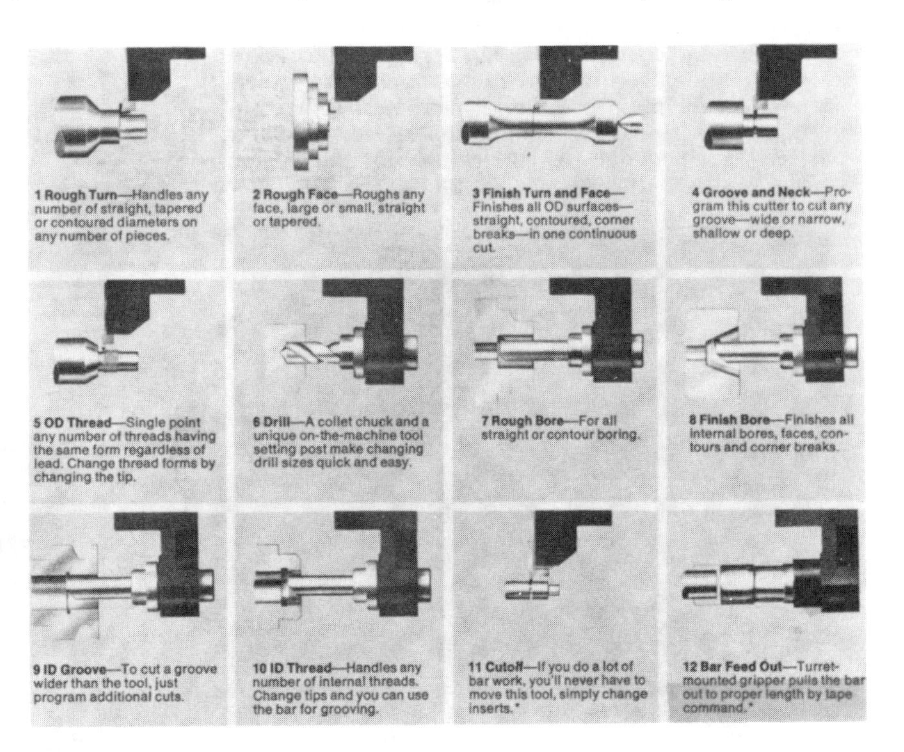

the type of versatility required from a work part standpoint.

Tooling versatility, as mentioned above, requires many tools that can be presented to a wide variety of work parts, with fast setup capability. By having a basic permanent tooling setup on the machine, the programmer and operator need only deal with the individual tool changes required for tooling that becomes unique for an individual work part. The Permanent Tooling concept offers this versatility. (See Figure 3)

SUMMARY

It is difficult to discuss the design of an NC Turning Machine without specifically referring to an individual machine. Machine tool selection should include a progressive evaluation of the required features as mentioned above. These features, when weighed against the specific application of ar individual manufacturer's requirements, will bring about the best decision for a NC Turning Machine that will offer the best of all of the many individual considerations that must be made.

SECTION IV

APPLICATION & WORK PART CONFIGURATION

An NC Turning Machine with the versatility to handle the wide variety of work part configurations that are presented to the machine, require holding equipment that maintains this same versatility. By mounting a thru-hole chuck on the spindle, the machine immediately has the ability to handle bar work, as well as chucking work, without removing the holding device. A hollow hydraulic cylinder at the opposite end of the spindle, maintains an open passage through the spindle for full length bar applications. With the thru-hole chuck, collet pads can be mounted on the end of the master jaws for bar work. To change to chucking, it is simply a matter of removing the collet bushings and mounting chuck jaws. For shaft applications, the NC Turning Machine should have a tape controlled tailstock that is available when needed to support the work part and yet can be out of the way when not needed. (See Figure 5) Let us look at the parameters of the various work part configurations and the effect of work part configurations on the cutting ability of the machine.

SIZE

Size itself is an important consideration in machining a work part. Very small work parts normally require very high spindle speeds to obtain normal surface footages. Traditionally smaller parts require closer tolerances to be maintained. Primarily on internal machining the number of tools required increase as the work part gets smaller, due to the rigidity requirements of small boring bars that must be very close to the bore size of the work part.

Another peculiarity of small diameter work parts is the frequency of the requirement to drill very small holes. In one case Warner & Swasey on its standard 1-SC size machine was required to machine extrusion tips, (used in the flow coating of insulation onto wire), whereby holes as small as .012 in diameter had to be drilled. (See Figure 4)

At the other end of the scale are very large parts which create their own specific problems. Loading and work part support is the first problem encountered. Aligning a large part can require a machine mounted loading device, which can assist in holding the work part against the locating surfaces while it is being gripped. If the part is large in diameter and long in length, it may be necessary to support the work part at both ends, using a turret mounted chuck along with the spindle chuck to support the part while a steadyrest surface is prepared. A steadyrest can then be brought

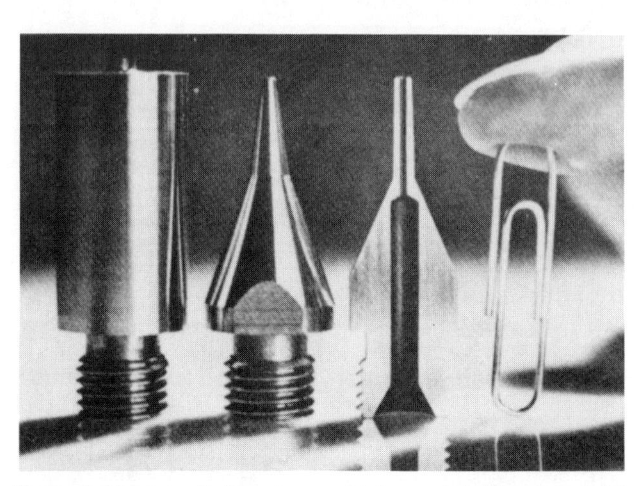

Figure 4. Very small diameter machining and drilling on a 1-SC machine.

Figure 5. In and Out positions of the retractable tailstock.

Figure 6. Swallow long shafts in spindle to reduce deflection and torque effects.

into play to support the work part after the chuck, in the turret, is removed from the work part, so that internal and external machining can be performed.

An example of this can be found in the large SC-45 Machine furnished by Warner & Swasey to a large oil equipment manufacturer in Southern California. This type of work part lends itself extremely well to 4-Axis cutting capability, whereby two separate slides are driven independently to provide combined cutting on the internal and external surfaces simultaneously. It is important to note that a long large part in the 30″ plus diameter by 120″ length involves a great deal of time-in-cut on each operation. This creates additional demands on horsepower consumption and potential overload on the servo drives, since the cuts are operating close to the continuous operation category of machining. (See Figure 7)

BAR MACHINING

Machining a work part from bar stock falls into a general category of "screw machine parts", which are usually thought of as various types of screws, bolts, nuts, washers, sleeves, etc. Usually bar parts are of a diameter between 1/2″ and 4-1/2″ bar stock. The usual length to diameter ratio is not more than 3 to 1.

Figure 7. SC-45 - 100 HP - 42″ chuck - 200″ maximum part length shown with 30″ diameter x 125″ long - work part for oil industry.

Parts machined from bar stock offer many advantages over chucking and shaft type parts, in that bar parts are usually held in a collet which offers the greatest gripping condition for the work part, and also the surface that is used for gripping is separate from the actual part being machined, which means that in many cases a bar part can be completed in one operation. The obvious disadvantages of bar parts are that since the part is in a long bar, the bar must be rotated with the spindle, which creates potential vibration and bar whip problems. This becomes an increasing factor as the diameter of the work part gets larger, in that bar stock over 3″ is usually hot-rolled material and is not perfectly round in diameter or straight throughout the bar length.

The problem of cut-off and parts catching can be a problem as well with bar parts. The cut-off operation is very similar to a very deep grooving operation and is not only time consuming, but can be expensive from a tooling standpoint. Following the basic philosophy of machining any work part on the least expensive machine in relation to the operation to be performed, it is often advantages to saw cut larger diameter parts and handle them as chucking jobs. This eliminates a parts catcher problem for large heavy parts, the high cost of cut-off, and reduces the cycle time. The exception to this is on parts that are made from tubing or where a relatively large diameter thru bore is required. In this situation the bar application is an excellent method of machining this type of work part, since most of the cut-off problems are eliminated.

Another advantage of machining parts from bar stock is that the machine can run unattended since loading and unloading is eliminated, due to bar feed and parts catching. Multiple machine operation is also a practical opportunity on bar machine work.

MACHINING CHUCKING PARTS

Chucking parts are generally larger parts made from castings, forgings, slugs or burnouts. This category of parts if often irregular in shape and may require a fixture for locating the part. A 3-Jaw power chuck is the usual holding device for chucking parts. When a fixture is required, it can be mounted on the face of the chuck. Conditions that must be dealt with in chucking applications are quite different than bar conditions. Work part loading and unloading is an operator function, and becomes a variable part of the cycle time. Chucking type parts usually require two or more operations. Chucking parts are more expensive, due to several factors. Cost of chuck jaws or fixtures are much higher than collet bushings used for bar work. Due to the irregular shape, or actual interrupted cuts of many chucking parts, speed and feed must be reduced to protect the cutting tool. Hard outside surfaces on castings and forgings also require reduced cutting rates. Out of balance work part conditions force reduced speeds. Centrifugal force on the chuck jaw can also cause reduced speeds unless a countercentrifugal chuck is used.

Although the basic reason for using castings and forgings in work part design is to reduce the amount of metal to be removed, it is actually often more advantageous to use a saw cut slug of bar stock and remove the metal at higher rates on an NC Turning Machine. This is also beneficial when considering the smaller lot sizes run on NC Turning Machines, which can reduce inventory costs of storing quantities of castings or forgings.

SHAFT MACHINING

This third category of work part is the least complicated and eliminates most of the problems encountered with bar and chucking parts. Shaft parts are classified as bar stock parts in sizes of 1" to 6" in diameter with length to diameter ratios of from 4:1 to up to 20:1. In diameters over 6", forgings are often used. The higher the length to diameter ratio is, the greater the machining problems. None of the holding problems of bar and chucking parts exists on shaft parts, since they are usually held between centers with a center driver on the spindle and a tailstock center at the other end. This is a balanced condition and, therefore, can utilize maximum cutting speeds. Also since the cuts are long, tool life can be a problem which offers a good application for ceramic inserts which offer longer life.

Lengths with over a 6:1 length to diameter ratio usually require a steadyrest support to reduce deflection of the work part. At a 10:1 or greater ratio, two steadyrests are required. Even though the part is now well supported and running in a balanced condition, another condition must be considered. When taking heavy cuts on long shafts, the torque of the cut will cause wind up of the shaft, which can result in chatter or vibration.

An excellent way to solve the torque problem and the need for expensive steadyrests, is to swallow one-half of the shaft in the spindle and grip the shaft in the middle with collet bushings or chuck jaws. (See Figure 6). This reduces the length to diameter ratio by one-half, and also offers the opportunity to face and center the part in the NC Turning Machine, thereby, eliminating a separate operation. A shorter bed length machine requirement can also be a plus factor in this concept.

One other consideration in shaft machining is that frequently lefthand turning or behind-the-shoulder turning is required. This cutting condition reverses the loads on the spindle and tailstock bearings with the tailstock bearings, then absorbing the thrust of the cut. A strong rigid tailstock with large bearings are required for this application.

SUMMARY

This paper presents the considerations required in determining speed, feed, and depth of cut as applied to NC Turning Machines. Heavy duty, rugged equipment is needed to obtain the maximum productivity when measured in cubic inches of metal removed per minute. Proper evaluation of an NC Turning Machine's features will result in the best selection of machine for the specific requirements. Recognizing the additional variables imposed by the work part configuration itself, will enable Method Engineers to do a better job in establishing cutting rates, Manufacturing Engineers to have more insight into machine selection, and Product Designers to design work parts for more economical manufacture.

ACKNOWLEDGMENTS

Acknowledgments —
This paper is the result of the joint cooperative efforts of the following Warner & Swasey personnel:

Ann Manix - Field Engineer, Los Angeles
Mark Mathis - Field Engineer, Los Angeles
Richard Anderson - Branch Manager, San Francisco
Derek Treherne - Branch Manager, Seattle
B.K. Srinivas - Section Head-Research Engineering Warner & Swasey Research Division, Solon, Ohio

Presented at the SME East-Central Tool Exposition, November 1975

The Development of a Heavy Duty Tooling System

by James W. Heaton
Kennametal Inc.

INTRODUCTION

What exactly is heavy duty machining.

Heavy duty machining is a relative term that is difficult to describe. There are no industrial standards defining it and every shop has their own interpretation due to the infinite number of cutting applications. In a plant making precision instruments, a "heavy" cut may be one that is taken with a 1/4" inscribed circle insert; while in a steel mill where roll forgings are roughed out, a "light" cut may be one that is taken with a 3/4" inscribed circle insert.

Many large shops, doing a great variety of work, take very substantial cuts on heavy duty N/C lathes (40 plus H.P.) with TNMG-543, 564, 666 and CNMG-866 type inserts. However, in some instances these inserts are used simply because a very small portion of the cutting operation requires a long cutting edge. Examples of this condition would be turning up to a large pre-formed or pre-machined shoulder; or boring to the bottom of a cast counterbore, etc. In these cases, the chip load imposed on the insert during most of the cut may be one of moderate roughing (0.250" - 0.300" depth of cut and perhaps a .020 - .025" feed rate) as opposed to "heavy" roughing.

Therefore, an attempt will be made to define what is meant by heavy duty roughing as the term applies to this paper.

We are primarily talking about work that normally requires an insert with an inscribed circle of at least one inch and/or a cutting edge length of at least one inch. Examples would be 3/4" inscribed circle triangles, 1" squares, 1" diamonds, 1" x 5/8" and 1" x 3/4" rectangles, 1" inscribed circle triangles, and 1-1/4" and 1-1/2" squares.

If the principal cutting conditions involve, as a <u>minimum</u>, a 0.400" depth of cut coupled with at least a .030"/rev. feed rate, the operation could qualify as heavy roughing. However, the size of the above chip load alone is not necessarily enough to establish the low end of heavy duty cutting. We must be cutting materials with yield strengths that develop fairly high cutting forces in the shear zone and extrusion region (friction on the rake face). And we must be travelling at a speed where a reasonable amount of work is being done (at <u>least</u> 18-20 cubic inches removed per minute).

The minimum conditions that qualify as heavy roughing would then be defined as:

0.400" depth of cut
0.030"/rev. feed rate
125 sfm
160 minimum BHN
(On 160 BHN material, this type of cut requires approximately 14 horsepower using conventional inserts, and puts a 3700 pound tangential load on the rake face.)

For other higher strength alloys, certain tool steels, and stainless steels, the minimum depth of cut and feed rate may be reduced and the chip load could still qualify as heavy roughing - simply because of the high work being done in the shear zone and the substantial amount of heat which is generated.

As regards these minimum conditions - for plain carbon and conventional alloy steels, a cut of 0.300" depth coupled with a 0.040" or 0.045"/rev. feed rate produces a chip load that would also qualify as heavy roughing.

The upper limits of heavy duty work are more difficult to define. However, we do observe feed rates up to 0.125"/rev. on some materials; and we recommend a maximum depth of cut of no more than 80% of the length of the cutting edge. Speeds vary widely at these extremely high chip loads but are usually low, and are most often dictated by workpiece size and shape and available horsepower.

These high load conditions assume that there is no premature insert breakage. They naturally require high horsepower (from 75-150) and may apply loads of up to 35,000 pounds on the insert.

The minimum tool holder shank size to qualify for this work is 1-1/2" x 1-1/2" x 6" long. The largest shank size can easily be as massive as 5" x 5" x 30" long, although the great majority of H.D. cutting tools have 2" x 2" shanks. When going up to shanks of 3" x 3" or larger; or very high beam type support blades (typically used on roll lathes) it is advisable to apply "cartridges".

BACKGROUND

Heavy duty carbide tooling has, historically, been developed through a slow evolutionary process beginning with large brazed tools. These brazed tools usually with rectangular shaped blanks were relatively effective because they had no mechanical clamps or superstructure to impede chip flow. But while chip flow was relatively free of superstructure type obstructions, there was a shelf-type or groove-type breaker normally ground into the rake face of the tool. As we now know, these configurations increase the feed force and at times horsepower consumption with little gain in productivity since they merely present a back pressure to work being done in the shear zone and offer high frictional resistance to the chip as it leaves the rake face. However, chip control was necessary and the grooves were an essential part of the design.

To overcome braze strains and chipping, relatively soft carbide grades were used and they further limited speed and chip load. Extremely high chip loads even caused conventional braze materials to soften and the blank would often move. Special braze material could overcome this problem but they increased the possibility of tool failure through braze strains.

The next development was a regrindable insert held in place with a top clamp. This insert had a chip control geometry similar to the earlier brazed tools but could be advanced after each use in order to regrind the cutting edge. The drawbacks of this tooling were: (1) Regrind time and cost, (2) very large expensive ground inserts, (3) clamp erosion from the chip, and (4) resistance to chip flow caused by the chip groove and clamp. A low profile spring clamp was next developed to hold the rectangular "LNU" style indexable -- non-regrindable insert. These superstructure components are often washed out with resulting holder damage. They also impede chip flow and decrease cutting efficiency. Spring clamps with carbide chip breaker plates were also developed to hold "L"-shaped inserts which had negative back rake and positive side rake. It was an early approach at limited or controlled contact cutting and we did have a freer cutting insert with this geometry. But again, it was expensive since it was ground all over and provided only two cutting edges.

DEVELOPMENT

Two currently popular inserts are the large squares (SNMA and SNMG-866 and 1066) and large rectangles (LNU style) because they have long cutting edges and are thick enough to handle heavy chip loads. However, with the low to moderate speeds and heavy loads involved in heavy roughing, carbide inserts are subjected to a series of conditions which cause cutting edge damage as follows:

(1) Edge chipping and depth-of-cut notching from load, runout, scale, hard spots, chatter, etc.

(2) Heat checking due to temperature changes.

(3) Thermal deformation or thermal bulge in the vicinity of the nose radius due to extremely high loads, high temperature and long, dry cuts.

(4) Extreme crater due to load and heat.

(5) Heavy flank wear because this kind of cutting permits more edge wear before it is necessary to index an insert for surface finish, loss of size, etc.

(6) At times the non-cutting edge directly below and opposite the cutting edge in use is damaged by the chip.

With this kind of damage it is highly questionable as to whether squares and rectangles can be turned over and still provide an effective cutting edge. An insert must be well supported by a flat, strong, dimensionally stable shim or seat. And when a badly damaged edge is turned over to press intimately against that shim, poor seating is the result. Unit loads of up to 150,000 psi are possible when taking these heavy chip loads and a smooth, clean, flat insert face must press against the shim. If this edge has heat checks, thermal bulge, or chips and crater, it makes a poor edge or surface to act as a support for the "new" cutting edge which was turned over and is now in use.

Therefore, the "land-angle" geometry, which will be discussed later, is placed on only one side of the insert which ensures proper tool support to handle these extremely heavy loads.

The latest developments have been pin type holders supporting indexable inserts with holes or holders employing a bolt and nut to hold the insert down. These holders can jam with scale or metal slivers; and because of the high torque which can normally be applied to the bolt or pin and the high cutting temperature, removal of the insert may be extremely difficult.

Bolts do not ensure that the insert is positioned and held back in the pocket - only down; and the pin types may not prevent flutter during extreme runout since they mainly hold the insert back.

Current molded inserts such as SNMA (1" square with a flat rake face) or SNMG (1" square with a molded chip breaker groove) present some chip control problems.

Just because the depths and feed rates are heavy in this kind of work, there is no guarantee that chip control will automatically occur; or be easy to facilitate.

It is often assumed that chips from heavy cuts are naturally deflected and curled against a negative rake geometry, even one that is smooth such as an SNMA style insert. This can happen (and we now believe that many materials want to curl but current chip grooves and deflectors do not seem to allow this free formation), but not necessarily as can be seen in the upper chip in Figure 1. This stringer type 4340 steel chip (Rc 28) was produced by a 1" square-smooth rake face, negative rake insert (SNMA-866) cutting at 200 sfm, 0.500 depth of cut, and a feed of 0.050"/revolution. The chip shows no inclination to curl, break or behave itself. Figure 1 also shows chip control with the land angle geometry under identical cutting conditions.

Therefore, even on very heavy cuts, we must often provide for chip control. This can mean a molded groove and the inherent danger of "over control" if we want to increase the feed rate; or the option of a superstructure to support a mechanical chip deflector. In that case, heavy chips, within a short period of time, often wipe off the chip breakers and clamp. The land-angle design provides natural chip control, without superstructure and often with reduced cutting forces. By using inserts with molded grooves we do get some chip control but if we want to increase the feed rate in order to be more efficient and productive, the chips may be badly crowded and deformed. This results in increased pressure, temperature and often chatter and breakage. Figure 2 shows chip formation with a groove type geometry. The groove will often slow down chip flow and may produce a thick chip (t_2).

The land angle geometry is very simple and was developed to allow the chip to escape as quickly and gently as possible from under the workpiece with minimum retarding of chip movement. Figure 3 shows the basic design. The land is considered to be the primary rake face and the floor the secondary rake face but still parallel to the plane of the land.

The forces on the land (Figure 4) are probably just as high on this design as on the land in the groove design. And chip interface velocity in this stagnant zone appears to be slow as little wear is detected on the land after long periods of cutting. The normal forces (F_n) between chip and insert along the angle appear to be lower than with the groove design. This angle may be assisting to move the chip downstream (perhaps similar to the land relief angle on an extrusion punch) rather than hinder its advance. This can be evidenced by the wear pattern being heavier on the floor near the angle than on the angle itself. Crater wear does not work its way back toward the cutting edge very rapidly. The land and angle appear to act as a barrier in this regard.

The chip leaves the rake face very rapidly and its initial radius of curl is small. This would seem to indicate a "crowded" chip but such is not the case. We feel that the interface velocity of the chip after it leaves the land is simply higher, in relation to the mean chip velocity, than a chip travelling through a groove or hitting a plate, and a natural curl is produced. In some cases, the chip has actually opened up or at least curls with the same radius as feed rate is increased (refer to Figure 5) which is opposite the conditions normally found with chip deflector devices and other geometries.

This land angle geometry has been placed on large heavy duty inserts (Figure 6) and is married to a clear top holder design in order to maximize productivity and minimize downtime.

The clear top holder, used with this insert geometry now combines some of the best features of the early evolutionary designs. We have an insert with a 72° conical countersink and no obstructions such as grooves, shelves, chip breaker plates or clamps (see Figure 7). The cutting is somewhat cooler, feed forces are reduced, lower horsepower may occur and yet the chip is controlled over a broad range of feed rates. High productivity is realized because land-angle can accept higher feed rates and still give a controlled, uncrowded chip.

The insert and shim are held in place by an anchor pin. These three pieces are quickly snapped together as an assembly through the use of a retaining ring located on the anchor pin.

This means as the anchor pin is retracted from the pocket, the three pieces come out together as a unit. The pocket can be easily cleaned of rust, scale, steel slivers and other material associated with extremely heavy duty cutting. The 72° countersink in the insert allows the anchor pin, with its unique head design, to hold the insert down and back simultaneously. We have the benefit of a cam or tilt pin pulling the insert back through the apex of the holder and the benefits of a secondary top clamp for holding the insert down and yet it is incorporated into a single anchor pin. This is the best possible lock-up configuration as far as ensuring a secure hold on the insert and power consuming obstructions have been eliminated.

When a hot insert assembly or package consisting of the insert, shim and anchor pin is elevated from the holder, it can either be indexed by spinning the insert to a new corner very quickly, or by removing and replacing it with a completely new cool assembly. The operator can then inspect the used insert and shim (while the machine is running) and make sure there is no severe damage. He also has the benefit of being able to handle a cool package to drop into a freshly cleaned pocket. The anchor pin has a lead-in pilot or nose which drops through a floating lock nut. After rotating the anchor pin approximately 2-1/2 revolutions clockwise, we have locked up the entire unit against a spring washer pack consisting of four Belleville washers. When the floating dovetail nut bottoms on the floor of the cavity in the underside of the holder, a 150 pound preload locks up the entire unit including the insert. This positive stop can be felt very easily and it is not necessary to apply any more torque to the anchor pin since this 150 pound vertical load on the insert is more than adequate to do the job of retaining the insert pack. The 150 pound preload on the insert in turn produces a 206 pound horizontal force back through the apex of the insert to hold the insert back in the pocket. These two forces prevent insert flutter during entry into the cut, through interruptions and runout; and when coupled with the dynamic cutting forces ensure that the insert is well seated and stays in position.

The 150 pound preload also ensures that there will be no backing out of the anchor pin under vibration during cutting.

While cutting in our Lab, we have observed, through the use of thermocouples, extremely high temperatures. For example - when cutting 4140 - Rc28 at 200 sfm, .500" depth of cut, and a 0.050 feed rate, we have measured insert and workpiece interface temperatures as high as 1800° F and insert temperatures (at the furthest point away from the cutting edge) of 575° F, pin temperatures of 475° F, shim temperature 450° F and holder temperature (0.5" below the shim) of 315° F.

These high temperatures coupled with potential "breathing" of the insert away from the holder walls during interrupted and runout type cutting cause concern when designing heavy duty tools. Because there is a severe difference in the coefficient of thermal expansion between carbide and the steel components, this breathing can be exaggerated and the spring washer pack is used to handle this expansion difference. It ensures that a nearly constant load will be applied to the insert even though the assembly is expanding and contracting as the insert, shim, tool holder, and pin heat up and cool down during and after cutting.

To remove the cutting unit, the anchor pin is turned 2-1/2 revolutions counter-clockwise and the threads act as an elevator (when coupled with the retaining ring) to lift the pin, insert and shim up and out of the pocket as a unit. The lead-in pilot can then be pushed up further by the operator's free gloved hand and the entire unit is removed or indexed very quickly. This is an advantage to the operator since this pack will be in the neighborhood of 400-650° F.

Figure 8 shows the application of a round land-angle insert and performance comparison to a groove type insert.

The clear top system provides a strong, firm lockup which accurately positions the insert and shim into the pocket and yet provides the resiliency to withstand the extremely high temperatures and pressures which are developed during heavy duty cutting. The head of the anchor pin is below the secondary rake face or floor of the insert which allows for free cutting with no possibility of pin erosion. The two outer corners of the holder have been relieved or notched. This provides a place for used insert corners, which may have mushroomed flank wear, to escape without being brinelled into the holder wall. This means that every time an insert is indexed it will always seat flat against both walls.

This tooling system will allow more productive and predictable machining when doing heavy roughing.

ACKNOWLEDGEMENTS

The author wishes to express his appreciation to Mr. W. L. Kennicott for his encouragement and assistance, Mr. J. F. McCreery for his unique contribution to this design, Mr. J. F. McCreery and Mr. J. S. Bator for their assistance in lab and field evaluations, to Mr. D. L. Burick and Mrs. Madeline Haines for preparation of illustrations and to Mrs. Cheryl Kondrich for the preparation of the manuscript.

BIBLIOGRAPHY

1. KENNICOTT, W. L., Recent Improvements in Metal Cutting Tools, 1970.

2. JONES, D. G. and MCCREERY, J. F., A Study of Preformed Chip Control Devices in Throwaway Carbide Inserts, 1973. SME Paper MR73-215.

3. HEATON, J. W. and ALBRECHT, A. B., The Development and Application of a New High Performance Cutting Insert. SCE Paper at SME 1975 International Tool and Manufacturing Exposition and Engineering Conference, Detroit, Michigan, April 7, 1975.

4. HENRIKSEN, E. K., Chip Breaking - A Study of Three-Dimensional Chip Flow, A.S.M.E. Paper No. 53-S-9, 1953.

5. HENRIKSEN, E.K., Findings and Directions in Chip Breakers Research, A.S.T.E. Paper 23-T-4, 1955.

6. CHAO, B. T. and TRIGGER, K. J., Controlled Contact Cutting Tools, A.S.M.E. Paper No. 58-SA-42, 1958.

7. TAKEYAMA, H. and USUI, E., The Effect of Tool - Chip Contact Area in Metal Machining, A.S.M.E. Paper No. 57-A-45, 1957.

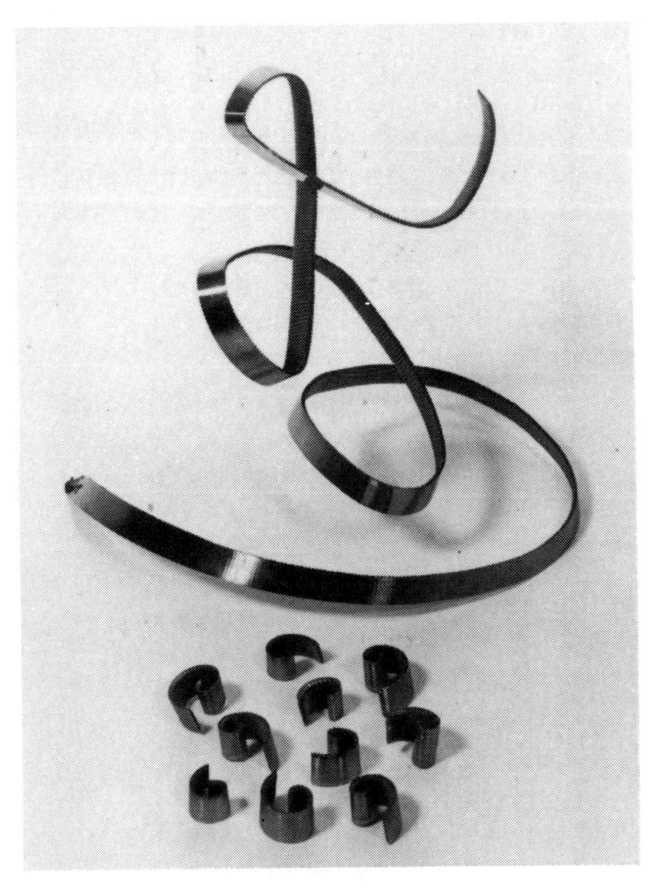

(Top)
Snarled 4340 chip

SNMA-866 insert
200 sfm
.500 depth-of-cut
.050"/rev. feed
32 horsepower
.078" thick chip

(No chip obstructions in order to find minimum power requirements)

(Bottom)
The natural chips from a land-angle geometry insert under identical cutting conditions. SNMH-866 insert, 30 horsepower, .072" thick chip.

FIGURE 1

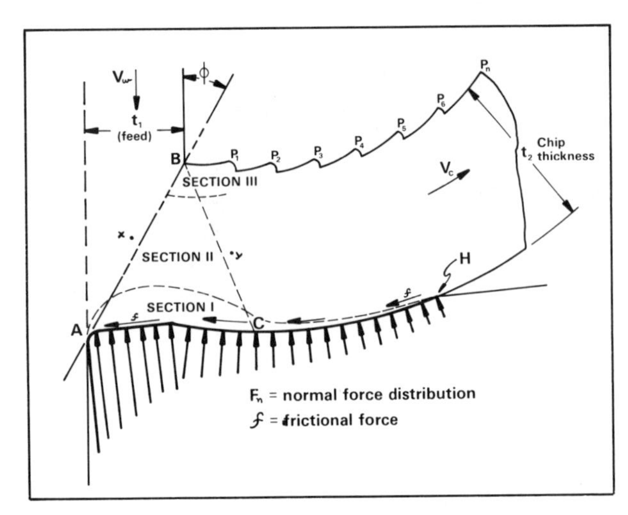

FIGURE 2

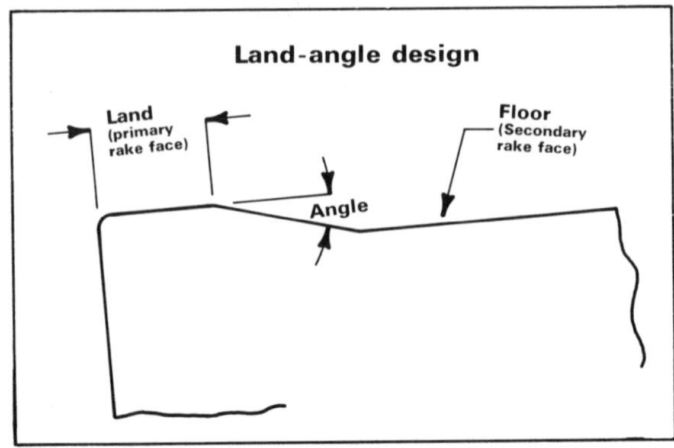

FIGURE 3

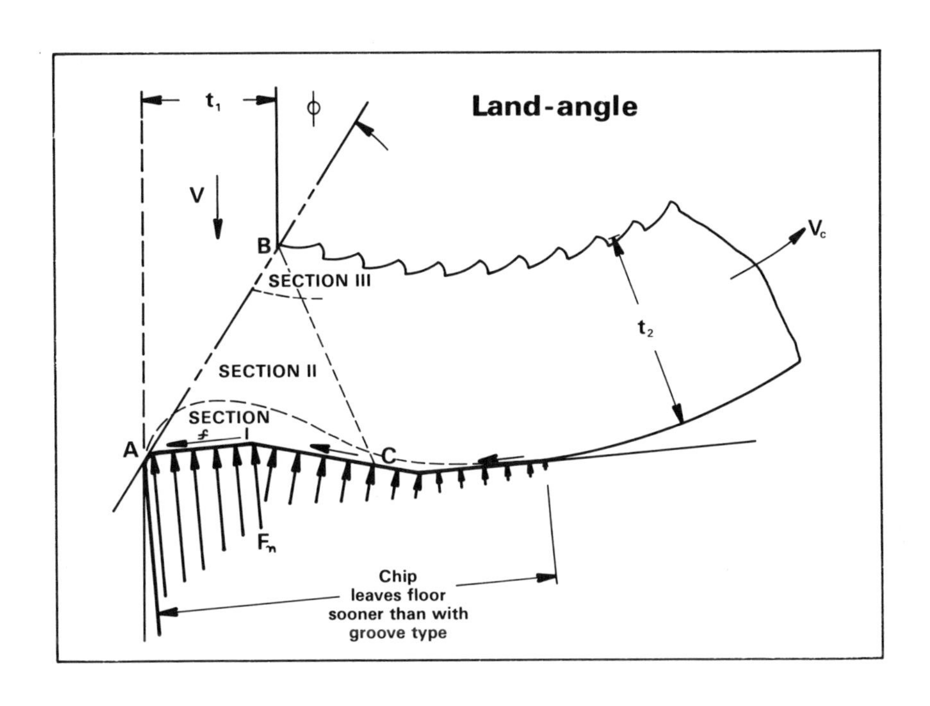

FIGURE 4

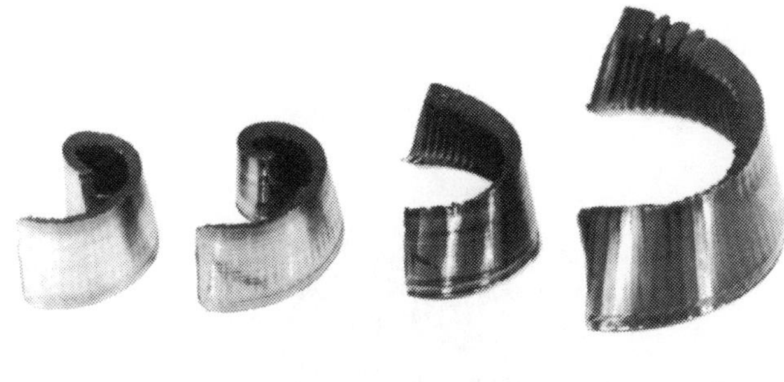

.060"/ rev.   .070"/ rev.   .080"/ rev.   .100"/ rev.

Four different feed rates applied to the same land-angle geometry while turning an alloy steel roll forging at 170 sfm and .600" depth-of-cut, no severe "crowding" was observed as feed rate increased.

FIGURE 5

1" i.c. insert and large chips

FIGURE 6

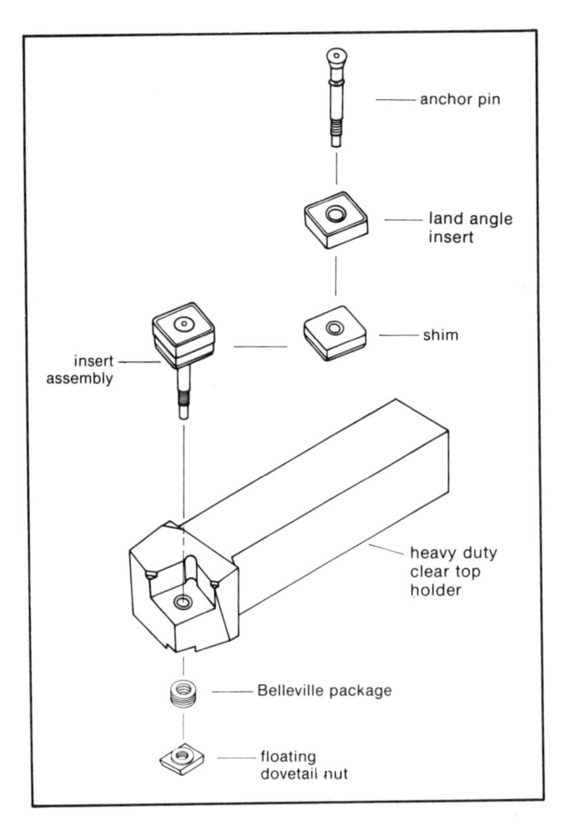

FIGURE 7

Large steel roll 250 sfm .300 depth-of-cut

RCMG-106 insert showing heavy oxidation from high temperature and deep cratering

feed=.040"/rev. (max.)

RCMH-106 (land-angle) No signs of high temperature

feed=.070"/rev.

FIGURE 8

Presented at the SME Chicago Area Conference, October 1977

Evaluation of Chucking System Needs for High Speed Turning

by Gerald E. Mueller
Universal-Forkardt Workholding, Houdaille Industries, Incorporated

The trend toward high speed machining has brought about an influx of changes among machine builders, chuck manufacturers and tooling suppliers. These changes have brought about new concepts with relatively unknown technical areas. To the end user these areas can be problematic. The most troublesome topics dealing with workholding, are the loss of gripping force and the distortion of thin walled objects at elevated speeds. Employing methods of graphical solution and applying formulas, help to reduce trial-and-error engineering. Different jaw counter-balancing systems are discussed, pointing out methods of approach. Machining methods are discussed to reduce manufacturing problems and improve performance.

INTRODUCTION

The trend of new high speed NC lathes is toward increased production, improved accuracy, surface finish, and more rapid changeover capabilities. These are typical demands industry has faced over the years. However, now more emphasis is placed on their significance.

New machines rapidly face obsolescence when concepts improve. The end user must make a precise analysis prior to his investment, to determine the best choice of capital equipment for present and future needs.

Increased production, means more returns per investment dollar. A variety of approaches to increased production are available. These approaches vary in significance and also in their applicability to a broad range of machining circumstances.

A production increase is accomplished by three major means; achieving higher cutting speeds and increased metal removal rates while improving uptime. Stress hasn't been so widely placed on improved uptime, except at the manufacturing engineering level.

Quick-change devices are available as options at the O.E.M. level to drastically reduce changeover times. Since NC machines provide low to medium production, changeover is an important factor of the efficiency spectrum.

Increased feed rates and higher speeds are the major topics of discussion among machine builders, chuck manufacturers, tooling suppliers and end users.

Machine builders are supplying machines which can perform not only at higher horsepower peaks, but also at elevated rpm. Higher rpm often require improved machine base structures to resist vibrations, new bearing designs to withstand increased peripherial speeds and refined bearing lubrication systems to withstand resultant heat gains. Improved bar-feed designs allow for increased rotational speeds while reducing vibrations. Constant surface speed rates have progressed to meet advanced tooling necessities.

Cutting tool manufacturers are continually improving existing carbides, coated carbides and ceramic tools. Tooling is more rigid and sturdy to overcome vibration problems, interrupted cuts and increased removal rates. Speeds are steadily rising to provide optimum metal removal.

Tooling using carbide and carbide bases allow for heavy depth of cuts and an extension of rpm limits. This approach uses brute horsepower to boost removal rates. It proves most effective for large amounts of metal removal but causes excessive power consumption and in extreme cases, causes system fatigue from exorbitant loads. Machines using this type of tooling are generally not classified as high speed machines, except when used in conjunction with ceramic tools for finishing operations.

Initially ceramic tooling concepts with high cutting speeds reduced depths of cut, and increased numbers of machining passes met much opposition. Its benefits have finally been realized and accepted on a broad scale. Advancements with respect to strength and durability have made increased chip loads and longer life possible.

Ceramic tools and the high speed concept, lend themselves to certain types of applications. Smooth surface finishes are readily achieved, along with accurately held dimensional tolerances. Parts manufactured from forgings and castings or slugged bar stock, are most effectively machined with the high speed approach.

Less stock removal is required on pre-formed work materials. Generally a single roughing and one finish pass is possible to complete the part surface. Carbide tooling requires the same number of passes for this type of application, but is accomplished at much lower speeds.

Slugged bar stock parts can be readily machined with the high speed concept for several reasons. Traditionally, bar cut-off operations have been very slow due to tool configuration and chatter. This time consumption can be very expensive due to NC machine and operator costs. Recently, the concept of slugging parts has received a great deal of analysis.

A steel handler can load a bar on the cut-off, facing and centering machine set on automatic cycle, while he performs other functions. This can be done at low machine costs and virtually no operator expense. Besides the reduction of cut-off costs, the most important savings can be attained from machining.

Slugs are easier to handle and can be machined at higher speeds. Rotation of bar stock at high rpm is not practical due to vibrations. Bar feed mechanism designs do not perform well at elevated speeds. Slugged workpieces are easily supported by a center, for stable working conditions.

Although this technique does not satisfy everyone's needs, it often means improved performance, greater safety, and less maintenance.

THE WORKHOLDING ELEMENT

The machine, workholding device and cutting tool are considered separate entities, but their performance relies on the ability to function in harmony. The workholding element requires particular attention since its performance is critical.

An improper chucking system can be hazardous, restrict performance or limit the machine's use. Its performance becomes particularly important in high speed applications where gripping force losses could mean tool breakage due to part slippage, or an accident from the workpiece escaping the jaw's grasp.

The chuck's rating is important to insure satisfactory operation, while maintenance is a key factor to preserve that rating. Since no concrete formula is available to determine a chuck's rating and standardization among chuck manufacturers doesn't exist, a significant difference may result from one chuck to another, although the manufacturers have similar ratings.

Parameters which play an important role in determining the rating are the maximum draw-force, clamping factor, lubrication and anti-friction constituents, jaw weight and position of its center of mass, and the percentage jaw force loss acceptable by a manufacturer.

The most common method of chuck rating is to specify the maximum clamping force. At low rotational speeds, the dynamic gripping force approximates the static force. However, at higher speeds this no longer holds true and the effects of centrifugal force must be considered. At elevated speeds it is possible to completely loose the jaw force with a conventional chuck. To achieve the maximum static clamping force, the chuck's input, mechanical advantage and frictional losses must be taken into account.

THE MAXIMUM DRAWFORCE

The chuck's input only varies slightly since rotating air and hydraulic cylinder forces are standardized to a degree. Cylinder designs and safety features differ significantly and deserve special attention at high speed operation. A cylinder should have features which maintain cylinder force even when the supply of the force media is lost.

CLAMPING FACTOR OR MECHANICAL ADVANTAGE

The clamping factor or mechanical advantage can vary from less than 1:1 to a ratio of 4:1 or more. A high input to output ratio can be advantageous if frictional losses don't deteriorate the gains. Low ratios require larger and more cumbersome cylinders which add weight to the revolving components. Constant speed changes on NC machines are required for optimum performance. This added weight can be detrimental to the drive system and braking elements. Weight (inertia) increases power consumption and accelerates drive system wear and reduces brake life.

LUBRICATION AND FRICTION

Costly frictional losses can be a result of inadequate lubrication or poor chuck design. To maintain a chuck's factory performance, it is necessary to grease the chuck at regular intervals. In severe cases, disassembly may be required, when the mechanism becomes overly contaminated with grit and chips. Poor lubrication canals restrict the proper flow of grease. New greases should force the old broken down lubricants out. New greases should not ooze out of the chuck, except in cases of over lubrication. Poor lubrication systems have a tendency to fling grease. Lack of grease enhances premature wear and can reduce jaw force up to as much as 50%.

In the presence of coolant environments, lubricants tend to wash away. Grease compounds containing molybdenum disulfide provide good lubricative and adhesive characteristics, and errosion resistance.

A good chuck design encloses the chuck mechanism completely. The base jaws and jaw guides should be shielded as much as possible to eliminate coolant and chips from entering, and grease from escaping.

The difference in performance between laboratory and actual working conditions is determined by the consequences of these variables. They affect the chuck's function of achieving maximum performance.

JAW WEIGHT AND POSITION

The jaw performance is an important parameter in determining the efficiency rating of the chuck, since it directly relates to clamping force at elevated speeds. Jaw weight, geometry and radial location affect the force at high speeds. The jaw weight, geometry and radial location of the center of gravity are a product in determining the moment of inertia for rotating elements. This inertia factor (P_c), is a function of the square of the rpm. The jaw force loss can be easily calculated by the following equation:

$$\text{(jaw force loss)} \quad P_c = (.112)(G)(r)\frac{n^2}{100} \qquad \text{equation \#1}$$

Where P_c (Kp) equals jaw force loss, G (Kg) the weight of base and top jaws plus T-nuts and bolts, r (cm) the radius of center of gravity of the jaw unit and n (rpm).

The clamping force of a rotating three jaw chuck at n (rpm) is given as the sum of all three jaws:

$$\text{(jaw force @ n)} \quad P_{sp} = P_o - 3P_c \qquad \text{equation \#2}$$

P_{sp} (Kp) represents the total jaw force at n spindle speed, P_o(Kp) indicates the total static gripping force and P_c(Kp) the jaw force loss due to centrifugal effects at n (rpm). P_o can be determined by using a static gripping force gauge or by calculating the drawforce input and multiplying this quantity by

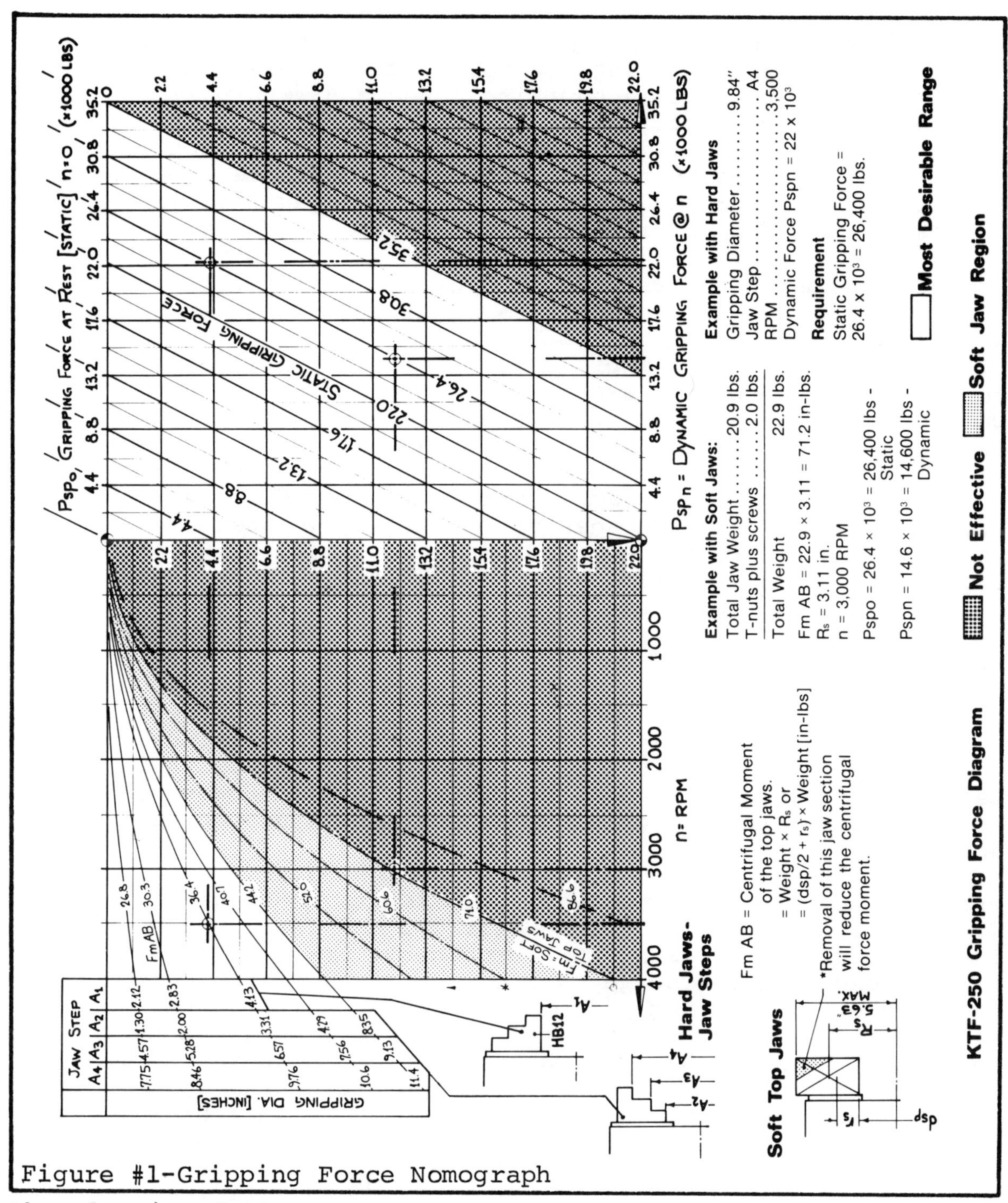

Figure #1-Gripping Force Nomograph

the clamping factor.

Jaw force losses become critical at high speeds since losses increase with the square of the speed. These calculations can become overbearing when constant chuck use is involved and different speeds and clamping diameters are desired. To alleviate this problem, nomographs are available from certain chuck manufacturers to aid in the calculation of dynamic jaw force. Figure #1 shows such a graph. This sample shows the Universal-Forkardt model 3KTGF-250 (10") countercentrifugal power chuck.

This nomograph is particularly helpful for several reasons. It dispenses with lengthy exponential calculations. It gives you the security of knowing how your chuck should ideally operate throughout the entire chucking range and with different top tooling. Maximum force values insure proper limits for workpiece rotation with solid or sturdy geometry. For workpieces of thin walled construction, it gives force limits which help to eliminate distortion problems. For soft jaw calculations, an easy to use formula is given to indicate safety areas of chuck performance.

It is easily understood that for optimum operation, the lightest possible jaw assembly is required, along with a jaw position which is located close to the chuck's center of rotation. Figure #1 shows this principle with jaw steps A4, A3, A2 and A1. These hard stepped top jaws can be positioned or reversed to achieve better dynamic gripping characteristics. Gripping a 4" diameter could be accomplished by either of two means, jaw step A2 or A1.

When looking at the schematic of jaw geometry showing steps A2, A3, and A4 it can be seen that the jaw's center of gravity is between steps A3 and A4, and nearest to A4. Gripping a 4" diameter on step A2 would make the jaw weight at the outer most position making r of equation #1 the greatest.

Gripping on step A1 in this instance would reduce the (r) dimension and bring the center of mass of the jaw close to the center of rotation. This makes gripping in this position most efficient because jaw losses are less at elevated speeds.

Gripping on step A4 would be most effective in the case of gripping an O.D. of 7.75 inches through 9.13 inches. The center of mass would be brought nearer to center. However, it is not always possible to grip on the most advantageous positions, due to workpiece geometry.

Since these options exist, it is imperative that the programmer indicate the proper jaw position to the operator. High speed applications become critical and failure to notify the operator could result in an accident.

Also note that the removal of a section of the soft top jaw as shown, would not affect jaw strength but reduces weight and moves the center of gravity (CG) closer to the center of rotation. These techniques can be helpful in achieving greater performance and insure a higher gripping force safety factor.

The example shown in Figure #1 with hard jaws at 3,500 rpm maintains 83% gripping efficiency. The variables have not been so critical, to allow for excellent efficiency. The example with soft jaws is not quite so effective. An efficiency of 55% at 3,000 rpm is realized since the jaw weight and (CG) have not been so favorable.

Various chuck manufacturers advertise concepts using jaw surfaces with serrations, machined directly onto the base jaws. Another technique allows for the mounting of collet pads onto the base jaws. These concepts attempt to reduce the inertia effects while reducing the bending moment of typically designed top jaws. These approaches are helpful to some degree, with bar stock workpieces which fit within the chuck bore.

They are attempting to approximate the concept of a collet type bar chuck. These approaches vary in applicability and significance of achievement. These chucks generally have large bore sizes with respect to outer diameter. Collet type chucks do not require such large O.D.'s since the method of guiding the collet is not done radially.

Collet type chucks are not affected by centrifugal force, since the pads are contained by the chuck body. Most bar machines are designed for use with collet chucks. They offer large bar sizes with small outer diameter restrictions. The diameter restrictions are a result of turret design. For optimum performance jaw chucks can not compete with collet chucks having small O.D. sizes and large bores. Jaw chucks do however have more versatile characteristics than collet chucks.

Chucks with large bores and small outer diameters generally have friction problems, due to jaw guide length reductions. The lack of sufficient jaw guide length causes the jaws to bind during movement. The above mentioned chucking approaches usually leave the jaw guides unprotected, and invite chips and grit to enter from lack of proper shielding. Earlier during the discussion of friction, we mentioned friction can consume up to 50% gripping efficiency.

The common fault of most jaw chuck collet pad approaches is the lack of peripheral support. Too often, this lack of support allows the insert to spring back at the outer edges. In practice, no finite accomplishments are realized. Rough stock is not perfectly round. A pad with surrounding circular geometry will not make even continuous contact. This can be seen by looking at the gripping pattern of the serrations on a rough workpiece. At the center, near the support, the serrations are embedded deep within the material's surface. The marks fade as the edges are approached. A collet type chuck has 360° peripheral support.

The effective driving force of both concepts shown in Figure #2 may be identical. Force is a product of stress and area. If the area is decreased, the stress or pressure is increased. Due to the decrease in area, the cobblestone serrations of Figure #2a will bury deeper into the material. The shear stress of a material may be greater than the coefficient of friction. Most of the contact area shown in Figure #2b will be driven by friction rather than by interlocked material to provide positive drive.

Parts having premachined gripping surfaces must be analized in a different manner. The premachined surfaces are almost perfectly round and can be considered as such. When pad forms or soft jaws are used, the gripping nature assumes a different pattern. The gripping characteristics are six contact surfaces instead of three. This can be seen in Figure #2 when assuming a three jaw chuck.

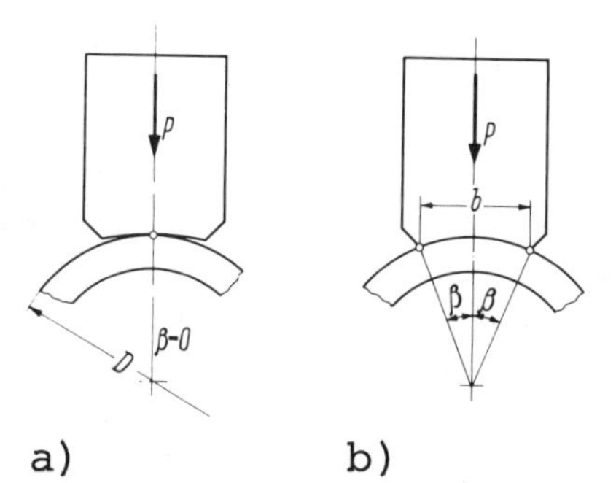

a) b)
Figure #2-Jaw Design

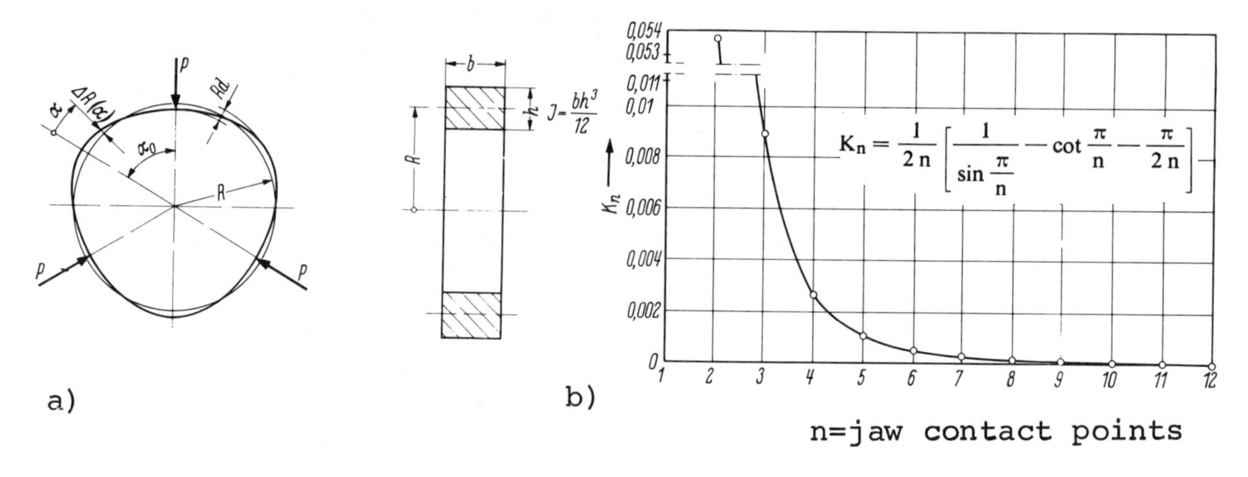

Figure #3-Distortion of Thin Walled Objects

Gripping marks on premachined surfaces are generally not acceptable. Therefore, smooth jaw surfaces are used and friction is the only driving force. The increased contact area in this case drastically increases the drive, while giving a wrap around effect on the workpiece. This is especially helpful in eliminating distortion of thin walled parts.

DISTORTION OF THIN WALLED OBJECTS

The distortion of thin walled parts is a very complex subject. To the manufacturing engineer, this can be a troublesome topic as well. Machining of such parts requires precise analysis due to their fragile nature. Gripping these components at high speeds can also be troublesome.

A high gripping force will distort the part, while too low of a force invites accidents. Forkardt has undertaken in depth studies to determine traits of commonality among thin walled objects and the results of gripping with numerous jaw arrangements. Figure #3 shows a segment of their research findings.

Figure #3a shows the distortion which occurs with a three jaw chuck and how it applies to the neutral axis of a ring shaped object. Figure #3b shows the effects of increased jaw contact as it relates to the K_n factor. K_n is a constant used to determine the total distortion (R_d) of the neutral ring axis. R_d is easily calculated from the following formula:

$$\left[\text{distortion of neutral ring axis}\right] \quad R_d = (K_n)(P_{sp})\frac{R^3}{(J)(E)} \qquad \text{equation \#3}$$

R_d(cm) is the total distortion of the neutral axis of the ring material. K_n is a dimensionless factor which is determined by the number of jaw contact points, P_{sp}(Kp) represents the sum of the jaw force, R(cm) is the neutral axis of the ring material, J (cm^4) is the polar moment of inertia, and E(Kp/cm^2) is the modulus of elasticity of the ring material.

K_n is given graphically and is determined by the equation given in Figure #3b. Additional information pertaining to Δ R and α are available from the chuck manufacturer. This topic

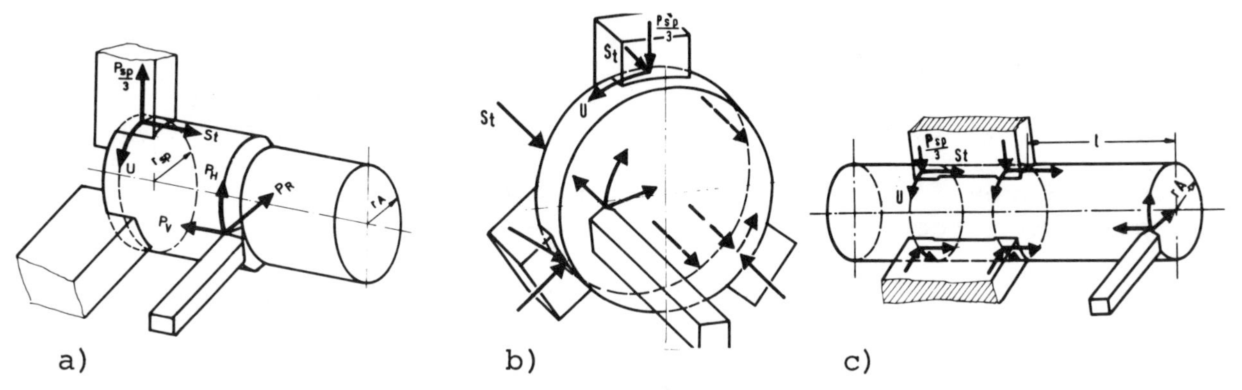
a) b) c)

Figure #4-Various Machining Exercises

becomes so complex that it cannot be covered in this text in any further detail. Equation #3 is often referred to as the Kohlhage equation after the man who did extensive research on this topic.

To eliminate distortion, the maximum permissible clamping force on a thin walled workpiece can be determined. We must also know the minimum possible clamping force to insure proper driving of the component. In critical cases, this is required so a force can be chosen within the extremes for optimum performance.

Figure #4 shows typical types of chucking examples. Figure #4a is a vector force diagram illustration. Figure #4b is of a flat workpiece where a bending moment from the cutting tool does not exist. Figure #4c is such a case where the moments (P_R x l) and (Pv x rA) exist.

Three main cutting tool forces are involved in the force diagram: P_H the main cutting force, P_V feed force and P_R passive force. P_V and P_R have counter reactions at the jaw by force St or by having the part placed against a tool stop. Stops such as a jaw step or the chuck body are common.

Drive is created by friction between the jaws and workpiece. This counter reaction must over come the main cutting force moment (P_H x r).

[1]This subject becomes complex due to the variety of different possible chucking procedures. A simplified equation giving the clamping force requirements for a specific situation is shown below. Equation #4 applies only to the machining of flat and tailstock supported components. Provisions for the effects of tilting and pullout are not included.

$$\left[\begin{matrix}\text{Force}\\ \text{Requirements}\end{matrix}\right] \quad P_{sp} = S \cdot \frac{P_H \cdot dA}{\mu_{sp} \cdot d_{sp}} = S \cdot \frac{N \cdot 716 \cdot 2}{n \cdot \frac{d_{sp}}{2} \cdot \mu_{sp}} \qquad \text{equation \#4}$$

1) portions of the given information were extracted from the Forkardt Workholding-Chucking techniques manual. Paul Forkardt KG, Duesseldorf, West Germany, 1974.

KEY TO SYMBOLS

P_{sp}=clamping force (sum)
U =circumferencial force
St =supporting force
P_H =main cutting force
P_V =feed force
P_R =passive force

S =safety factor
dA=cutting diameter=2r
d_{sp}=clamping diameter=$2r_{sp}$
N =power rating of machine
n = spindle speed
μ_{sp}=friction value

VARIOUS FRICTION VALUES WITH DIFFERENT JAW FORMS

μ_{sp}=.15 with smooth jaw surfaces
=.25 with cobblestone serrations on jaws
=.35 with sharp jaw serrations

VARIOUS FRICTION VALUES WITH DIFFERENT WORK SURFACES

μ_{sp}=.26 on unmachined component
=.17 on rough machined component
=.10 on finish turned component

The main cutting force can be calculated by the simplified formula below:

(main cutting force) $P_H = a \cdot s \cdot K_m$ equation #5

(a) indicates depth of cut in mm,(s) feed in mm/rev, and (K_m) a dimensionless material constant with dull tool factor included. This factor can be found in almost any technical journal where recommended cutting tool speeds and feeds are indicated. These formulae were derived by H.J. Warnecke for his dissertation at the Braunschweig Technical college in 1962.

ACCEPTABLE JAW FORCE LOSSES AND RATINGS

The reason for a lack of standardization among chuck manufacturers'ratings is quite obvious. The numerous variables coupled with the multitude of machining exercises make standardization almost impossible.

Different manufacturers accept different jaw force loss percentages. Some manufacturers rate the maximum rotational speed when only one fourth of the maximum static jaw force remains, while others accept one half of the static force. These percentages are typical for the rating of conventional chucks.

One half the static rating should give an adequate safety factor. However, little may be accomplished at so low a speed. One fourth the static rating leaves little regard for safety. Using a principle of one third the maximum static force for the rating of standard chucks is practical.

Chuck manufacturers attempt to supply devices which will function adequately for general purpose use. Exotic machining applications deserve additional attention. All high speed applications should be given critical scrutiny, since some designs do provide greater performance and safety features than others.

CHUCK DESIGNS

A multitude of chuck designs are available to industry. This magnitude is quite clear, when it is understood that each chuck manufacturer offers several models. This variety can be broken down into three major catagories for mode of operation. Of these three types, the lever and wedge-hook models are most popular and operate in similar fashion. The wedge-block version, functions under a completely different principle.

Lever and wedge-hook type power chucks have served industry for many years. Development of high speed chucks, made the adaptation of these two principles with counter-centrifugal balancing methods the most obvious approach.

Lever Chucks

A typical lever chuck is shown in Figure #5. The diagram depicts the mode of operation. Push and pull forces created by axial movement of a hydraulic or air operated cylinder, induce an "in and out" radial jaw movement. Generally the cylinder is located at the opposite end of the spindle. The chuck and cylinder are connected by means of a drawtube or drawbar. This transfer of motion is accomplished by rotation of levers, located by cross-pins at each jaw. A 90° transformation of motion is accomplished by rotating the lever to change axial movement into radial motion. The difference in link length of the lever accomplishes different clamping factors or mechanical advantages which were discussed earlier.

Wedge-hook Chucks

The wedge-hook chuck shown in Figure #6 also uses a similar principle to change the axial movement of the cylinder into "in and out" radial jaw motion. Wedges are used instead of levers. The wedge has a Tee configuration and is located at a 10°-20° incline. The change in incline produces a different clamping factor. Action of the cylinder causes the Tee section to slide on the puller ramp, causing "in and out" radial jaw motion. The term wedge applies, since the small angle acts to provide a wedging action which serves as a semi-sticking surface. This sticking helps to a degree, to insure against jaw force loss even if the cylinder force is diminished.

Both of these approaches have a limited jaw stroke. To achieve various clamping diameter extremes, the top tooling must be relocated. This is a time consuming effort.

The inner components require precision machining to insure equal jaw movement of all three jaws. Figure #6 also shows that the piston rod can be ground and guided within the spindle bore. This prevents the puller from tilting. The puller is the part connected to the wedgehook or lever. Tilting of this element causes unequal jaw movement and loss of accuracy. Guiding this element reduces run out and increases repetitive accuracy.

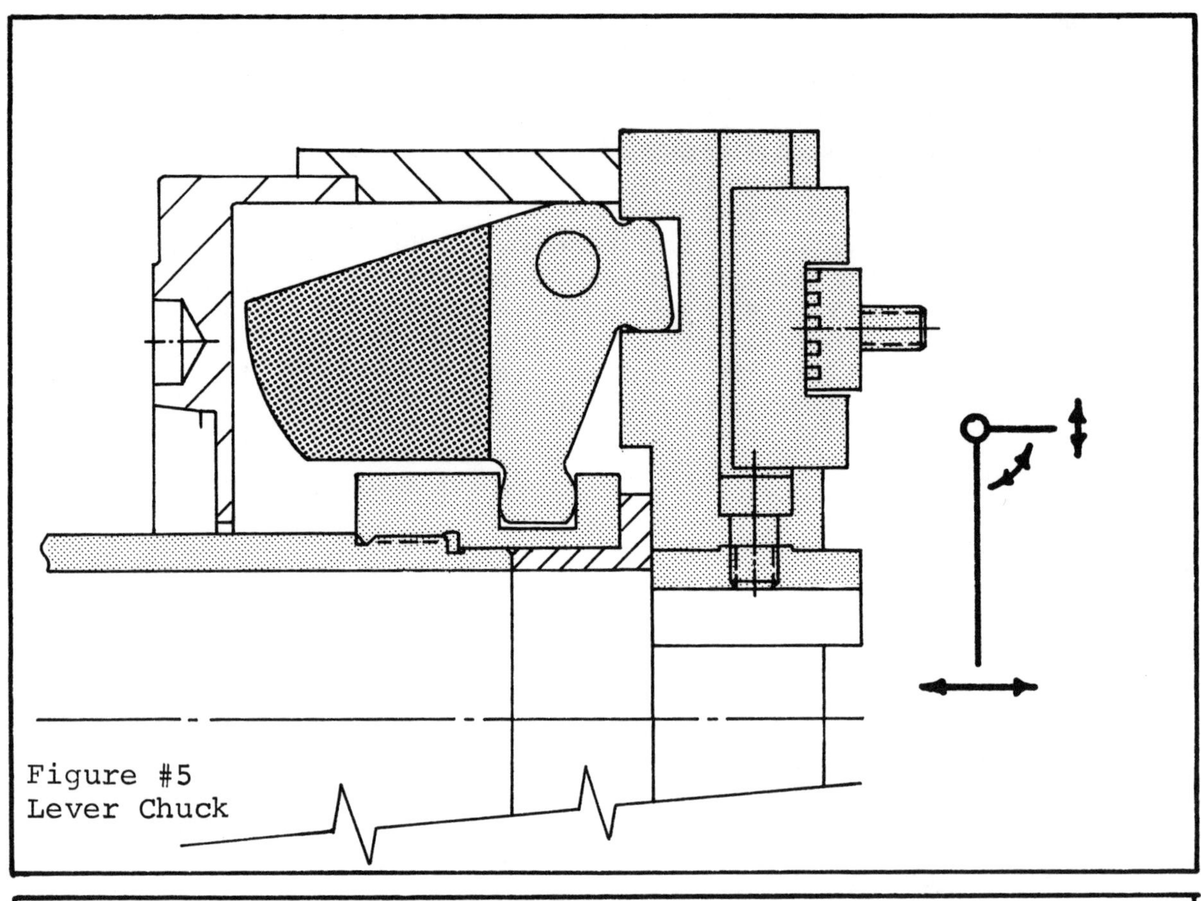

Figure #5
Lever Chuck

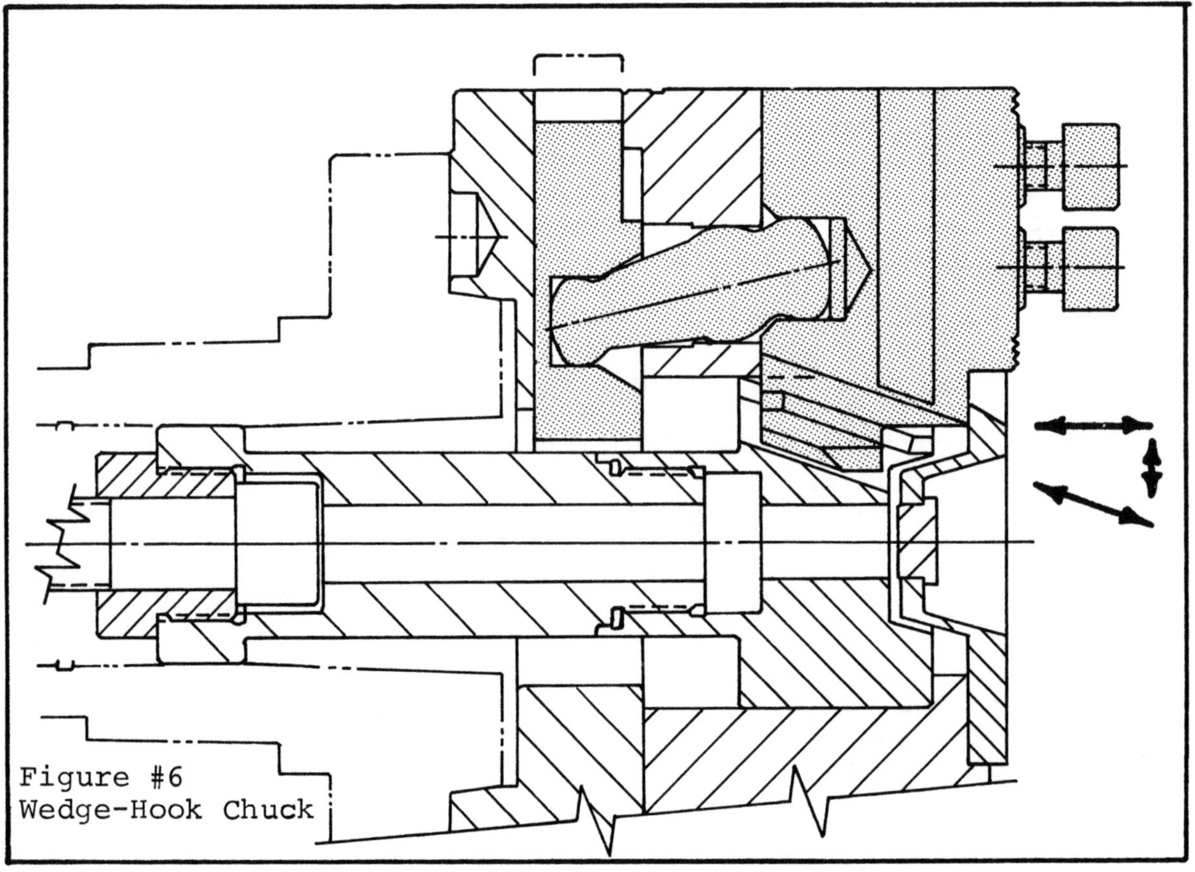

Figure #6
Wedge-Hook Chuck

COUNTER-CENTRIFUGAL BALANCING

The application of counter-centrifugal balancing methods (flyweight compensation) is done in a variety of ways. Each manufacturer applies a slightly different technique, which they feel provides the best solution.

Figures #5 and #6 show different techniques of applying this flyweight action. The most common method used on lever chucks is shown in Figure #5.

The shaded portion located at the rear of the lever shown in Figure #5, is the mass which counter-reacts the combined mass of the base jaw, bolts, Tee-nuts and top jaw. This shaded portion is a solid connection between the cylinder, drawrod and the base jaw.

Figure #6 shows a slightly different method where the flyweight mass is located independently from the drawrod. Only a solid connection between the counterweight and the base jaw is offered. This was done since the centrifugal force directly relates to the jaw and counterweight, not the cylinder drawrod. Connection to the drawrod was felt to function as a dampener.

Normally the counterweight mass is designed for a "mean" jaw position with standard top tooling. This counterweight is of a specific weight. The inertia of the top jaw can vary drastically, due to the range of positions and reversal features. The base jaws are generally used in a normal mean position of the limited jaw stroke. The base jaw therefore causes no problems for the calculation of centrifugal force. The difference of top jaw position and inertia makes the counterbalancing difficult.

Nomographs such as shown in Figure #1 show the effects. Small gripping diameters can be overly compensated for and thus, an increase in gripping force can be accomplished at higher speeds. Large gripping diameters or rather large inertia factors are under compensated for, and greater force losses are encountered.

Some manufacturers offer interchangeable weight packages which can be mounted on the counterbalancing system. This works fine for extreme cases. An operator must be very careful not to forget to change the balance weight, if the next set-up operation calls for an increase in inertia. This could very easily become a safety hazard.

Wedge-block Chucks

Figure #7 shows the wedge-block type power chuck. It uses the principle of a rotary actuator instead of a cylinder with axial movement. The actuator is connected to the chuck by means of a torque tube which is splined on both ends. The mating spline can be seen on the chuck's back face, shown as the innermost portion of Figure #7a. Rotation of the torque tube moves the helical gear. The gear is responsible for moving the three wedge-blocks simultaneously. The wedge-blocks have inclined rack teeth on two sides. One side mates with the helical gear. The other rack tooth side mates with the base jaw. A

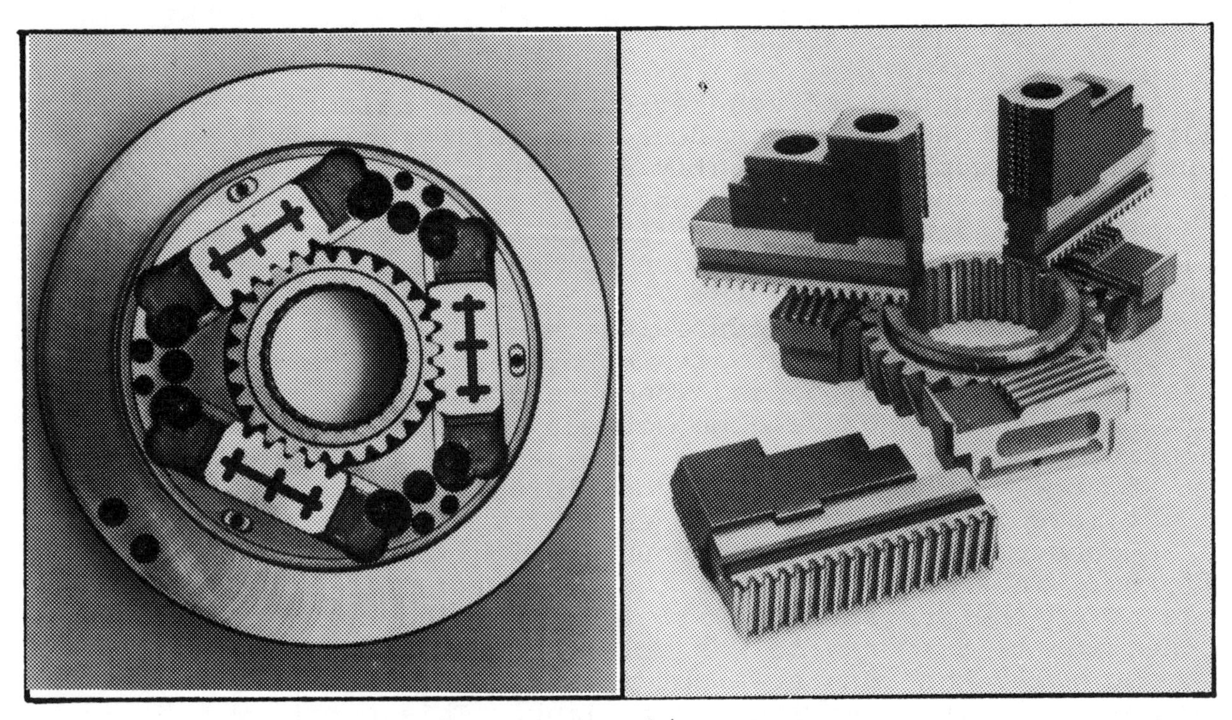

a) b)

Figure #7-Wedge-Block Chuck

portion of this rack on the base jaw side has been removed as seen in Figure #7b. Normal rotational movement of the actuator operates the jaws in a fully engaged rack tooth stroke. Over-riding of the actuator allows for disengagement of the mating rack teeth between wedge-block and base jaw. This allows for rapid disconnect and changeover of jaws or jaw position. This feature provides a substantial time savings over conventional power chucks, while also offering a larger jaw stroke. The mechanical advantage of the wedge-block chuck is so great that jaw forces in the range of three times the magnitude of conventional chucks are realized.

This chuck does not have centrifugal counter-balancing since, even if two-thirds the jaw force is lost, its force is greater than static flyweight chuck maximum readings. This chuck can be used for high speed operation where solid or sturdy workpiece geometries are to be machined. Pressure reduction allows for reduced clamping force on thin walled objects. A variation of this wedge-block design has been made, which allows for ultra-high speed operation, of a 6" diameter chuck at 10,000 rpm.

REFERENCES

Chucks for High-speed Turning, R.N. Stauffer, Manufacturing Engineering, March 1977

Counter-Centrifugal Chucks for High Speeds, Tooling & Production, April 1977

Einfluss des Futters auf die Verspannung verformungs-empfindlicher Werkstuecke, H. Antoni, Werkstatt und Betrieb, Heft 12, 1976

Faster Chucks with N/C, J. Thornton, AMM/MN-Tooling Section, May 16, 1977

Forkardt Workholding--Chucking Techniques, Paul Forkardt KG, 1974

High-Speed Chuck & Cylinder Combos, Metlfax, July 1977

New Chucks Meet High-speed Needs, B.A. White, American Machinist, McGraw-Hill, June 1977

Power Chucking for Profit in NC Applications, W.A. Mossner, Tooling & Production, September 1976

The Quick-Jaw-Change Chuck...A New Way to Cut NC Downtime, W. A. Mossner, Modern Machine Shop, December 1976

Time to Come to Grips with NC Chucking Problems, W.J. Reed, Machine and Tool Blue Book, July 1975

Presented at the SME Precision Machining Workshop, May 1977

Diamond Turning and Flycutting for Precision

by P. Donald Brehm
Pneumo Precision Inc.

Diamond turning and flycutting of precision non-ferrous metal and plastic parts is a method of generating a high degree of form and finish accuracy at low cost. The basics of diamond machining for precision are presented in this paper together with examples of diamond turning and flycutting.

INTRODUCTION

Precision machining may be defined as the production of surfaces by material removal processes to geometry (form) accuracy of .000,020 (0,5um) or better and surface finish of 4.0 microinch AA (0,1um AA) or better. Although not all workpieces may need this level of accuracy, the machines and tools involved should display this level of capability.

Precision machining at this perfection level is accomplished with specialized machine tools using single point diamond cutting (for non-ferrous metals and plastics) and grinding (for ferrous and harder materials. Although tools other than diamond (ceramic, borazon, coated carbide, etc.) can be used, only diamond tools can consistently produce work to finishes of 4.0 microinch and better - thus this paper will discuss turning and flycutting with diamond tools only.

DISCUSSION

Diamond as a cutting tool - single crystal diamond has a number of unique physical properties (1) making it the ideal choice for a cutting tool material for precision machining. Diamond has the highest known wear resistance of any cutting tool material so there is little edge wear to cause size changes in the work. Diamond has the lowest known coefficient of friction of cutting tools so that chips slide across the tool face without causing any built up edge. Diamond also has the lowest coefficient of expansion together with the highest thermal conductivity. This means any heat generated in cutting (very small anyway because of the sharp edge and low friction) is carried away from the cutting edge without causing tool expansion. Diamonds have extremely low compressibility and thus do not deflect under cutting forces. Diamond tools can be sharpened, using specialized edge lapping techniques, to an edge quality capable of microinch level depth of cuts and surface finish such that optical reflectance and scatter of cut surfaces are near the best optical polishing results. The best diamond tools require crystal orientation by X-ray diffraction techniques since diamonds have directional hardness and must have the cutting edge oriented to the proper crystal axis. The mounting of the diamond is critical as is the final low vibration level edge lapping with diamond micro abrasive. Diamond tools can be relapped a number of times to restore edge sharpness lost by cutting wear.

<u>Cutting speeds</u> - we routinely operate diamond tools at cutting speeds up to 12,000 surface feet per minute, both on continuous cuts (turning) and interrupted cuts (flycutting). Surface finish appears to be relatively independent of tool speed. We

use the high speeds to minimize production time and reduce temperature effects since work and machine temperature do not have as much time to change during the shorter cutting cycle.

Cutting feeds and finish - we normally perform diamond turning and facing and diamond flycutting at feed rates in the range of .0004 to .004 inches per revolution. (0,01 to 0,1 mm/rev) using a round nose diamond tool with a typical nose radius of .200 (5mm). Using the approximate equation for theoretical finish with a round nose tool [2] (see figure 1)

$$\text{peak to valley finish} = \frac{(\text{feed per rev})^2}{8\,(\text{tool radius})}$$

these feeds result in theoretical peak to valley finish values of 0.1 microinch to 10 microinch. Since peak to valley finish is normally about 3 to 4 times A.A. finish these values fall within our definition of precision machining. Since finish is determined not only by feed rate but also by machine movement errors, machine internal and external vibrations and tool edge quality, we have found that round nose tools can give exceptional results even at production type feeds. Where still higher production feed rates must be used, flat nose (facet type) diamond tools can be used. Flat nose tools have the disadvantage of requiring precise alignment to the work but once adjusted can produce parts at exceptionally high production rates.

Diamond turning - including diamond facing and boring is one of the best applications of diamond tools in precision machining. Turning can be used for precision machining of simple geometric forms or for computer controlled contour turning. We have found diamond turning ideal for machining workpieces such as computer memory discs, printing gravure rolls, photo copy rolls, metal mirrors, plastic lenses, lense mounts, guidance system components, ordnance parts, and other parts where the cost of lapping and polishing is prohibitive or where the part shape (cones, contoured or internal surfaces, etc.) or material (soft metal or plastic) does not lend itself to lapping or polishing.

A typical production example is the facing of aluminum memory discs at 2000 rpm - see figure 2. These parts are 14.000 dia x .075 thick (350 mm x 2 mm) and require the use of a special vacuum chuck. Depth of cut is typically .0005 per side with a .200 radius tool and feed of .002 (50um) per revolution. Finish obtained is better than 2.0 microinch AA. In this application we have used flat nose tools also at feed rates up to .0015 (40um) per revolution, giving a finish of 0.8 microinch AA. Another example of diamond turning on a MSG-300 lathe is a thin walled copper part requiring finishing of a conical I.D. and O.D. Finish requirement was not critical (16.0 microinch AA) but roundness and wall thickness variation was. The sharpness and wear life of a diamond tool allowed the thin wall to be cut without deflection. The part was held in special conical vacuum chucks, depth of cut was .001 (25um) for roughing and .0005 (12,5um) for finishing.

A third industrial diamond turning example involves the finishing of a thin shell aluminum tube approximately 12.000 (300mm) long. Using the MSG-300 lathe with air bearing headstock and tailstock and air bearing tool slide. figures 3 and 4 - it was cut at 2400 rpm and a feed of .004 (0,1mm) per revolution with .001 (25um) depth of cut. Surface finish achieved measured under 2.0 microinch AA. The same part was finished to 0.3 microinch AA with slower feed and .0005 depth of cut.

The above examples were achieved in a normal air conditioned shop environment. Special precautions were taken with chip removal, and special chip extractors were used for each operation to remove the continuous chip without allowing it to touch and thus mark the cut surface of the work. Both tool geometry (rake angle and clearance) and cutting fluids have been found to have an influence on tool life and work finish in the above examples. In all above cases coolant was applied directly at the tool tip by air/coolant mist.

Diamond Flycutting - also known as single point diamond milling is an ideal method of producing flat, optically smooth surfaces. A single diamond tool is rotated at high speed (up to 12,000 feet per minute) while the workpiece passes by. The tool takes one continuous chip across the work per revolution. With a rigid machine and with the sharpness of the diamond tool there is very little surface edge effect at the beginning and end of this interrupted cut (typically 1.0 microinch or less). In conventional grinding and lapping of flat surfaces, control of edge flatness can be a common problem.

To obtain tool relief so that cutting occurs only during the leading 180° of each revolution the trailing edge of the flycutter is normally set slightly higher by about .0001 (2,5um) by tilting the spindle axis. The concave error that this can cause in the workpiece can be expressed by: (see figure 5).

A typical flycutting example is the precision machining of the aluminum face of a part to a flatness of one light band (11 microinches) with a surface finish of 5 microinches peak to valley. Part size is 3.000 x 9.000 (75 mm x 225 mm). Parallelism to the opposite face must also be held to close limits and the surface to be cut has a series of holes. Flatness must be maintained to the very edge of these holes. This part was processed with a 6.000 diameter (150mm) flycutter, revolving at 6000 rpm. Feed rate was .0006 (15um) per revolution with a depth of cut on the final cut of .000,250. The diamond tool had a .200 radius (5mm), was adjusted for slight positive rake in a special adjustable tool holder, and air/coolant mist was used. Chips were removed by a vacuum shroud round the cutter. The machine used was a Pneumo MSG-500. The surface produced is optically reflective with no feed lines discernable to the naked eye.

We have used flycutters up to 12.000 diameter (300mm). For higher production, multiple tool flycutters can be used with stepped tools although the finish will not be as good since the roughing tools can effect the finishing tool. Dynamic balance is very critical and tools must be weighed to within 0.1 gram and replaced with a tool of the same weight. Our work has been done with round nosed tools since the adjustment of a flat nosed tool in a rotating flycutter would be very time consuming. Flycutting can also be done with the workpiece on a rotary table rather than a slide and by tilting the worktable axis spherical surfaces can be generated.

Workpieces that have been successfully cut using the diamond flycut process include multi-facetted optical scanners - see figure 6. Ideal since angles, flatness, and parallelism to an axis must be held), flat hydraulic motor and pump plates and other flat surfaces. In the case of hydraulic plates, the diamond flycutting leaves a surface flat and smooth enough to provide a metal to metal seal under high pressures.

General Considerations - for successful precision machining with diamond tools, a number of criteria must be satisfied:

1. Compatible workpiece material - diamonds cannot cut all materials.
2. Precise machine movements - spindle, slides, etc. - accuracy and smoothness.
3. Lack of vibration in machine and isolation from external vibrations.
4. Proper quality diamond tools - edge sharpness and tool geometry.
5. Chip removal.
6. Temperature control.
7. Material condition - stress level, cleanliness and purity.
8. Operator training and attitude.
9. Proper coolant.
10. Proper gaging - non contact type if surface scratching

is to be avoided.

11. Machine controls, programming, servo-resolution, etc.

This list is not in order of importance since each criteria can be of greater or less importance depending on the workpiece size, shape, and level of perfection desired. It is most important to consider the total diamond machining system since each portion of it can spell the difference between success and failure in precision machining to millionths.

CONCLUSION

The demand for precision parts that cannot be made by lapping or polishing techniques is becoming greater each year. Precision diamond turning and flycutting is a way of producing many of these parts both on a laboratory and production basis. The specialized machines and tools needed for such precision machining are now available.

REFERENCES

1. Tolansky, S., "The Strategic Diamond".
2. Donaldson, R. R., "Nomograms for Theoretical Finish Calculation with Round Nose Tools" UCRL-76047.

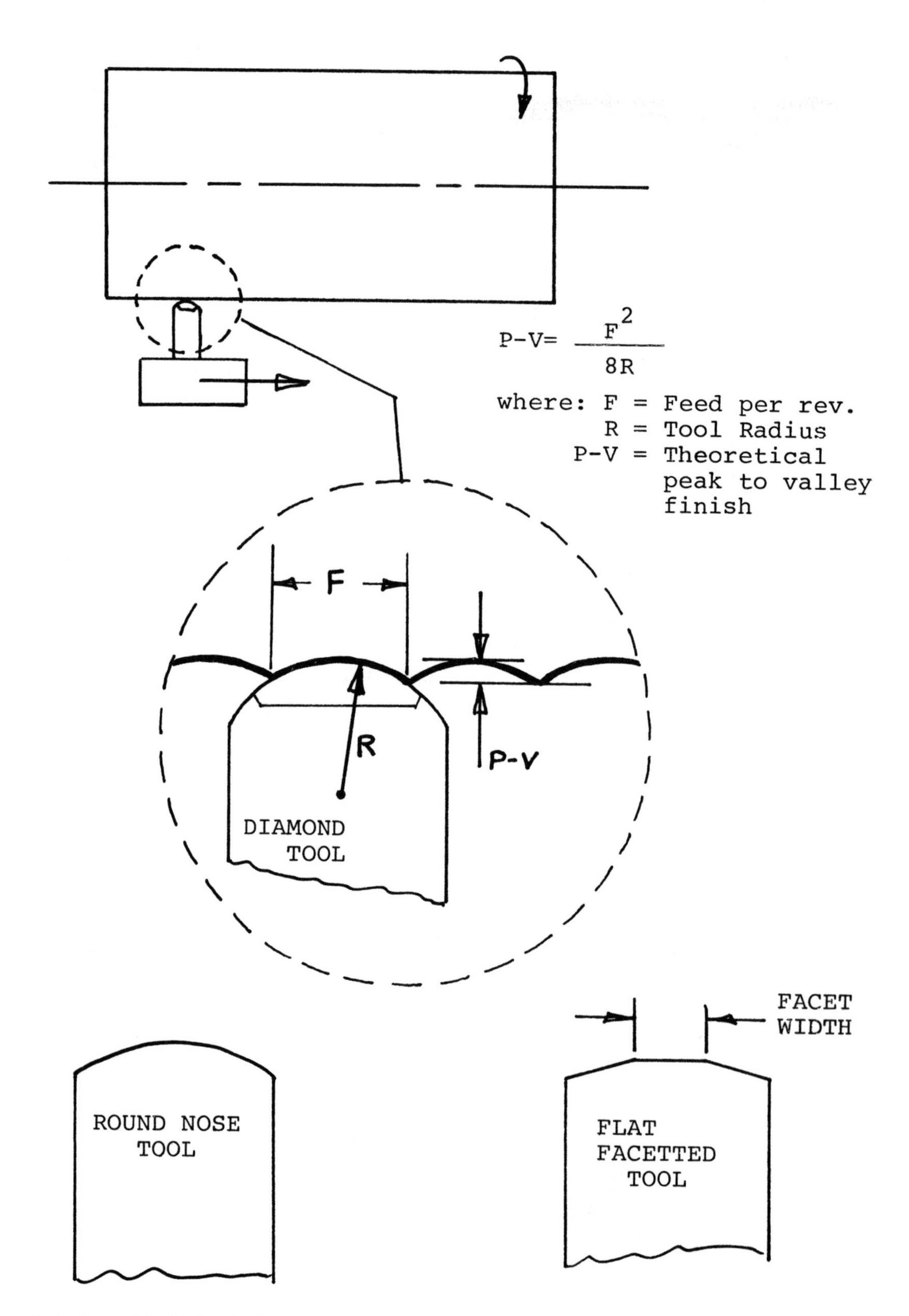

FIGURE 1. THEORETICAL FINISH WITH ROUND NOSE TOOLS

Facing aluminum memory disc at 2000 rpm. Part held on vacuum chuck. Toolholder with chip extractor tube shown at top of fine infeed slide.

FIGURE 2. DISC FACING

FIGURES 3. & 4. Diamond Lathe set up with air bearing headstock and tailstock and air bearing tool slide for turning of tubes to 0.3 microinch AA finish.

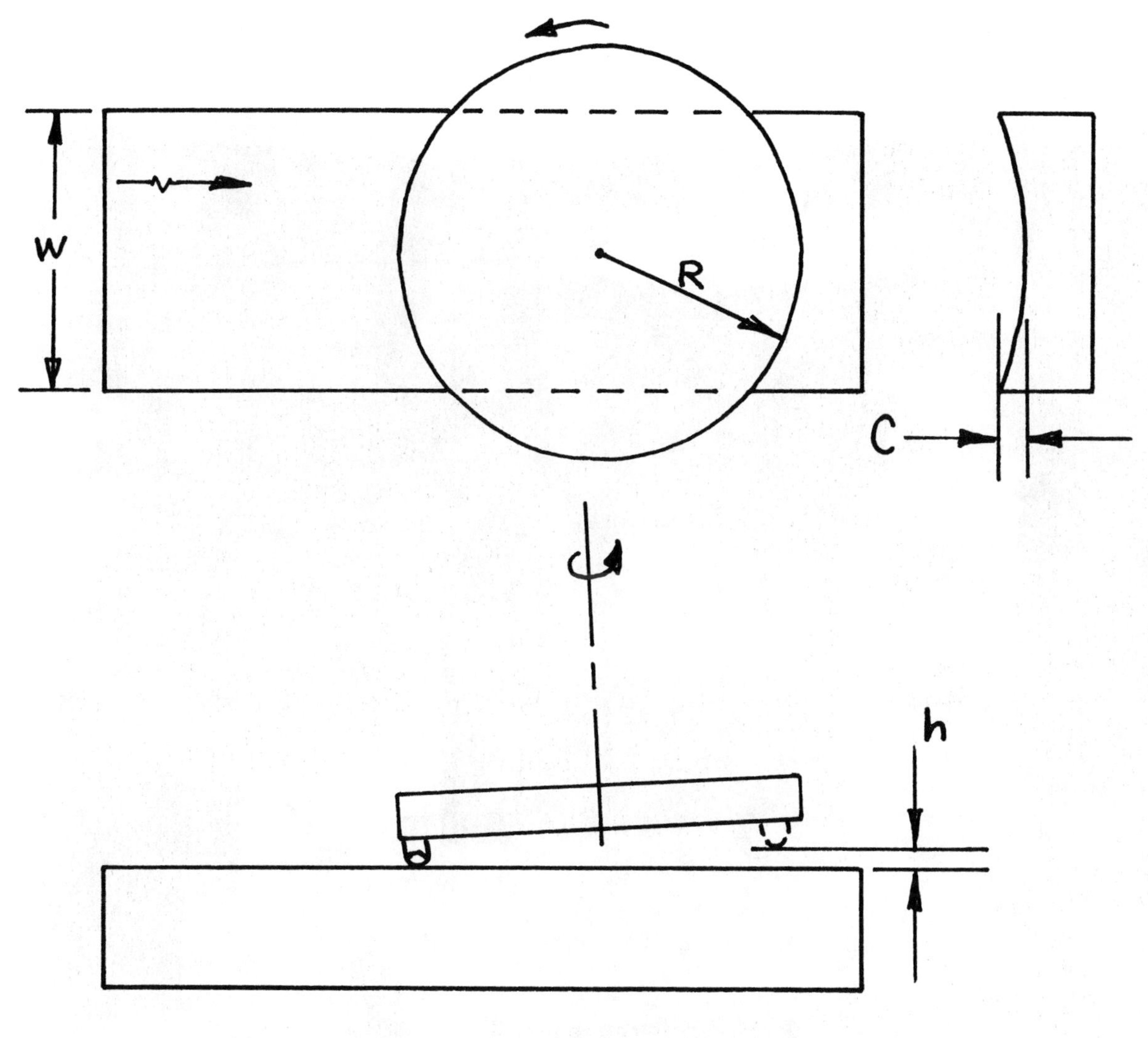

$$\text{CONCAVITY (C)} = \frac{h}{2} - \frac{h}{4R}\sqrt{4R^2 - W^2}$$

where:

h = Amounts trailing edge is lifted
R = Flycutter radius to cutter
W = Workpiece width

FIGURE 5. Concavity resulting from tilting of flycutting axis.

FIGURE 6. Diamond Flycutting of multi-facetted aluminum laser scanner mirrors. Parts held in indexing fixture (up to 10 cut at one loading). Flatness held to 1/8 wavelength per facet. Cutter speed 12,000 feet per second.

CHAPTER 2

TURNING

Reprinted from *Machine and Tool BLUE BOOK*, February 1983

New Insert Designs Update Turning Technology

by Jack Lynch
Sandvik Metalworking Products Division, Fair Lawn, New Jersey

Higher standards on tolerances, surface finishes and productivity have prompted development of a new generation of positive rake inserts.

By JACK LYNCH
Product Specialist, Turning Tools
Sandvik Metalworking Products Div.
Fair Lawn, N.J.

Other than the improvements in grades and coatings, positive rake turning has been accomplished with much the same basic tools for 20 years. For example, flat top inserts with mechanical chipbreakers are still the most popular choice when positive rake tooling is desired. However, this design has always had inherent disadvantages. The top clamp used with loose chipbreakers can hinder chip flow, resulting in surface finish damage, machine downtime and reduced productivity. This can be especially troublesome during boring operations, a common application for positive rake tooling.

To overcome these disadvantages, new positive rake inserts and clamping systems have been developed.

With recently perfected grinding and pressing techniques, very close tolerance cutting edges can be produced. New inserts are produced with pressed-in chipbreakers without loss in cutting edge strength. In addition, the land width around the cutting edge of the inserts can now be more accurately controlled which results in consistent chip control. In fact, a Sandvik round insert contains 72 individual chip formers or dimples built into the cutting edge.

Clamping system developments include a design in which a Torx screw, which clamps the insert, fits well below the surface of the insert to allow an unobstructed flow of chips. The socket-type design of the screwhead provides maximum contact between screw and screwdriver, decreasing the tendency of slippage or stripping.

In boring operations, the relatively small clamping area needed to secure the insert means greater access to small diameter holes where tool clearance is a problem. This feature combined with insert performance makes this design ideal for special tooling.

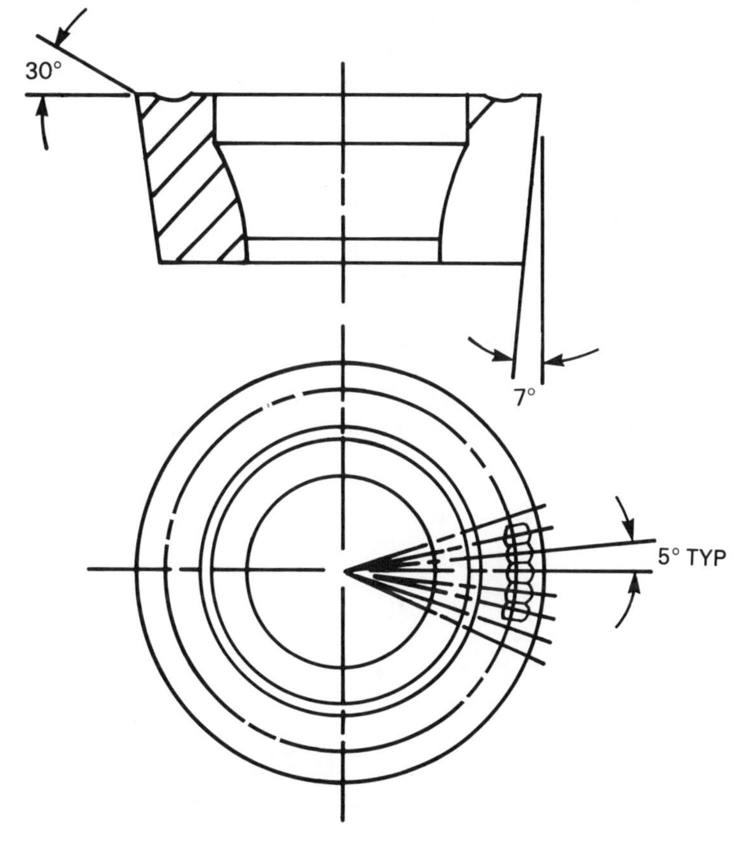

Through improved pressing techniques, greater accuracy and detail around the cutting edge are now possible. In this area, Sandvik's 5 mm round insert with 72 individual chipformers provides users with an insert designed to optimize chip control.

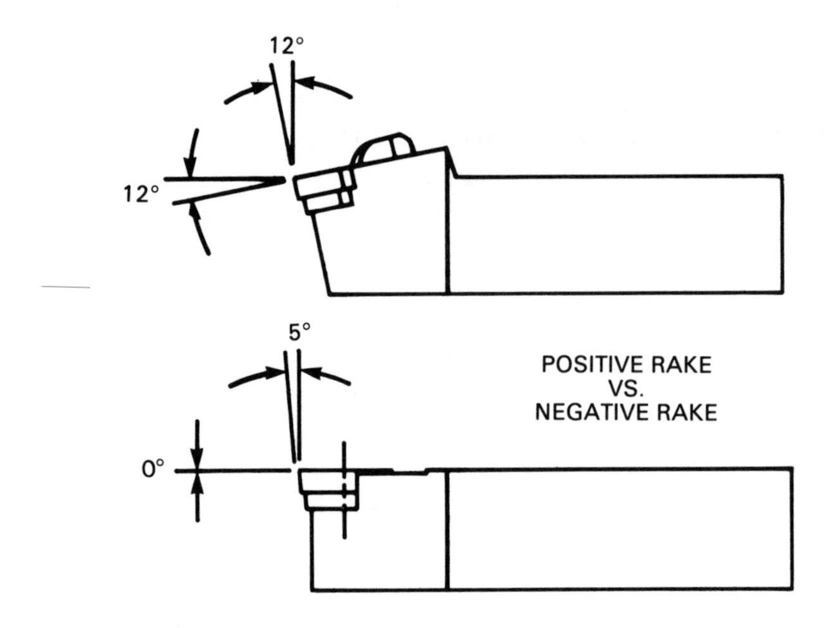

A positive rake insert does not have to be inclined to obtain clearance. Lower cutting forces are generated which result in less chatter and superior surface finish.

In general, these new positive rake systems have increased performance considerably by offering a free-cutting, high-strength cutting edge. Geared to meet the tooling needs of small, high-precision components, their most common application will be in single and multispindle automatics, smaller sized NC and CNC machines, and center and toolroom lathes.

A Sandik T-Max U 35-degree profiling tool provides an example of the performance that is possible from these new positive rake configurations. The tool provides consistent chip control with lower cutting forces over a wide range of applications.

When compared to a standard 35-degree negative rake VNMG copying insert, a stronger single-sided positive insert proved to have less insert breakage and vibrations. Additionally, lower cutting forces helped produce a part with better utilization of machine horsepower. Because of this, an overall increase in productivity was achieved with fewer scrap parts.

On the other hand, VBMM inserts are held in a neutral position in the toolholder with the top face sitting at 0 degree. Side clearance required for cutting is built into the insert. A standard VNMG insert, with its basic negative design, must be tilted in the toolholder to as high as minus 10 or minus 12 degrees in order to obtain the correct cutting clearances. This negative rake creates much higher cutting forces on the relatively weak 35-degree point.

In plunging operations, the additional clearance on the sides of the new center-screw insert systems means less rubbing or drag on the insert. This is most evident on the trailing edge as the depth of the plunging operation increases. In general, test reports show that a new 35-degree copying insert can take up to double the depth of cut when compared to a similarly sized VNMG insert.

Insert breakage is greatly reduced at these higher productivity levels due to the support area offered by the single-sided design. This consideration is extremely important in automated systems which give a high priority to cutting edge security.

When productivity is considered, the new inserts have proven successful in materials ranging from stainless steel and cast iron to heat-resistant alloys and aluminum. • • •

Presented at the SME Increasing Productivity
with Advanced Machining Concepts Clinic, August 1982

The Forces in Turning

by Bala K. Srinivas, CMfgE, PE
Beekay Associates

INTRODUCTION

In any metalcutting or chip making process, the tool accomplishes the cutting action by overcoming the shear strength of the workpiece material. In so doing, the cutting tool and the toolholder transmit the resistance encountered at the tool/workpiece interface back to the machine tool which supports both the tool and the workpiece. The development of workpiece materials with higher and higher shear strengths in recent years has tended to increase the load on various machine-tool components. Also, the shear strength of the workpiece material is not the only parameter that affects the loading on the machine tool. Machinability parameters themselves—speed, feedrate, and depth of cut—also affect the forces in metalcutting. The development of new cutting-tool materials with higher hardness and/or transverse rupture strength enables them to be used at higher speeds, feedrates, and depths of cut for higher productivity. Such high machinability parameters impose greater loads on the machine tool than heretofore. It is essential, therefore, to know what levels of forces are being generated in today's and tomorrow's machining processes.

This paper describes an experimental procedure to measure the forces in one metalcutting operation—axial turning—and concludes by developing a mathematical model of the forces in turning. The mathematical model can then be used to predict the forces in any axial turning operation.

THE MECHANICS OF METALCUTTING

The general relationship

$$F \ \alpha \ \tau_w^a \ v^b \ f^c \ d^g \qquad (1)$$

where F = Cutting or Feeding force, lbf

τ_w = Shear strength of the workpiece material, psi

v = Cutting speed, sfm

f = Axial feedrate, ipr

d = Depth of cut, inches

and a, b, c, and g are exponents to be determined

expresses the dependence of the cutting and feeding forces in axial turning, on three machinability parameters. The shear strength of a material is proportional to its tensile strength which, in turn, is proportional to its hardness. Manufacturing engineers, NC programmers, and machine-shop personnel are more familiar with the hardness of a workpiece than its tensile or shear strength. Furthermore, hardness is more conveniently measureable in a machine shop. Hence, equation (1) can be rewritten as:

$$F = K\ (Bhn_w)^a\ v^b\ f^c\ d^g \qquad (2)$$

where K = a constant

and Bhn_w is the hardness of the workpiece on the Brinell scale

In an axial turning operation, the three components of force are:

1. The tangential cutting force, F_c
2. The axial feeding force, F_f
3. The radial thrust force, F_r

The relationship of these three forces is shown in figure 1. Data from many researchers and from many experiments has shown that, of these three forces, the tangential cutting force, F_c has the highest value, followed by the axial feeding force, F_f, and the radial thrust force, F_r. As a rule of thumb, the value of F_f is one half of that of F_c, and F_r has a value equal to one half of that of F_f. This is best remembered as the "1, 2, 4 rule."

EQUIPMENT AND INSTRUMENTATION

The experimental procedure described in this paper covers a series of test-cuts in which two of the three components of force were measured and recorded. The force component not measured was the least of the three, F_r. The equipment and instrumentation used for the tests consisted of:

1. A Warner & Swasey SC-13, 4-axis NC turret lathe with 30-hp motor, variable-speed spindle drive, W & S 4-axis contouring control, and a horsepower meter which showed horsepower input to the spindle-drive motor.

2. Kennametal's style KSBR-164 toolholder. This is a negative-rake toolholder with 1-in. square shank, 6 in. long. Tool signature was: -5, -5, 5, 5, 15, 15, 1/32. The length of the toolholder was reduced to 4 in. to minimize overhang. The toolholder itself was converted into a force transducer by bonding resistance-type strain gages, type EA-06-062TT-120 to form two half-bridges as shown in figure 2.

3. Workpieces were cylindrical bars of various steels and various hardnesses, preturned to 7.9 in. dia x 24 in. long.

Both ends were faced and centered. The workpiece was held in a 10-in., 3-jaw, self-centering scroll chuck and the outboard end was supported by a "Rohm" revolving center on the end turret.

4. The cutting- and feeding-force recording system consisted of two Bridge Amplifier Meters (BAMs), one for each set of strain gages, and a two-channel "Brush" paper-tape recorder. The strain gages and other circuitry are detailed in figure 2.

TEST PROCEDURE

Axial turning cuts were made at various speeds in the 300 to 2,000 sfm range, depending on axial feedrate, depth of cut, and workpiece hardness. The feedrate range was from 0.005 ipr to 0.025 ipr; the depth of cut ranged from a low of 0.050 in. to a high of 0.250 in. Workpiece hardness ranged from 204 Bhn to 453 Bhn. The machine-tool control was set in the Manual Data Input (MDI) mode. The length of a cut varied from 3 in. to 12 in.

Because the toolholder itself was a force-transducer, all screws and bolts holding the cutting insert in the tool-holder and the toolholder in the turret station were tightened with a torque-wrench. The force-measuring and -recording system was routinely calibrated once a week by applying known loads on the toolholder through a calibrated hydraulic jack. Also, the system was calibrated after catastrophic tool failures, as and when they occured during the test period.

For a typical test-cut, the desired speed, feedrate, and depth of cut were manually input into the machine-tool control system and the BAMs and the "Brush" recorder were turned on. At the end of the cut, all instruments were turned off and a notation made on the paper tape of the "Brush" recorder indicating the workpiece material, its hardness, the tool material, speed, feedrate, and the depth of cut. The horsepower consumed at the motor during a cut was also noted and recorded. All force data obtained through experimentation was grouped according to workpiece hardness, feedrate, and depth of cut. Data in each group was averaged to give the "Analysis of Forces" shown in Table I.

MATHEMATICAL MODELING

The work of many researchers, including this author's, has shown that, of the three machinability parameters—speed, feedrate, and depth of cut—cutting speed has the least effect on cutting and feeding forces. At low cutting speeds there is a slight drop in the cutting force as the cutting speed is increased but, in the normal operating range of most cutting tools, the cutting force can be assumed to be virtually independent of cutting speed. Therefore, the exponent for v in equation (2) becomes zero and that equation can be rewritten as:

$$F = K\ (Bhn_w)^a\ f^c\ d^g \qquad (3)$$

In the above equation, the values of K, a, c, and g are determined as follows:

Cutting-force values for 0.015 ipr feedrate and 0.150 in. depth of cut for steels of different hardnesses (from Table I) are used as the "y" values for a regression analysis; the "x" values for the regression analysis are the hardness numbers, Bhn_w for the corresponding force values. Such a regression analysis for a non-linear power curve of the general form:

$$y = x^a \qquad (4)$$

yields a value of 0.43 as the exponent for Bhn_w. This means that

$$F_c = K\ (Bhn_w)^{0.43} \qquad (5)$$

Similar regression analyses done on the data from Table I yield the following values:

For F_c: Exponent for f = 0.83
d = 0.92

For F_f: Exponent for Bhn_w = 1.56
f = 0.45
d = 1.03

Thus, the mathematical models (equations) for F_c and F_f can be written as:

$$F_c = K_1\ (Bhn_w)^{0.43}\ f^{0.83}\ d^{0.92} \qquad (6)$$

$$\text{and } F_f = K_2\ (Bhn_w)^{1.56}\ f^{0.45}\ d^{1.03} \qquad (7)$$

Value of the constant K_1 in equation (6) can be calculated by substituting the appropriate values for F_c, Bhn_w, f, and d from Table I in equation (6). Similarly, the value of K_2 can be calculated by substituting the appropriate values for F_f, Bhn_w, f, and d from Table I, in equation (7).

Now, the final form of the equations for the tangential cutting force, F_c and the axial feeding force, F_f becomes:

$$F_c = 11{,}031\ (Bhn_w)^{0.43}\ f^{0.83}\ d^{0.92} \tag{8}$$

$$F_f = 3.23\ (Bhn_w)^{1.56}\ f^{0.45}\ d^{1.03} \tag{9}$$

DISCUSSION OF RESULTS

Equations (8) and (9) enable us to calculate the tangential cutting force and the axial feeding force for turning any STEEL workpiece. The inputs required are the Brinell hardness number of the workpiece, the feedrate, and the depth of cut. Equations (8) and (9) show that cutting and feeding forces are more sensitive to changes in depth of cut than to changes in feedrate. Also, feeding force is more sensitive than cutting force to changes in workpiece hardness.

With the tangential cutting force known for an axial turning operation, the horsepower consumed at the cutting edge can be calculated from:

$$hp_c = \frac{F_c\ v}{33{,}000} \tag{10}$$

where hp_c = Horsepower consumed at the cutting edge

The metal-removal rate in an axial turning operation is:

$$M = 12\ v\ f\ d \quad \text{cu in./minute} \tag{11}$$

From the horsepower consumed at the cutting edge, the horsepower required at the motor can be calculated from:

$$hp_m = \frac{hp_c}{E} \tag{12}$$

where hp_m = Horsepower required at the motor
and E = Efficiency of the headstock drive

The Jan 22, 1973 issue of "American Machinist" published a nomogram [1]* for calculating the metal-removal rate and the horsepower required in turning.

By simultaneously plotting metal-removal rate and tangential cutting force on a log-log grid against cutting speed and axial feedrate, Lambert and Taraman [2] have shown how

---

* Numbers in square brackets denote references at the end of the paper.

the metal-removal rate (productivity) can be increased significantly with hardly any increase in the load on the machine-tool components (cutting force). Figure 3, based on the Lambert-Taraman concept shows contours of cutting force and metal-removal rate at 0.150-in. depth of cut when turning C-1045 steel with a hardness of 204 Bhn.

Let us assume that a turning cut is being made at 360 sfm and 0.0125 ipr feedrate on a lathe with 15 hp available at the motor. This condition is shown by point A in figure 3. The tangential force corresponding to point A is approximately 500 lbf. By moving horizontally from point A, along the 500-lbf cutting-force line to point B, the metal-removal rate has been increased from 8 cu in./minute at A to 13.5 cu in./minute at B. The cutting speed corresponding to point B is 600 sfm. Productivity has been increased 69 percent without increasing the load on the machine tool. However, it must be remembered that increasing the cutting speed from 360 sfm to 600 sfm decreases the tool life. A more comprehensive economic analysis will have to be made to see if the increase in productivity is also cost-effective.

BIBLIOGRAPHY

1. Srinivas, B. K., "Nomogram Simplifies Turning Calculations," American Machinist, Jan 22, 1973, pp 94-95.

2. Lambert, B. K., and Taraman, K., "Development and Utilization of a Mathematical Model of a Turning Operation," International Journal of Production Research, Vol 11, No. 1, 1973, pp 69-81.

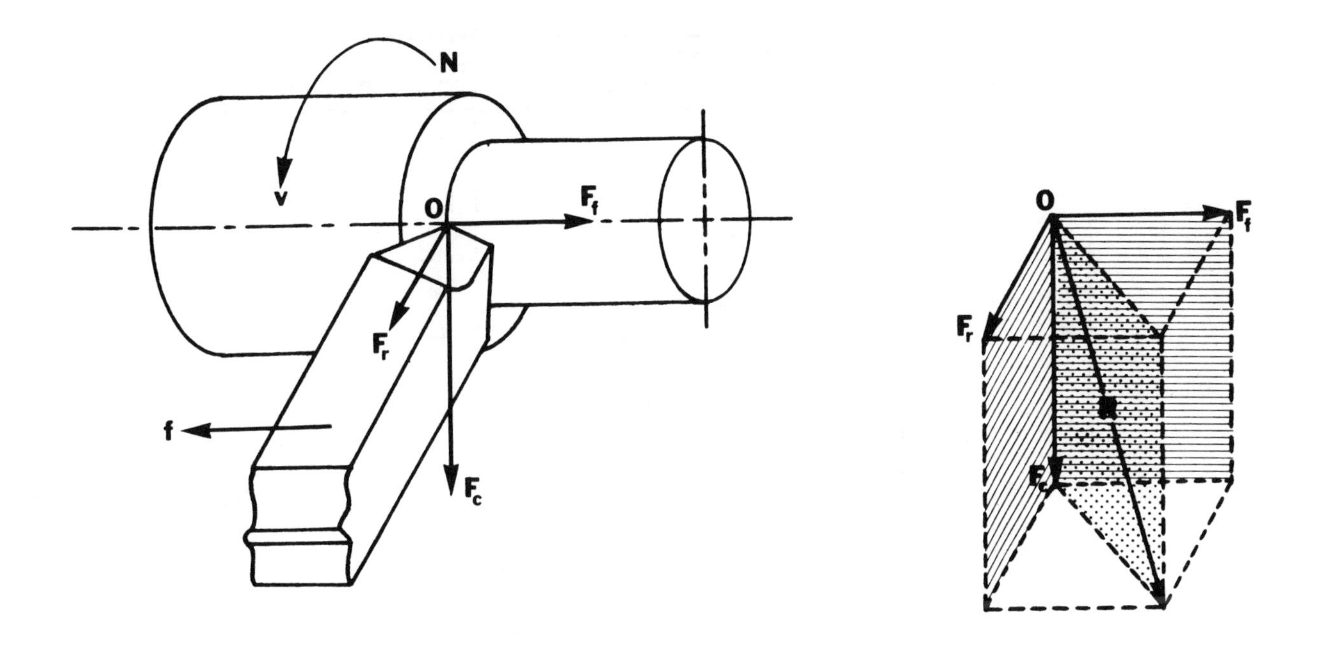

Fig 1. The Forces in Turning.

courtesy: S M E

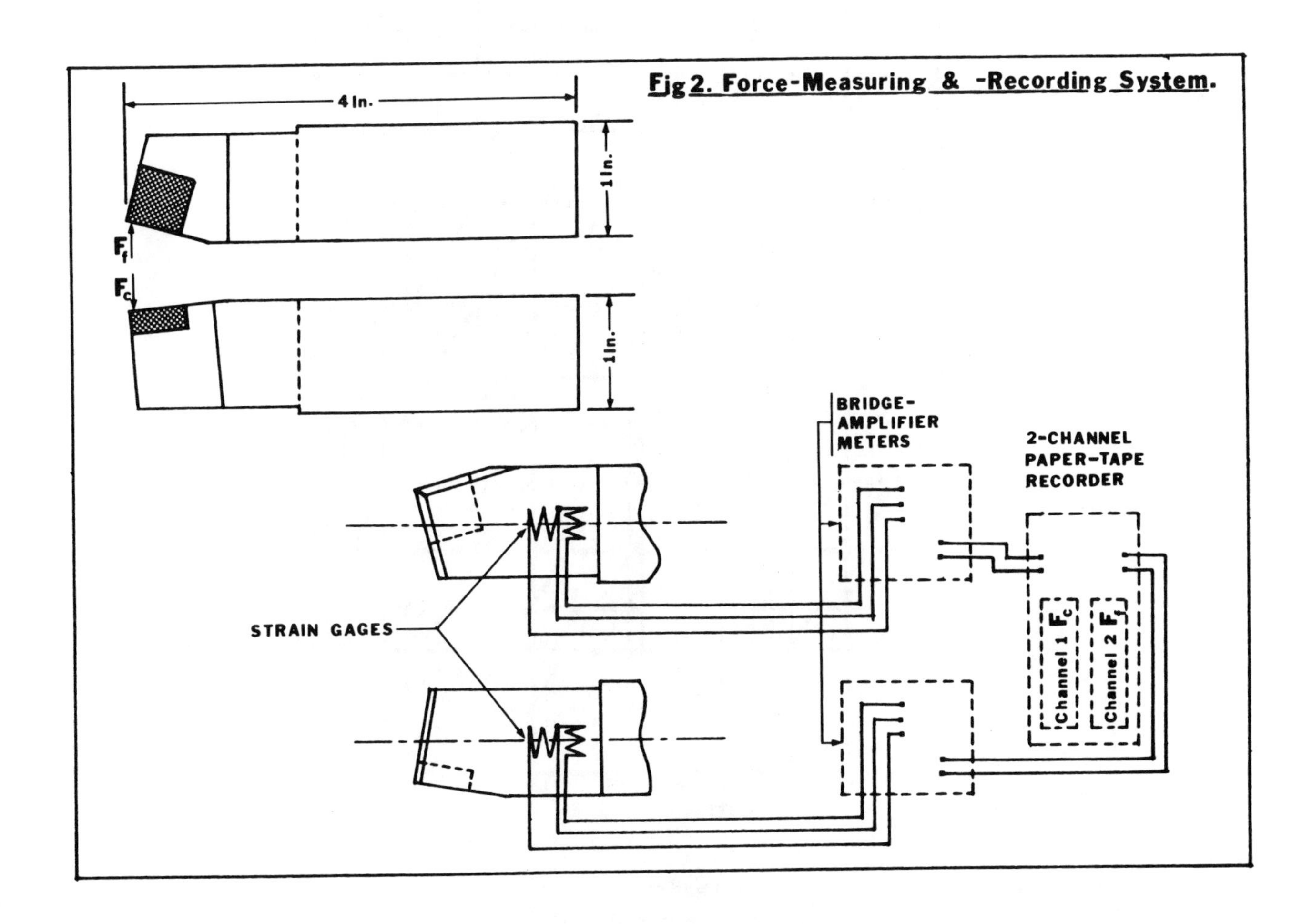

Fig 2. Force-Measuring & -Recording System.

TABLE I ANALYSIS OF FORCES									
Feedrate	Depth of Cut	C-1045 Steel 204 Bhn		A-4150 Steel 228 Bhn		A-4150 Steel 257 Bhn		A-4340 Steel 453 Bhn	
ipr	in.	F_c lbf	F_f lbf	F_c lbf	F_f lbf	F_c lbf	F_f lbf	F_c lbf	F_f lbf
0.005	0.050	90.00	45.30						
0.008	0.050	123.16	73.68	136.00	70.07				
0.010	0.050	156.67	69.33						
0.015	0.050	200.59	87.41	228.67	97.60				
0.020	0.050	280.00	98.67						
0.025	0.050	318.40	102.08	358.75	110.31				
0.005	0.100	172.50	112.00						
0.010	0.100	293.33	136.00					360.00	520.00
0.020	0.100	555.00	201.00						
0.008	0.150	337.32	215.37	369.13	208.26				
0.015	0.150	580.00	279.68	611.05	287.37	640.00	400.00		
0.025	0.150	887.14	353.43	964.00	389.00				
0.005	0.200	315.00	200.00						
0.010	0.200	540.00	293.33						
0.020	0.200	940.00	380.00						
0.008	0.250	544.67	357.17	572.00	371.50				
0.015	0.250	924.14	495.86	1014.00	534.00				
0.025	0.250	1445.88	600.00	1560.00	650.00				

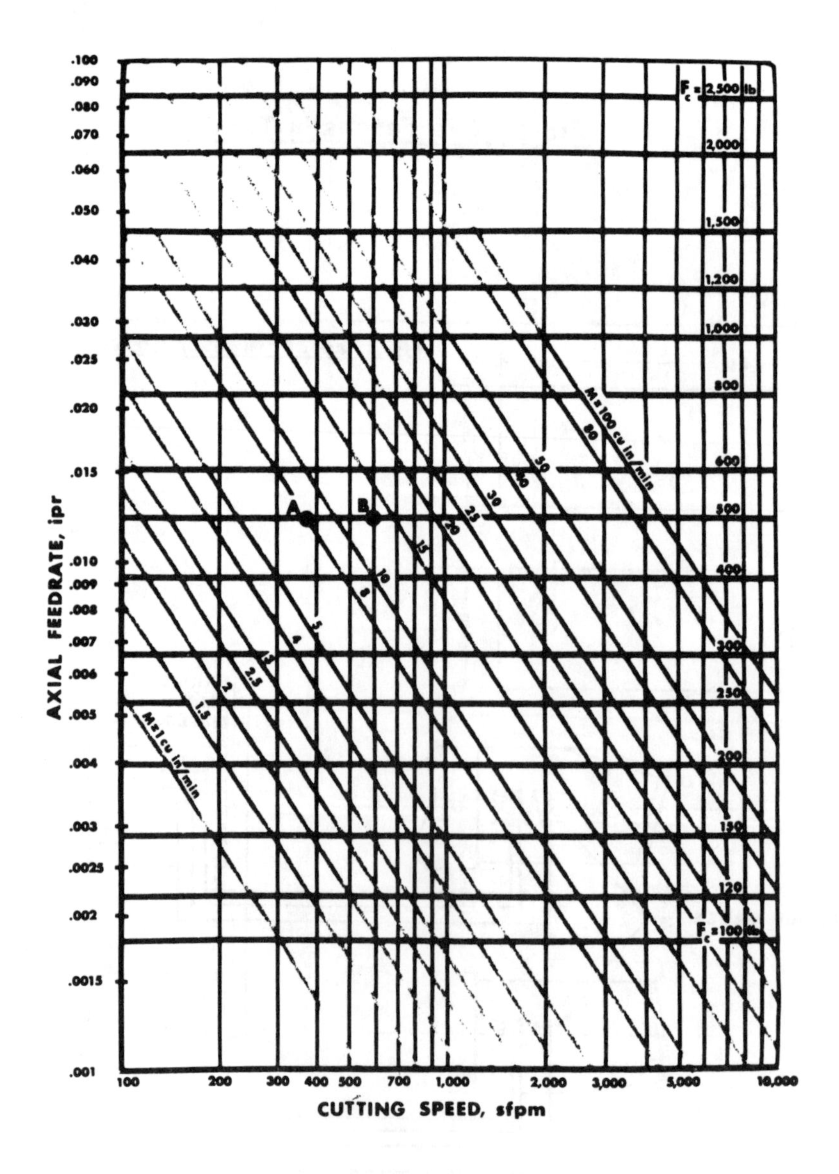

Fig 3. Contours of cutting force and metal removal rate at .150 in. depth of cut. (workpiece material : C-1045 steel)

Presented at the SME Westec '83 Conference, March 1983

Block Tool System—Key for Conventional and Unmanned Manufacturing

by James J. Nimphius
Sandvik, Incorporated, Bloomfield Hills, Michigan

HISTORICAL PERSPECTIVE

Since the beginning of time, civilizations advanced with development of tools, and in particular cutting tool materials. In support of this statement, we must only look to our history book to see how historians define the advance of civilization (i.e. Stone Age, Iron Age, etc.).

When we look at manufacturing in more contemporary terms, we can see from Fig. 1 that at the beginning of the 20th Century when carbon steel was the

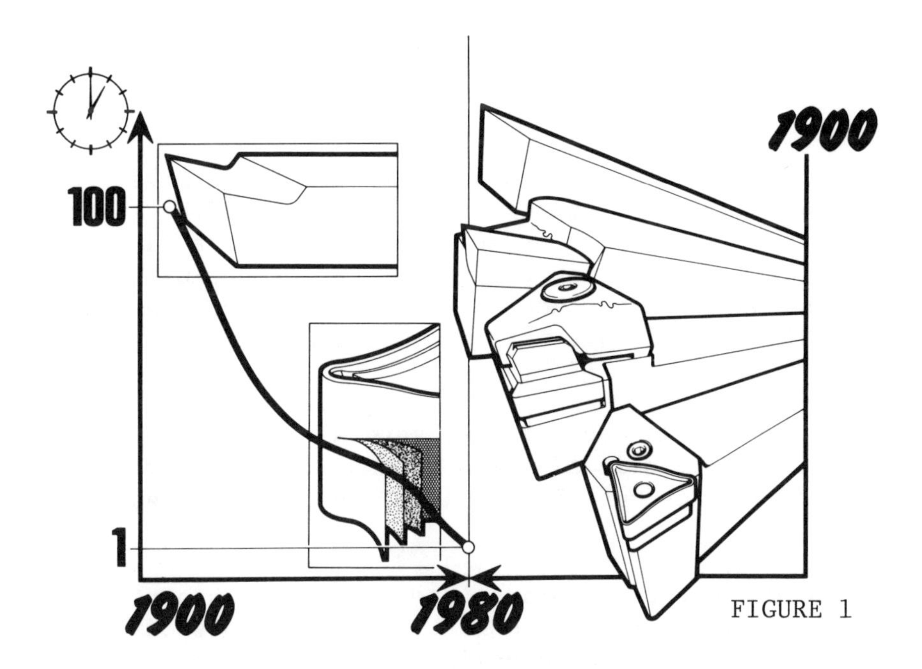

FIGURE 1

predominate cutting tool material, a given component took 100 minutes to machine. Through the advance in tooling materials, i.e. high speed steel, brazed carbide tools, indexable insert coated carbide, ceramic, and triple-coated carbide, the same component 80 years later now only takes one minute to machine. All this development by tooling manufacturers concentrated around the cutting edge and cutting tool materials has forced the machine tool manufacturers to develop faster and more rigidly efficient machines. The manufacturing world has benefited with higher productivity from faster machining rates. Thus, the actual metal cutting portion of production has become very efficient. However, as can be seen by Fig. 2; metal cutting only represents 6 to 22% of the overall production time - 6% being for general manufacturing and 22% for highly automated transfer line manufacturing such as in the automotive industry.

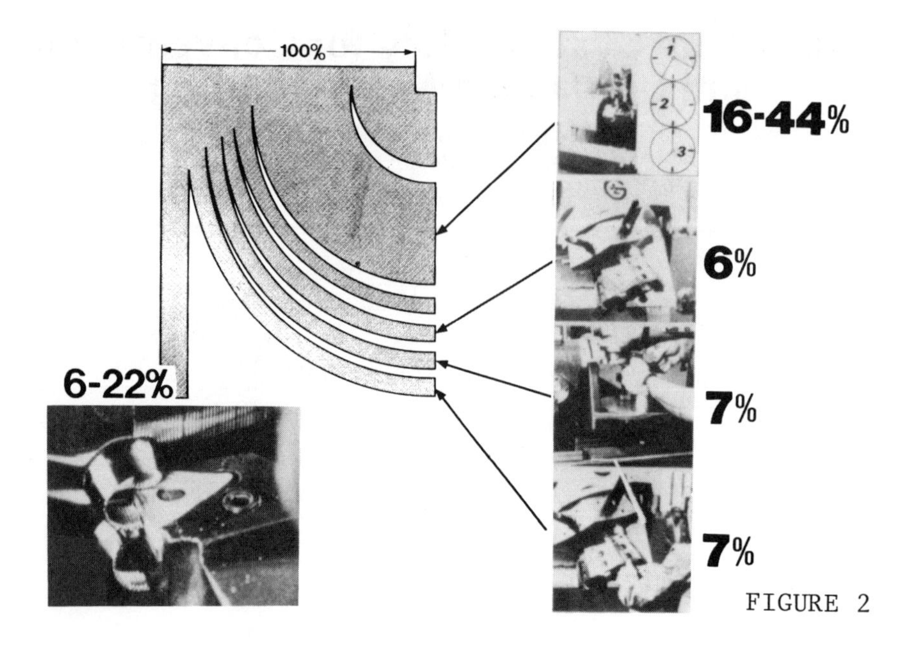

FIGURE 2

The balance of this time represents seven percent for tool change, seven percent for set up and gauging, six percent for tool related break down, and 16-44% from unutilized shift. As can be seen from this, more time can be saved in the production cycle by eliminating or reducing a percent of the time spent on these functions. We have now reached a point in time where gains as a result of improved tool materials are no longer as important as the other aspects mentioned above.

In order to gain savings in these areas, several things must be accomplished through the rationalization and automation of the manufacturing process. Several prerequisites must be fulfilled in order to reach significant improvements.

First of all, the components should be designed with automation in mind. They should have the capability for convenient gripping and should have the possibility to utilize automatic tool changing. It may be necessary for us to design more complicated components to eliminate manual assembly or a complicated component that is difficult to machine may have to be designed into several simplier parts that can be assembled automatically. The choice of material can be very important. Often if the right material is selected, critical surfaces can be locally hardened through induction instead of case hardening, thus the heat treat process can be integrated into the total manufacturing chain. This could eliminate several days or even weeks from the manufacturing process and reduce the need for large in-process inventory as buffer stock.

Also, more emphasis should be placed on the replacement of traditional materials with the use of new materials such as plastic and fiber which make it possible to drastically change product design and manufacturing methods. Cold forming of certain components along with laser technology for both cutting, heat treatment, and gauging must also be investigated.

AUTOMATION

It is an accepted fact that with mass production, high levels of automation are the most economical methods of manufacturing. Several factors today have made automation not only economically justified in smaller lot manufacturing, but in many cases necessary for survival. They are:

1) The increased cost of todays sophisticated CNC machines has required improved machine utilization.

2) Absenteeism in manufacturing has in some cases reached the 20% level and the reduced machine utilization resulting from this has made it difficult to justify such large capital expenditures.

3) The need to reduce in-process inventory through a decrease in set up time and the ability to interrupt production schedules without causing increased labor costs, in order to be more responsive to market needs.

4) To give management better control over the manufacturing process through better reporting and more current information.

Most of these things can be accomplished today with modern Flexible Manufacturing Systems and Automated Production Cells. These are for the most part today focused on singular problem areas and as yet have not been incorporated into the total manufacturing process. Several reasons for this exist around the interfacing of control and software with larger main frame computers. These are being worked on and solutions are expected before the end of this decade.

However, there are some basic tooling and tool changing problems that must be answered. This is what we will discuss.

NEED FOR AUTOMATIC TOOL CHANGING IN UNMANNED MACHINING

When we look at an automated production cell that consists of several machine tools along with inspection or gauging stations all serviced by one or several robots linked with material handling system, should one cutting tool require changing because of failure or normal wear, the entire cell would be put out of operation because someone would have to change the tool. The robot could not function while the man was changing the tool so not only would the machine he is working on be down, but all others as well because the robot can not perform its functions. This establishes the need for automatic tool changing.

There is yet another problem that we must face in machining and that is machinability. Many people talk about machinability as a synonym for tool wear but machinability is not just wear on the tool.

If a material causes a tool to wear more than another material, we say that material No. 1 has a low machinability and material No. 2 has a better machinability. If the cutting forces are higher for one material than another, the first is of poorer machinability than the second. When the surface quality of one material is worse than that of a second, then the first has poorer machinability than the other. When chip formation is poor, we say this material is of poor machinability.

To overcome these variables in machinability with automatic tool change, we must establish a very conservative tool changer frequency or use more sophisticated methods to be discussed on page 12.

BLOCK TOOL SYSTEM

As we have stated earlier, the time has come when tooling manufacturers must forsake their traditional area of development and look for inovative ways to reduce the time required in the manufacturing process. The BLOCK TOOL SYSTEM has done that by the development of a unique coupling. As shown in Fig. 3, two-thirds of the standard late tool is used for just holding the tool in the machine.

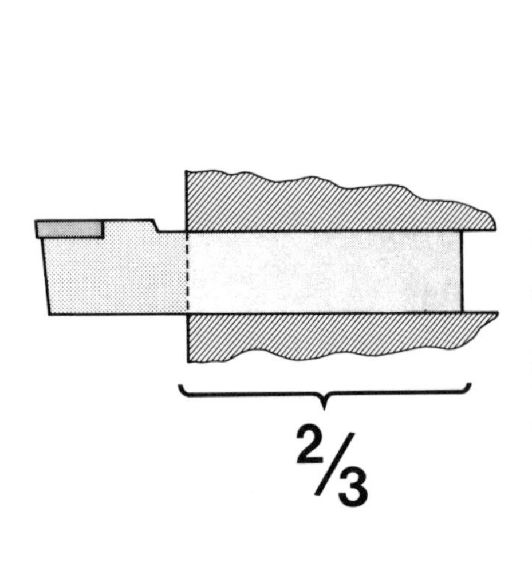

FIGURE 3

By detaching the cutting portion of the standard turning tool and providing a clamping unit that becomes an integral part of the machine, similar to what is shown in Fig. 4, we now have the basis for an automatic tool changer for lathes.

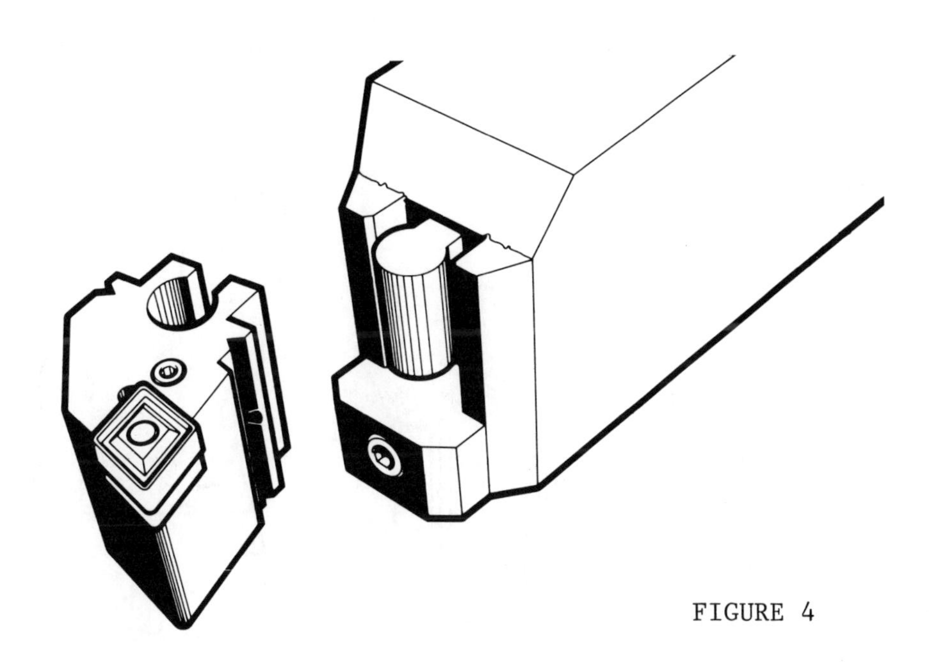

FIGURE 4

The Coupling

This new concept has been made possible through the unique new coupling. Basically, the cutting unit and clamping device have been recessed to fit together with a radial and axial location face on either side of the coupling. The mouth of the cutting unit is forced open by the centre drawbar of the coupling when this is pulled back. This pushes the four location faces of the unit against the corresponding faces that locate and support in the clamping unit (See Fig. 5). This provides an extremely rigid connection without possibilities of play even during rough machining. When the drawbar is released, the jaws spring back and sufficient clearance is obtained for the cutting units to be easily lifted out of the coupling.

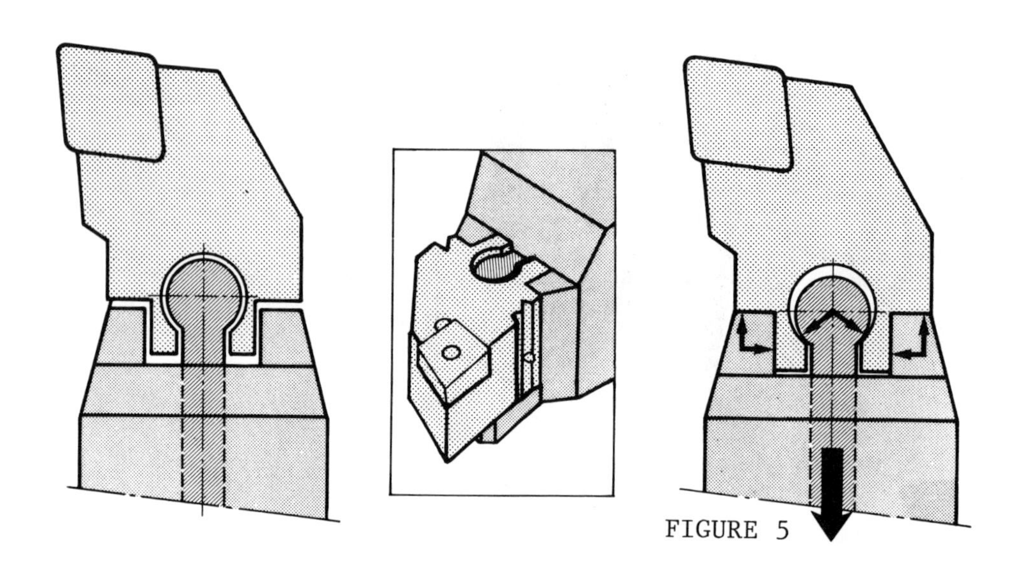

FIGURE 5

The cutting unit is inserted from above the coupling and rests firmly on a supporting face in the bottom. The design of the coupling and its position just behind the operating part of the tool means that the all important tangential support has been extended as far as possible (See Fig. 6), further than what is achieved for a conventional tool holder. Clean locating faces are the key to success in any accurate coupling and this has been given special consideration throughout the development and trials of the BLOCK TOOL SYSTEM. The design of the coupling is such that there are no dirt pockets only plane, open faces as well as provisions for jets of compressed air and cutting fluid around the coupling to prevent dirt or chips from building up on the locating surfaces.

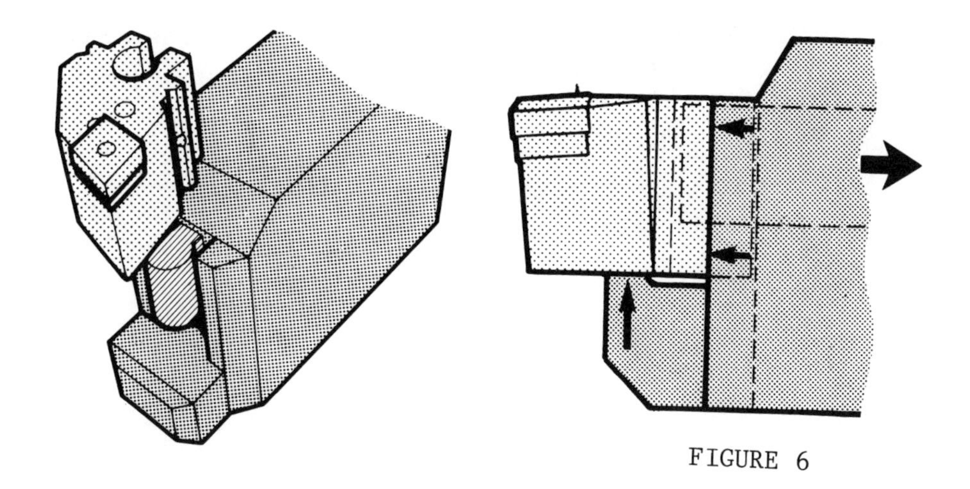

FIGURE 6

In addition to the excellent rigidity provided by this totally play-free coupling, the design also achieves a very high level of repeatability as can be seen in Fig. 7 (± eighty millionth on an inch is the total radial variation that can be expected from this coupling).

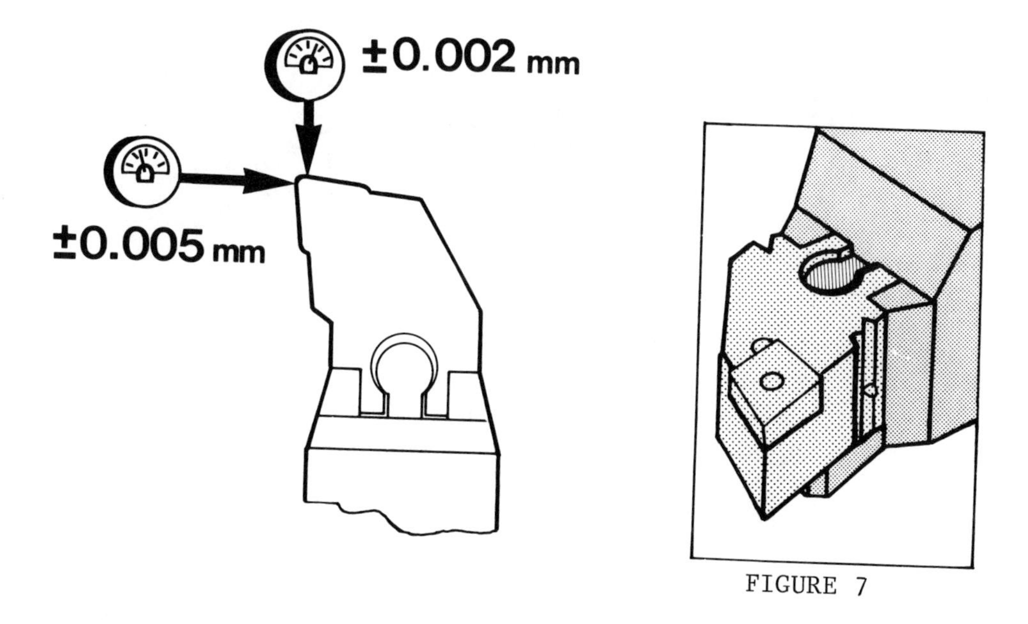

FIGURE 7

As in any coupled system, the success of its performance depends on the clamping force being correct. It is, therefore, extremely important that the drawbar be activated with a pull of 5,600 lbs. (25kN) of force as represented in Fig. 8. This can be ensured by loading the drawbar permanently with a set of spring washers. The force then can be overcome by means of a hydraulic cylinder for release of the tool. This then allows for quick manual change of the tool or completely automatic tool changing.

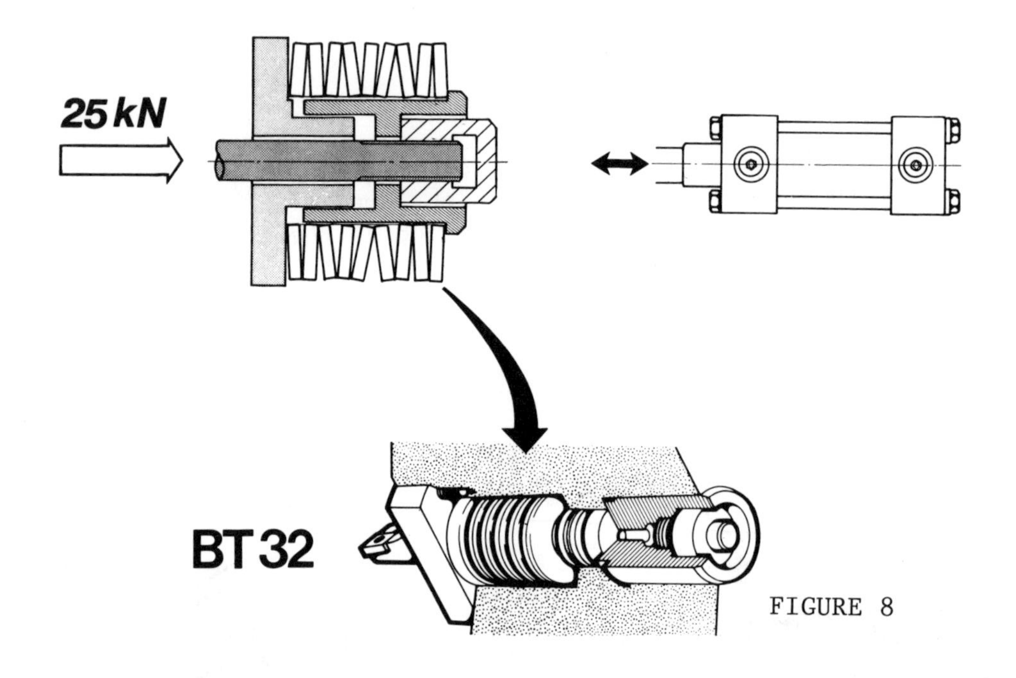

FIGURE 8

Tool Storage and Changing

The limited size of the coupling and clamping unit means that the system can easily be built into a machine turret or tool post. Therefore, with a relatively simple pick and place device; the tool can be automatically changed in the turret, as can be seen in Fig. 9. This then eliminates the need for a man to replace a worn or used cutting tool.

FIGURE 9

Also, by reducing the mass of the tool by two-thirds through the use of the Block Tool coupling, storage problems for back-up tooling are greatly reduced. Fig. 10 shows a typical storage system for the inventory of back-up tooling.

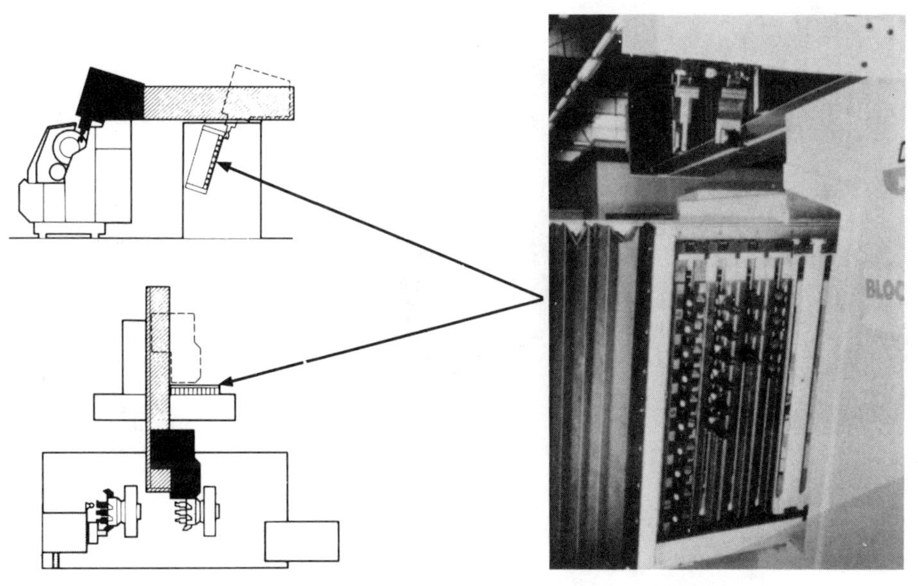

FIGURE 10

The Probes

Measurement of tools and workpieces in machines is efficiently carried out with a touch trigger probe. Its principle is simple in that an electrical circuit is broken when the stylus of the probe is deflected from its zero position. This triggers off an inductive signal for a dimensional value. The signal can be obtained from radial and axial deflection when brought into contact with the object at relatively high traversing rates and with considerable over travel. When withdrawn, the stylus returns to its zero position and can be used again immediately. The probe is connected to the CNC system of the machine which receives signals which either verifies a position or indicates that corrective movement is needed. Switching accuracy of this probe is in the region of 0.005 - 0.001 mm.

The touch trigger probe can be mounted on the headstock, machine bed, or a moveable arm for tool position setting and tool state inspection. The position setting is carried out by the cutting edge of the tool being driven into contact with the stylus. The deflection of the probe determines the tool off-set for the machine control. A cycle is written into the control to bring the necessary tools to the stylus. Each time the tool to be checked is put into the turret by the gripper arm, it is automatically set against the probe. In seconds this qualifies the tool precisely on the machine, eliminating any external tool setting. The tool can then be inspected as regards cutting edge condition by the same probe. Wear, damage, nose radius, and edge build up can be checked through program routines in the machine control unit.

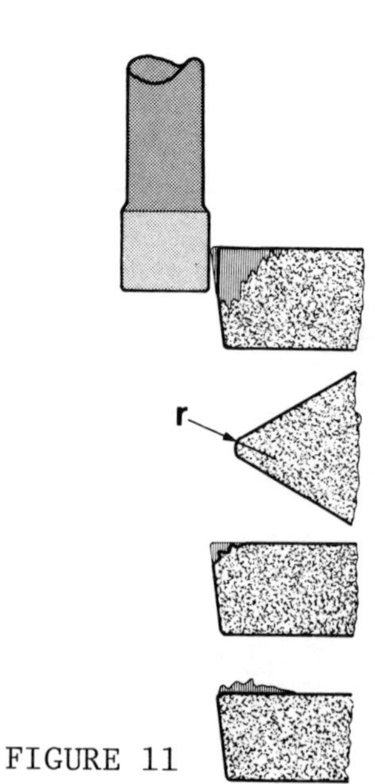

FIGURE 11

Feed Force Monitoring

To safeguard an unmanned machining system, some type of tool monitoring device must be employed. This is best accomplished by mounting sensors along the feed axis. The signals from the sensors are fed into a processor as illustrated in Fig. 12. When all these things are tightened together and can be monitored with large main frame computers, we will have the factory of the future today.

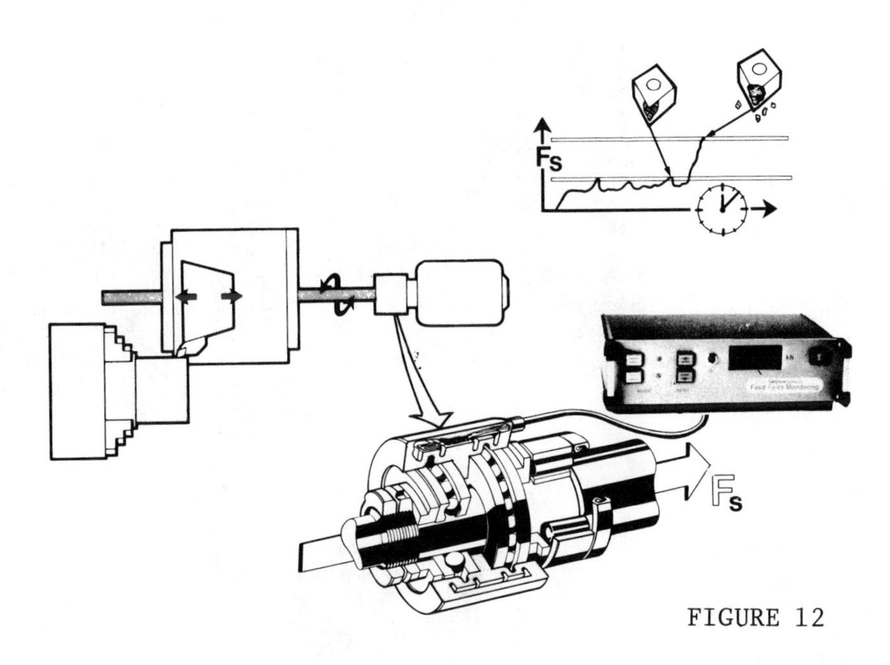

FIGURE 12

Reprinted from *Manufacturing Engineering,* January 1981

Quick Turning on Long Shafts

Turning long, thin shafts without the assistance of steadyrests typically means slow speeds. But not so with this tooling

Positive-rake, on-edge style insert in a 0° lead tool lessens cutting forces and so reduces chatter problems.

ONE OF THE ADVANTAGES of using positive rake tools is that the cutting pressure is generally decreased. This is of particular importance when doing shaft work where there are high length-to-diameter ratios and steadyrests aren't employed. By minimizing the cutting pressure, flexing of the shaft and chatter are both reduced, if not eliminated.

Case in point is found at Weldun International, Bridgeman, MI. Weldun designs and manufactures highly specialized test, assembly and metal-removal systems for the automotive industries, as well as intricate plastic and diecast molds, special fixturing and other products. Some of the shaft turning it performs is done on pieces that have length-to-diameter ratios of 25:1. Much of the work is done on a 15-hp (11-kW) Monarch engine lathe without the use of a steadyrest.

Conventional positive-rake tools were employed; satisfactory results were achieved. But the company needed more production. The answer was found in the selection of a tool with a different configuration: a 0° lead Ingersoll-Hertel Fix-Perfect tool with a 12° positive rake insert positioned on edge.

The tool, available from the Ingersoll Cutting Tool Div. (Rockford, IL), has eight usable cutting edges. Its clamping system utilizes a clamp stud that secures the insert in the direction of the main cutting force in the toolholder pocket. This design provides the strength and stability that facilitate higher metal-removal rates than would otherwise be possible with conventional positive tools.

According to Skip Linn, machine supervisor at Weldun's Special Machine Div., the new tooling has, in many cases, permitted metal-removal rates up to 50% higher than those of the previously used tools. He also reports improved chip control.

In Action

In one application of the Fix-Perfect tools at Weldun, a 20″ (508-mm) length on a 1½″ (38-mm) diameter CR 1018 bar is turned down to 0.800″ (20.32 mm) diameter. Roughing is completed in four passes within the feed and speed limitations of the lathe and without the use of a steadyrest.

The first cut is taken at 0.150″ (3.81 mm) depth, 0.028 ipr (0.71 mm/rev) feed, 720 rpm, and at 283 sfpm (86 m/min). On the second pass, the depth of cut is 0.100″ (2.54 mm), the feed is 0.018 ipr (0.46 mm/rev), and the rpm is increased to 914, which provides 287 sfpm (87.4 m/min) at 0.050″ (1.27 mm) depth of cut and 0.014 ipr (0.36 mm/rev). At this point, the length-to-diameter ratio is 22:1.

The final cut is taken at 1200 rpm, 282 sfpm (85.9 m/min), 0.050″ (1.27 mm) depth of cut, and at 0.0093 ipr (0.236 mm/rev) feed. The resulting shaft has a length-to-diameter ratio of 25:1. Metal-removal rates in the application are nearly double what they had been. ■

Reprinted from *Manufacturing Engineering*, January 1981

Focus on Face Grooving

Here's a look at a line of tools that can stand up to the demands of improved turning machines

CUTTING TOOLS

ROGER DUFFY
Operations Manager
Manchester Div.,
The Warner & Swasey Co.

A TURNING MACHINE that has increased capabilities cannot operate at its maximum efficiency unless it is equipped with a very rigid toolholder assembly that can take more. One application in which this can be readily observed is face grooving.

That is, much of the change currently going on in the area of face grooving involves tougher assignments for the toolholders and inserts. These assignments, in turn, require greater rigidity and higher quality tool construction so that the tooling system allows the machine to operate at its maximum efficiency.

Manchester Div. of The Warner and Swasey Co., Akron, OH, has developed a line of face grooving tools that accomplish this. The line includes toolholders that are suitable for use on turret lathes, chuckers, single-spindle automatics, boring mills, and any other type of turning equipment, NC, CNC or manually controlled. In addition to the extremely broad range of machines they can be used on, the tools are highly flexible in terms of applications. They can make face grooves (including O-rings) and they can also produce discs, slugs and rings.

A face grooving system has been developed by the company. It's based on the extension of a curved support blade, which provides depths of cut up to 2″ (51 mm) and widths of cut from ⅛ through ⅜″ (3.2 to 9.5 mm). Face groove diameters, depending on the specific components, range from 2 to 60″ (51 to 1524 mm). The system includes components for counterclockwise and clockwise rotation.

Increased rigidity means that carbide tools can be used in this face grooving system. Higher speeds and feeds result.

Innovations and Rigidity

Innovations in the area of face grooving contributed by Manchester include a tool that has the ability to combine face grooving and facing operations and so produce grooves that are much wider than the width of the insert. The line also features a tool that can groove as small as ½″ (13 mm) in diameter, a tool that can make a recessed groove at the bottom of a bored hole, and a tool that can be mounted on a rotating spindle.

These tools are well suited for face grooving because of their high built-in rigidity, an engineering accomplishment that is the single most important benefit of the company's unique four-piece component toolholder assembly. This assembly consists of the toolholder, the insert, the clamp, and the curved support blade.

The importance of this rigidity becomes evident in the case of a wreck during a production run. With single-piece construction, the entire toolholder must be replaced. With a four-piece assembly, only the damaged component need be replaced. This means a less expensive repair and less downtime.

Another important advantage of four-piece construction is that it permits heavy cuts and extremely close tolerances in ordinary metals as well as exotics, stainless steel, and various heat-treated metals.

The system is so rigid that the use of carbide inserts is feasible, permitting the much higher speeds and feeds that the inserts are designed to be used at. As a result, cutting is performed on all metals, including the exotics, with a high degree of efficiency, and so increases the productivity of any turning machine.

Toolholders in the line are available to handle a wide range of work. They are designed with rectangular or round shanks in various sizes to fit conventional, NC and CNC machines. Right and left hand configurations are included. ■

Reprinted from *Manufacturing Engineering*, November 1983

Machining Tough Abrasive Plastics

Abrasive, reinforced plastics pose a special machining problem. Cemented carbides and diamond cutting tools meet the challenge

ROBIN P. BERGSTROM
Managing Editor

MACHINING OF many common plastics and other nonmetallics is generally an easy task. Cemented carbide cutting tools, or even in some cases tool steel, will do a satisfactory job. However, engineers at Carboloy Systems Department of General Electric Co. say that the machining of rubber and plastic parts can be particularly challenging when the material is reinforced with abrasives or metals. Often the particles contained in these parts differ from those found in abrasive cast irons and steels.

When cutting extremely hard or abrasive materials at high speeds, the effectiveness of many conventional cutting tools is severely limited by rapid edge wear. As edges wear, quality suffers. It becomes increasingly difficult to maintain critical tolerances. Part size begins to vary, and surface finish deteriorates.

There are ways to overcome these problems, of course. Changing tools when they become dull obviously will help. So, too, will reducing machining speeds. Although both of these alternatives effectively reduce tool wear, they also effectively reduce production.

To overcome these problems, specific tough, abrasion-resistant grades of cemented carbide or diamond cutting tools, depending on the application, may be the answer.

Cemented Carbide Tools

Parts with difficult-to-machine contours requiring intricate form tools are best machined with carbide cutting tools. This is mainly due to the difficulty in the fabricating of diamond tools with anything more sophisticated than simple cutting shapes. Carbide form tools, on the other hand, are readily available. Also, when part size is very small and it is impossible to attain the necessary surface footage, carbide is again the best choice.

Carbide cutting tools are most often specified for machining of abrasive plastics when severe interrupted cutting conditions exist, or where depth-of-cut exceeds 0.150″ (3.81 mm). This is because most commercially available cutting tools tipped with polycrystalline diamond blanks have a maximum usable cutting edge length of 0.150″. Hence, diamond tools are most frequently specified for lighter finishing cuts at high speeds.

Cemented carbide cutting grades recommended for use on plastics and other nonmetallic materials have differing amounts of wear or abrasion resistance and toughness. Since no chips are formed in most plastics machining, crater resistance is not an important factor.

- Grade 883 has good abrasion resistance with exceptional toughness to withstand breakage from hard particles contained in the part material. Grade 883 is recommended for roughing operations on plastics.
- Grade 895 has good toughness and abrasion resistance for medium-duty to finishing operations.
- Grade 999 has moderate toughness but excellent abrasion resistance for use in high-speed finishing where resistance to wear is the primary consideration.

Diamond Tools

Certain of the more exotic materials, such as graphite-reinforced plastic, can only be machined successfully with diamond cutting tools. Machining operations performed at cutting speeds in excess of 1000 sfpm (305 m/min) as well as high-volume machining operations are also good candidates for diamond cutting tools.

There are two basic types of diamonds used as cutting tools. Single crystal natural diamonds were the first type of diamond employed in cutting. They have outstanding wear resistance, but they are not capable of withstanding high shock loading. Furthermore, their performance may vary by a factor of 10, depending on the orientation of the crystal structure.

The second type of diamond tooling consists of tiny Man Made® diamond crystals fused together through a high-temperature, high-pressure process and bonded to a carbide substrate. This entirely new material was developed by General Electric Co. and is marketed under the trade name Compax®. Carboloy Systems produces CarboPax™ cutting tools from Compax blanks. The advantage of this type of diamond tooling is that its carbide substrate and diamond surface crystal structure give it very good resistance to shock loading. The completely random orientation of the diamond crystal structure prevents cracks from propagating through the surface and provides uniform wear.

Carbon, ceramics (unfired), fiberglass composites, graphite, plastics, rubber epoxy resins—these are just a few nonmetallic materials which can be effectively machined with diamond tooling. On these

(A)

(B)

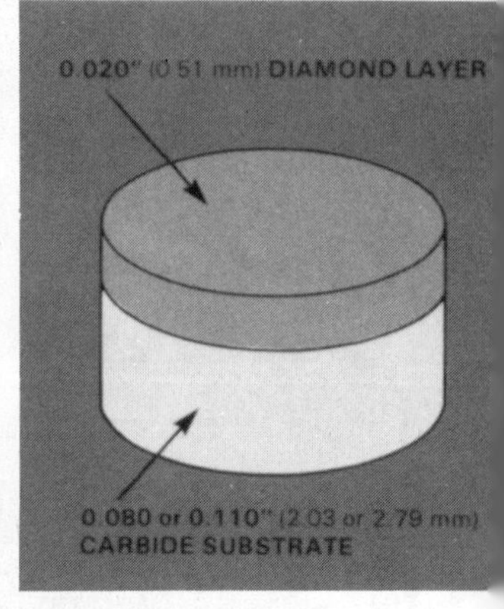

In 1957, General Electric was the first company to develop and commercially produce synthetic diamonds. The largest diamonds made by GE today are the size of granulated sugar. The diamond at right (A) is single-crystal natural diamond, shown for comparison. Compax, a laminated polycrystalline diamond product (B), was introduced by GE ten years ago. These unique tools are manufactured under temperatures in excess of 1000° C and pressures of more than 1 million psi (7 million k Pa), resulting in a metallurgical bond between the diamond and the tungsten carbide substrate. Due to the need for the pressures to be developed evenly, the blanks are made in round form, with a cutting surface that is uniformly hard in every direction.

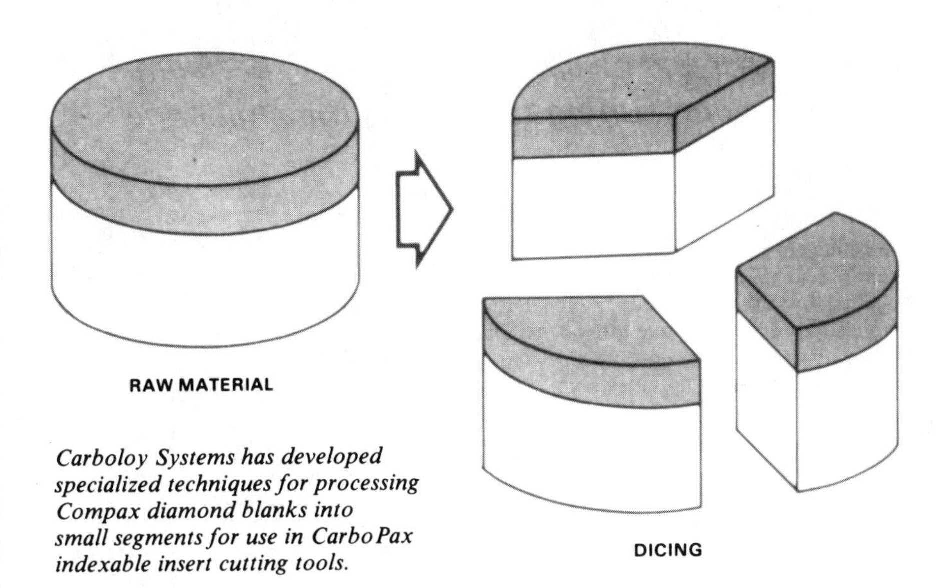

Carboloy Systems has developed specialized techniques for processing Compax diamond blanks into small segments for use in CarboPax indexable insert cutting tools.

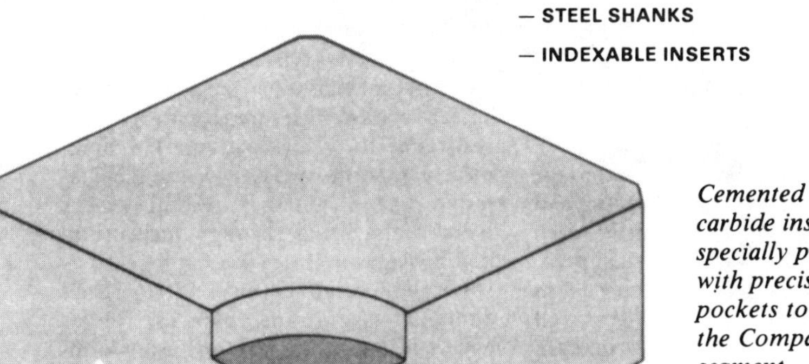

Cemented carbide inserts are specially prepared with precision pockets to accept the Compax segment.

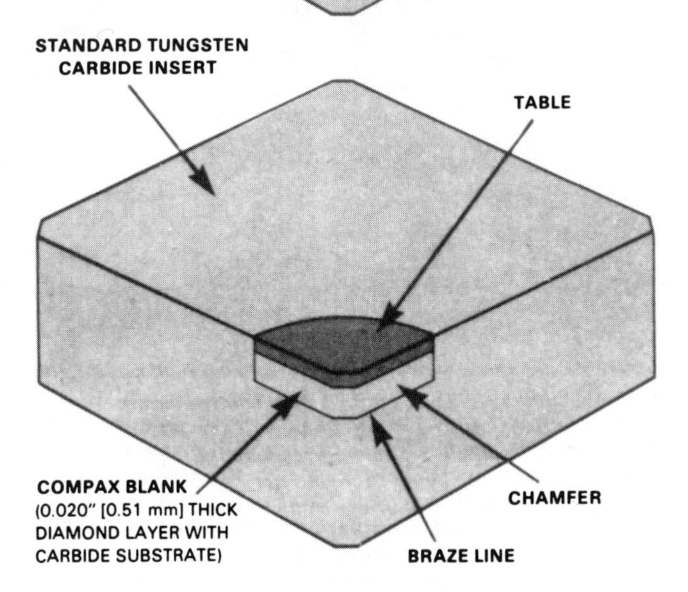

After the segment is brazed to the carbide insert, the CarboPax insert is ready for use.

materials, diamond tools offer dramatic performance improvements over carbide; tool life is often 100 times better. Diamond tools can also offer better control over part size, improved surface finish and surface integrity, and higher cutting speeds.

Generally, diamond cutting tools can be applied according to the same basic rules applied to cemented carbide tools. But there are a few exceptions.

- Positive rake tooling is recommended for the vast majority of diamond tooling applications. This is partly because diamond tools are normally used to machine soft, abrasive materials and partly because negative rake diamond tools have only one side and thus offer no economic advantage.
- Built-up edge should be avoided. If this is a problem, try increasing cutting speed and use a higher (more positive or less negative) rake angle.
- It is sometimes possible to control breakage and chipping by reducing feed rate.
- Generally, use of coolant is not necessary. However, since dust is often a problem with plastics machining, coolant may be prescribed to help reduce airborne particles.
- Regrinding diamond tools is recommended. Consult your cutting tool supplier for detailed grinding procedures.

Applications

In one automotive plant, tungsten carbide cutting tools have been replaced by diamond blank tools for machining glass-filled phenolic brake pistons. Tool life has been increased significantly, and faster cutting speeds and greater depths of cut have been accompanied by better surface finishes and lower scrap rates.

When using tungsten carbide tools, only 250 pieces were machined before regrinding was required. One regrind yielded approximately 100 additional parts. When Compax polycrystalline diamond cutting tools were substituted, output was increased to 15,000 pieces per tool before regrind. And six additional regrinds were possible, each producing up to 15,000 pieces. Downtime for tool adjustment and tool changes was reduced from 15 hours per week to less than one hour.

In another application, a manufacturer needed to turn parts made of silica reinforced ablative plastic. Silica reinforcement of the ablative produced a hardness far greater than plate glass and much more abrasive than ground glass. In thin sections, the silica phenolic was very brittle due to the high (70%) silica content.

Because of the hardness of silica and its abrasive nature, diamond and carbide cutting tools were necessary. Carbide

Glass-filled phenolic piston for disk-brake assembly is precision machined with polycrystalline diamond tools. Surface finish must be continuous and free of depressions and protrusions that might cause leakage of break fluid.

Finish-turning the outer diameter of a cam gear made of tough fabric-impregnated phenolic. The cutting tool (just out of photo beneath the gear) is a CarboPax polycrystalline diamond insert similar to the two shown just left of the gear.

tools provided the necessary toughness for roughing, while diamond, because of its exceptional hardness, was able to hold size consistently on long cuts.

In roughing, the carbide tool was able to withstand the shock of the lumps of silica phenolic—something the diamond tooling would not. However, because of the extreme abrasiveness of the silica, diamond cutting tools were specified for finishing to precise tolerances.

In yet another application, replacing high-speed steel cutting tools with polycrystalline diamond tools helped the manufacturer eliminate a problem in the machining of laminated aircraft parts. This job involved the machining of laminates which were reinforced with polyester, phenolic, and epoxy thermosetting plastics.

Machining these materials with conventional HSS tools resulted in the rapid deterioration of the cutting edges. Frictional galling caused binding of the cutter and workpiece, resulting in delamination. Replacing the HSS tools with diamond cutters caused little or no delamination, and burring of the parts was reduced.

In a final application, a major U.S. auto manufacturer machines cam gears from tough-to-work, fabric-impregnated phenolic. The gears are roughed out and finished in four operations on a twin-spindle Olofsson boring mill using CarboPax polycrystalline diamond tools. The switch to diamond tools has had significant impact. For example, a year's production at 4600 gears a day required 20,700 tool changes when using carbide cutting tools. That figure drops to a mere 60 tool changes per year with the use of diamond tooling. The manufacturer estimates that the switch in materials has resulted in freeing up nearly 350 hours of machining time.

These are just a few examples of how modern carbide and polycrystalline diamond cutting tools have been successfully applied to the machining of abrasive plastics and other nonmetallic materials.

In certain applications, the choice between cemented carbide and diamond tools is clear; the decision can be based on the relative hardness or abrasiveness of the part material. In other cases, there are economic considerations. Cutting tools produced from cemented carbide cost considerably less than polycrystalline diamond. Therefore, each job must be evaluated from a standpoint of part material, machining parameters, and manufacturing volumes. ■

Presented at the SME 1982 International Tool & Manufacturing Engineering Conference May 1982

Chip Control in Turning

by Sakayuki Nakamura
Daido Steel Company, Limited
Garry J. Wuebbling
Metcut Research Associates Incorporated
and John D. Christopher
Metcut Research Associates Incorporated

INTRODUCTION

Continuous chips produced in turning operations can damage the workpiece finished surface, chip the tool, and create difficulty in chip handling. In an unmanned manufacturing plant which is one of the ultimate goals of manufacturing, the control of chip shape is a key factor which will determine the successful operation.

A review of literature reveals that there are three primary methods of chip control as follows:

1. Selection of appropriate depth of cut, feed rate and cutting speed.

2. Application of chip breakers on the tool.

3. Selection of workpiece materials whose chips are easily broken, such as free-machining materials.

The influence of cutting conditions on chip breakage has been discussed in terms of chip flow diameter,[1-4] chip flow direction[5] and fracture strain of chips,[6] all of which have been studied based on the fundamental theory on chip formation. Storing the cutting conditions for well controlled chips which have been determined from turning tests under various cutting conditions is another aspect of this kind of study designed for more practical use.[7] Development of a chip-shape sensor for use in an adaptive control system has also been attempted.[8]

There are many reports on the relationship between the geometry of obstruction-type chip breakers and the radius of chips,[1,3] optimum configuration of the obstruction-type chip breaker,[1] and functions of the groove-type chip breaker,[4,9] and the non-groove-type (land-angle) chip breakers.[10]

One work material with good chip breakability (chip breakability being that characteristic of chip breakage from the viewpoint of the workpiece material) is free machining brass whose chips tend to be small fragments. Steels containing additional sulfur and/or lead show better chip breakability over a wide range of cutting conditions and can be machined without a chip breaker using normal cutting conditions.

Necessary information for engine lathe operators and NC lathe programmers includes suitable cutting tools and optimum cutting conditions. In order to provide such information, it is necessary to collect and store data in advance, which is the goal of this study. Since the combinations of workpiece materials, cutting tools, and cutting conditions are infinite, estimation of chip breakage by experimental or theoretical equations is useful toward saving man-hours and reducing cost.

This study consists of two parts: First, an experimental study on the effect of side cutting edge angle, cutting speed, workpiece material, and chip-breaker shape on chip breakage phenomena; second, a discussion of the ways to estimate chip breakage for the combination of workpiece material and cutting tool using the experimental constants which are obtained from the first step.

CHIP CLASSIFICATION

Several ways to classify chip shapes have been proposed. The primary difference among them is the number of groups into which the chips are classified. Similar definitions on acceptable and unacceptable chips are specified in all cases.

For this study, the chip-shape classification described by E. K. Henriksen(1) was modified to include snarling chips not exceeding 2" in diameter and compressed crescent-shaped chips which were suggested by D. G. Jones, et. al.(10) Figure 1 gives a representation of those different types of chips. Of these chips, No. 4 through No. 7 are judged acceptable, and No. 1 through No. 3 are considered unacceptable. No. 8 chips are not too harmful from the viewpoint of disposal. However, they are classified as unacceptable because they are accompanied by a high cutting force which reduces tool life.(10)

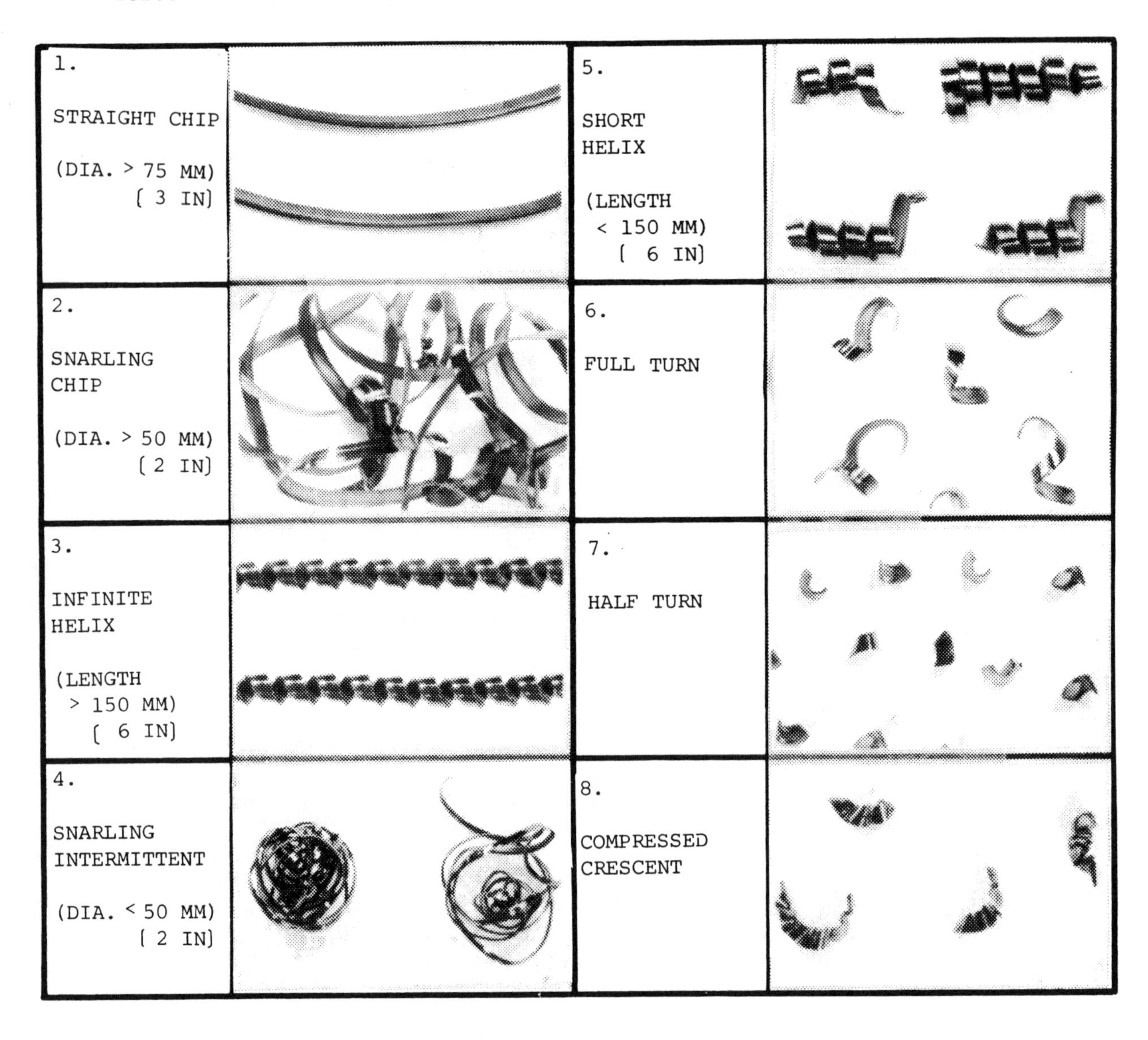

FIGURE 1. CHIP CLASSIFICATION

EXPERIMENTAL PROCEDURE

AISI 4140, AISI 8620, Inconel 718 and Ti-6Al-4V were used as workpiece materials for this study. Their chemical compositions are given in Table 1. The materials were procured as three-inch-diameter, hot-rolled round bars and were subsequently heat treated. Three levels of heat treatments were given to AISI 4140 in order to examine the influence of hardness on chip breakage phenomena. The Brinell Hardness after heat treatment is listed in Table 1. A multi-phase TiN-coated carbide was used for AISI 4140 and AISI 8620, but a C-2 grade carbide was used with Inconel 718 and Ti-6Al-4V as indicated in Table 2.

TABLE 1. DETAILS OF WORKPIECE MATERIALS

WORKPIECE MATERIAL	CHEMICAL COMPOSITION (%)								HEAT TREATMENT*	HARDNESS (HB)
	C	Si	Mn	P	S	Ni	Cr	Mo		
AISI 4140	0.41	0.20	0.83	0.010	0.010	0.05	1.10	0.25	ANN Q&T Q&T	179 341 415
AISI 8620	0.20	0.28	0.67	0.021	0.018	0.42	0.44	0.16	ANN	156
	C	Ni	Cr	Mo	Nb+Ta	Ti	Al	Fe		
INCONEL 718	0.05	53.17	18.07	3.11	5.16	1.02	0.67	BAL.	STA	376
	C	Al	V	Fe	O	N	H	Ti		
Ti-6Al-4V	0.006	6.02	4.12	0.13	0.18	0.006	0.002	BAL.	STA	302

* ANN: Annealed
Q&T: Quenched and Tempered
STA: Solution Treated and Aged

A set of cutting tests which was composed of a combination of five different tools (four tools for Inconel 718 and Ti-6Al-4V), five depth-of-cut levels, and five feed-rate levels was regarded as "a unit of test". In other words, 'a test' for a workpiece material, a cutting speed, and a side cutting edge consisted of all these combinations. Table 3 shows the detail of the combinations.

TABLE 2. CARBIDE GRADES APPLIED TO WORKPIECE MATERIALS

WORKPIECE MATERIAL	TOOL MATERIAL
AISI 4140, AISI 8620	MULTI-PHASE TiN COATED CARBIDE (KENNAMETAL KC850)
INCONEL 718, Ti-6Al-4V	C-2 CARBIDE (KENNAMETAL K68)

TABLE 3. A UNIT OF TEST

(A unit of test consists of all the combinations of tool, depth of cut, and feed rate.)

TOOL (INDEXABLE INSERT)	A: CNMA432 G: CNMG432 M: CNMM432* P: CNMP432 S: CNMS432
DEPTH OF CUT, MM (IN)	0.25, 0.51, 1.27, 2.5, 5.1 (0.01, 0.02, 0.05, 0.10, 0.20)
FEED RATE, MM/REV (IPR)	0.064, 0.13, 0.25, 0.38, 0.51 (0.0025, 0.005, 0.010, 0.015, 0.020)

* Tool M was not applied to Inconel 718 and Ti-6Al-4V.

All the tools were 80-degree diamond inserts with 1/32-inch nose radius. Figure 2 illustrates the cross-section of the tool describing chip breaker design. Tool A (CNMA432 as described in this paper) has no chip breaker, but a flat rake face. Tool G (CNMG432) has a groove-type chip breaker. Tool M (CNMM432) has a land-angle type, which was designed to reduce the cutting force at the high feed rates. This insert shape was not used for Inconel 718 and Ti-6Al-4V because the C-2 grade was not commercially available. Tool P (CNMP432) and Tool S (CNMS432) are obstruction types in which the geometry varies along the cutting edge, depending on the distance from the tool point. The side rake angle of the inserts set in the tool holders are also given in Figure 2. The values for chip flow radius will be explained in a later section.

The depth of cut and feed rate were selected to cover both roughing and finishing operations. Other cutting parameters such as side cutting edge angle and cutting speed will be explained later.

The workpiece material was machined until the chip shape stabilized. The chips were collected during each operation and were ranked as previously described.

DESCRIPTION OF ACCEPTABLE CHIP REGION

One of the ways to describe the effect of cutting conditions on chip breakage is to draw an acceptable and unacceptable region on a rate versus feed depth of cut diagram. Examples of such diagrams obtained in this study are shown in Figure 3. The boundary lines between under-control and good control in these diagrams were determined by fourth power linear regression analysis after log transformation. Since the chip shape rating number has no mathematical meaning, this regression method is not mathematically rigid. However, it helps us draw the smooth lines, and the lines drawn by this method are a good representation of the real data. On the other hand, the border lines between good-control and over-control could not be determined by this method because there are only a few data points belonging to the over-control region. These lines were drawn as a smooth curve linking the over-control data.

At the smaller depth of cut, the tool nose has a large effect on chip flow direction, chip curl radius, and eventually chip breakage. Most of the discussion in the following sections will be on the boundaries defining under control and good control at a depth of cut larger than 1.27 mm (0.05 inch) where there is little influence by the nose. The average feed rate along the under good/control control boundary for the depth of cut between 1.27 and 5.1 mm (0.05 and 0.20 inch) is defined as the lower limiting feed in this paper. The average value along the good control/over control boundary is the upper limiting feed.

In the following sections, the effect of side cutting edge angle, cutting speed, workpiece material, and cutting tool/chip breaker geometry on chip breakage will be discussed.

TOOL NAME	CHIP BREAKER SHAPE	CHIP BREAKER TYPE	SIDE RAKE	CHIP FLOW RADIUS
A (CNMA432)	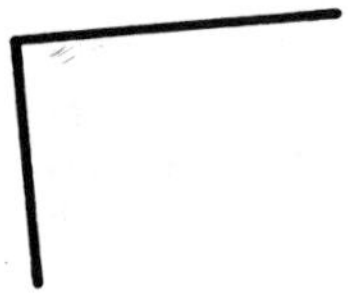	NO CHIP BREAKER	-5°	----
G (CNMG432)	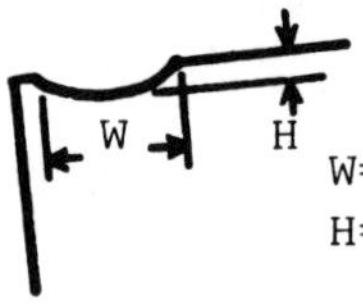 W=1.68 MM H=0.10 MM	GROOVE	-5°	3.6 MM (0.14 IN)
M (CNMM432)	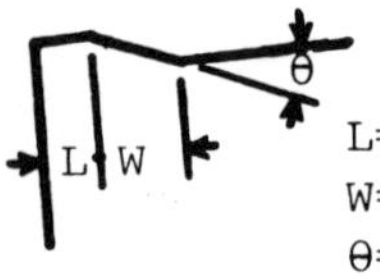 L=0.30 MM W=0.43 MM $\theta=16^{\circ}$	LAND-ANGLE	-5°	4.1 MM (0.16 IN)
P (CNMP432)	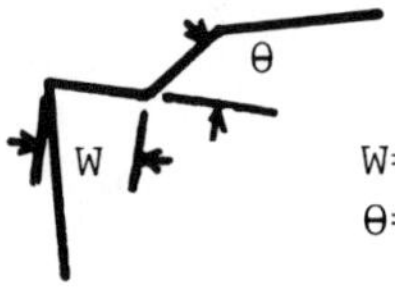 W=1.30 MM* $\theta=55^{\circ}$	OBSTRUCTION	$+5^{\circ}$	2.5 MM (0.10 IN)
S (CNMS432)	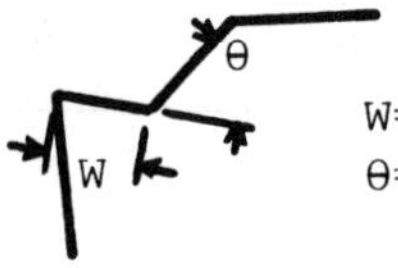 W=1.33 MM* $\theta=60^{\circ}$	OBSTRUCTION	$+15^{\circ}$	2.3 MM (0.09 IN)

*AVERAGE VALUE

FIGURE 2. GEOMETRY OF CHIP BREAKER CROSS SECTION

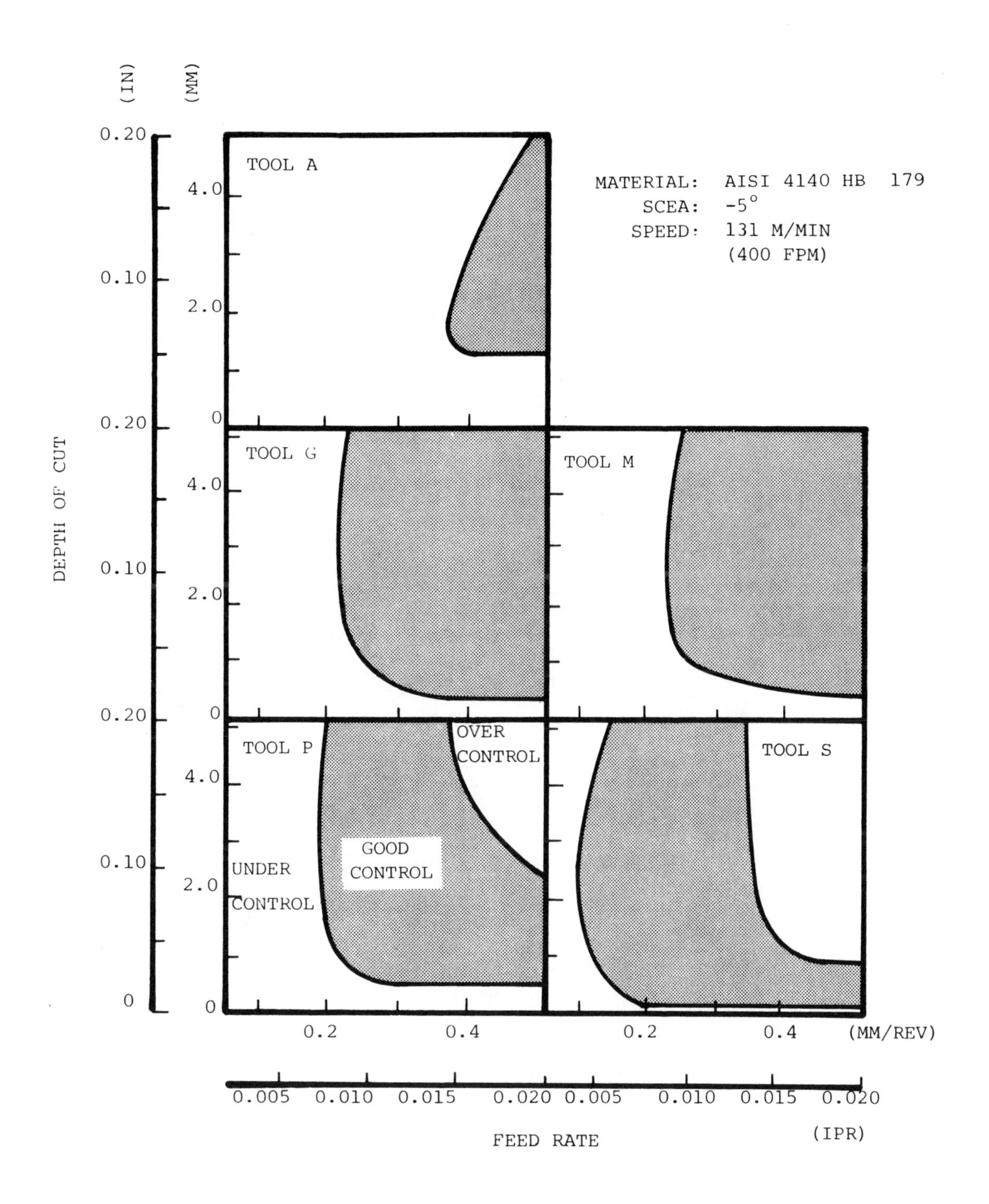

FIGURE 3. ACCEPTABLE CHIP REGION ON FEED RATE DEPTH OF CUT DIAGRAM

EFFECT OF SIDE CUTTING EDGE ANGLE ON LOWER LIMITING FEED

Lower limiting feeds are plotted against side cutting edge angle in Figure 4. Annealed AISI 4140 steel was used for these tests. The cutting speed was 131 m/min (400 fpm). This figure indicates that the side cutting edge angle has no major effect on lower limiting feed, but the increase in side cutting edge angle does slightly reduce the lower limiting feed. The reduction is about 12% over the side cutting edge angle range of -5° to +15°.

While the larger side cutting edge angle (absolute value) produces the thinner chips which are less easily broken, it also gives more opportunity for chips to hit obstacles such as the workpiece or the tool. The results shown in Figure 4 suggest that there is a greater tendency for chips to hit obstacles than for thinner chips to be produced.

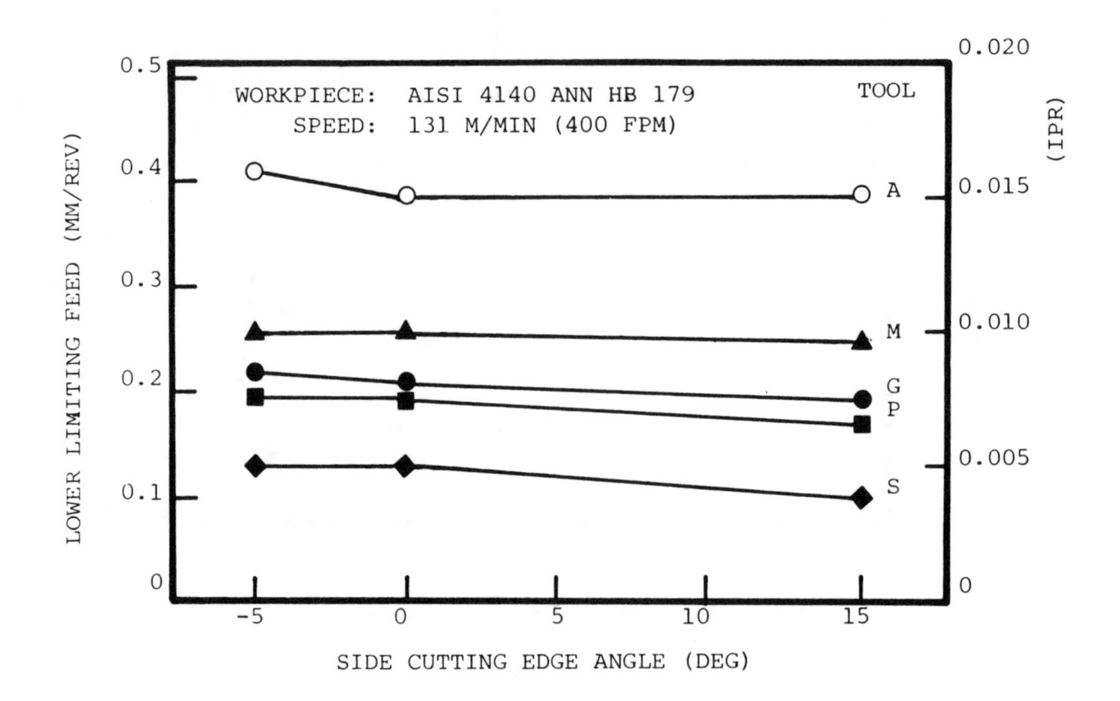

FIGURE 4. EFFECT OF SIDE CUTTING EDGE ANGLE ON LOWER LIMITING FEED

EFFECT OF CUTTING SPEED ON LOWER LIMITING FEED

The relationship between cutting speed and lower limiting feed for chip breakage is given in Figure 5. All tests were carried out with annealed AISI 4140. The side cutting edge angle was -5°. There was no built-up edge observed on the tool rake face during these tests.

The lower limiting feed becomes slightly higher when the cutting speed is increased. The increase of lower limiting feed caused by the increase in cutting speed from 66 to 197 m/min (200 to 600 fpm) is about 22%. The thinner chips are obtained at the higher cutting speeds. The change in measured chip thickness was about 25% within the speed range of this test. The coincidence of these two values indicate that the increase in lower limiting feed brought about by the increase in cutting speed is related to the reduction of chip thickness.

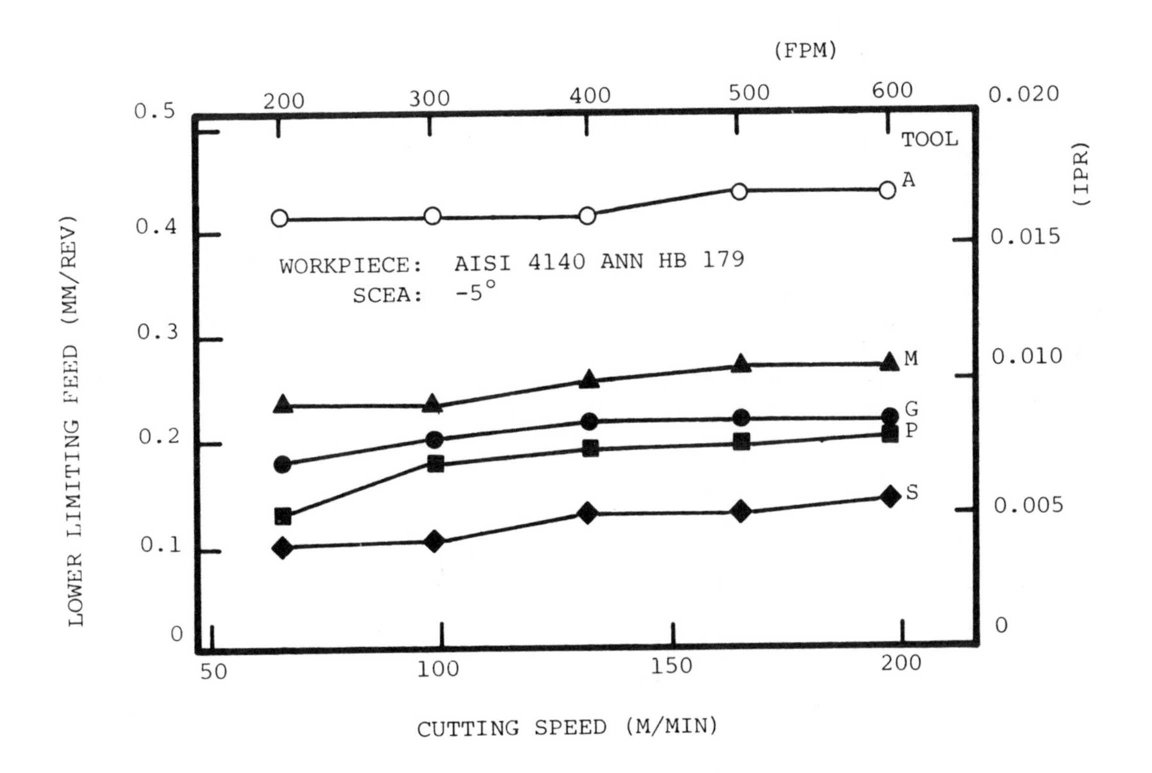

FIGURE 5. EFFECT OF CUTTING SPEED ON LOWER LIMITING FEED

EFFECT OF WORKPIECE MATERIAL ON LOWER LIMITING FEED

Lower limiting feeds for all tested workpiece materials are shown in Figure 6. The cutting speed for each material was selected from the Machining Data Handbook to avoid undesirably high tool wear rate. The values of cutting speed are given in Figure 6. The difference in cutting speed need not be seriously considered when comparing the materials because cutting speed causes little change in the lower limiting feed.

The workpiece material has a significant effect on chip breakage phenomena. Annealed AISI 4140, AISI 8620 and Inconel 718 have better chip breakability than Ti-6Al-4V and hardened AISI 4140. Since similar trends were observed in all the tests with five different insert configurations, these chip-breakage features can be considered characteristic of the materials themselves.

Lower limiting feeds are plotted against Brinell Hardness in Figure 7 which shows that harder material has poorer chip breakability. This trend is also well explained by the change in chip thickness: harder material produces thinner chips which are less easily broken. The harder material also produces higher strength chips which are more difficult to break.

EFFECT OF CHIP-BREAKER TYPE ON LOWER LIMITING FEED

The effect of chip-breaker type is shown in Figures 4 through 6. In all those cases, Tool S showed the smallest lower limiting feed, followed by Tool P, Tool G and Tool M. Tool A had the poorest chip control as expected from the rake face configuration. Note that the influence of the chip-breaker shape was as great as that of the workpiece material. From the fact that the chip control ranking of the tools does not change for different side cutting edge angles, cutting speeds, and work materials, the effect of chip-breaker type may be described by a parameter related to its geometry. Details of this consideration will be discussed later in this paper.

UPPER LIMITING FEED

Over-controlled chips were only observed in turning with Tool P and Tool S. Quantitative discussion is not possible because there are not enough data. However, it appears that the higher the lower limiting feeds of a tool or a material, the higher their upper limiting feeds.

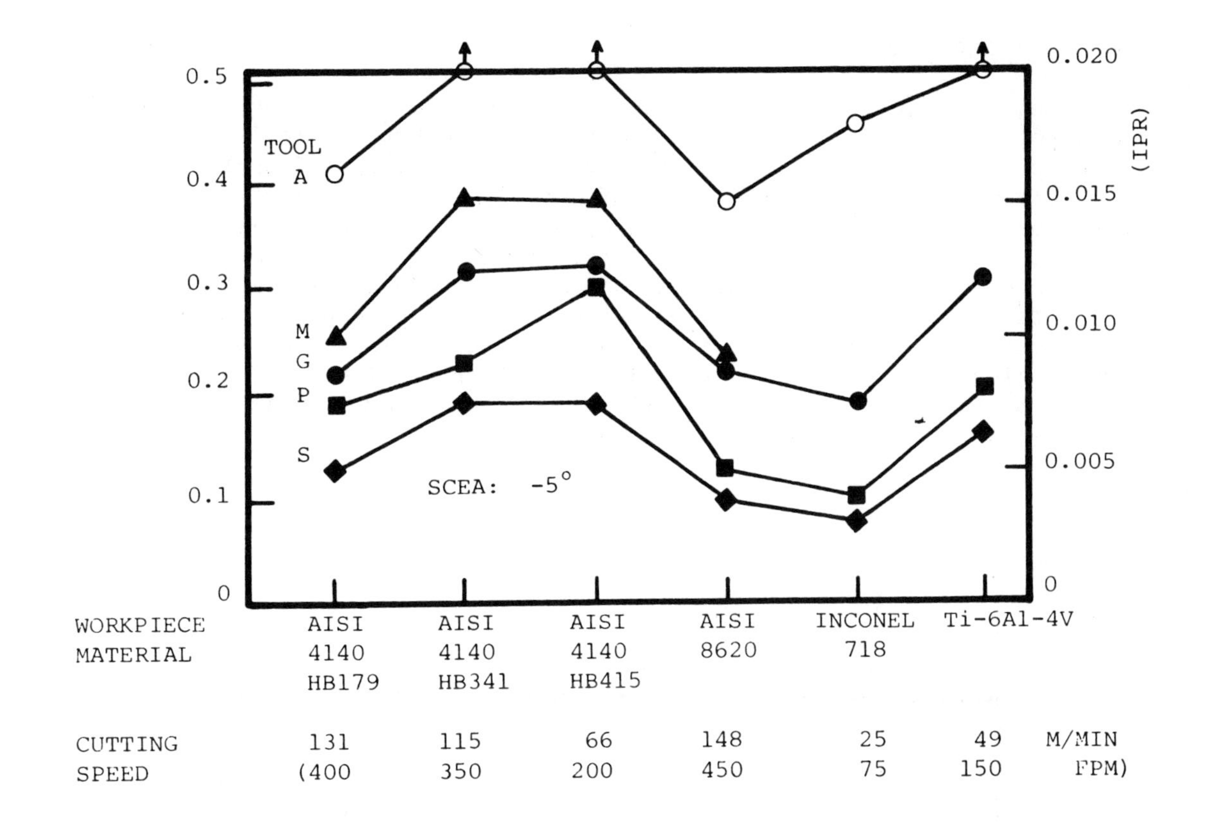

FIGURE 6. LOWER LIMITING FEED OF VARIOUS MATERIALS

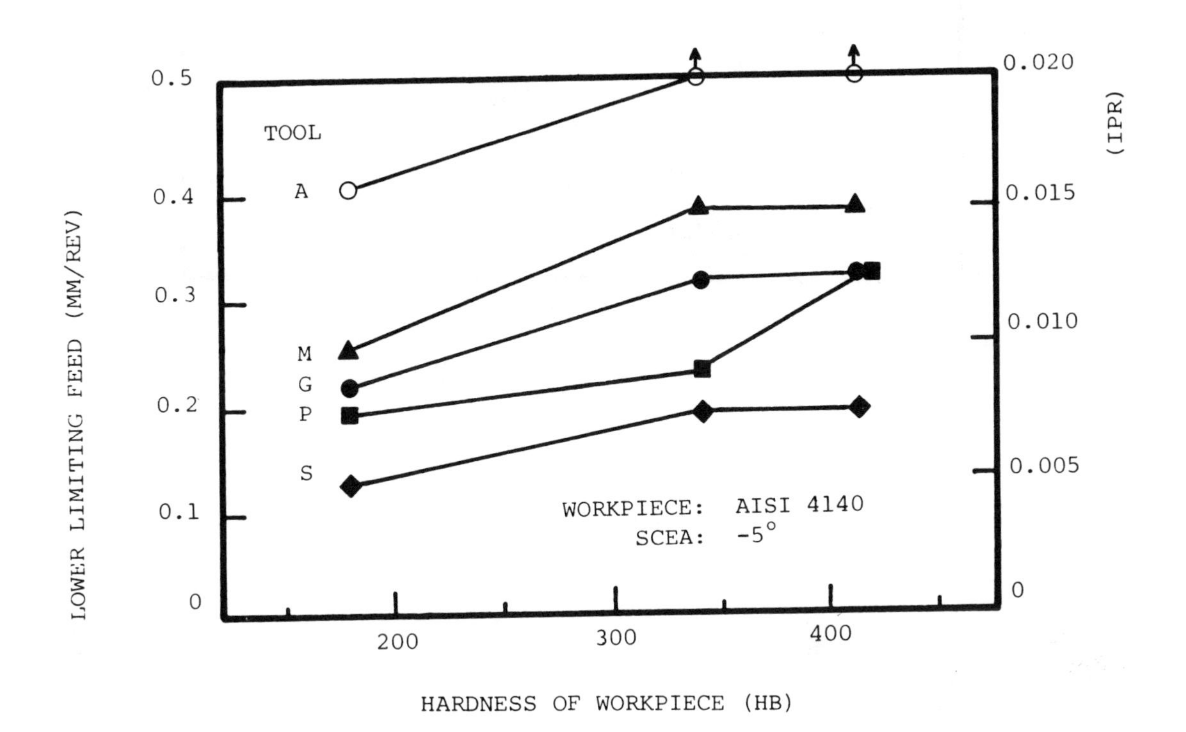

FIGURE 7. EFFECT OF WORKPIECE HARDNESS ON LOWER LIMITING FEED

LOWER LIMITING DEPTH OF CUT

When the depth of cut is smaller than the feed rate or nose radius, the direction of chip flow is perpendicular rather than parallel to the workpiece axis. The chip thickness under such conditions is strongly dependent on the value of the depth of cut. Therefore, the chip breakage phenomena at small depths of cut and high feed rates are similar to those at low feed rates and large depths of cut. Figure 8 shows the relationship between lower limiting depth of cut and lower limiting feed. There is a good correlation between them. Because the configuration of the chip breaker at the nose region is different from that along the side cutting edge, each tool has a different correlation between the lower limiting feed and the lower limiting depth of cut. These relationships are shown as different lines in Figure 8.

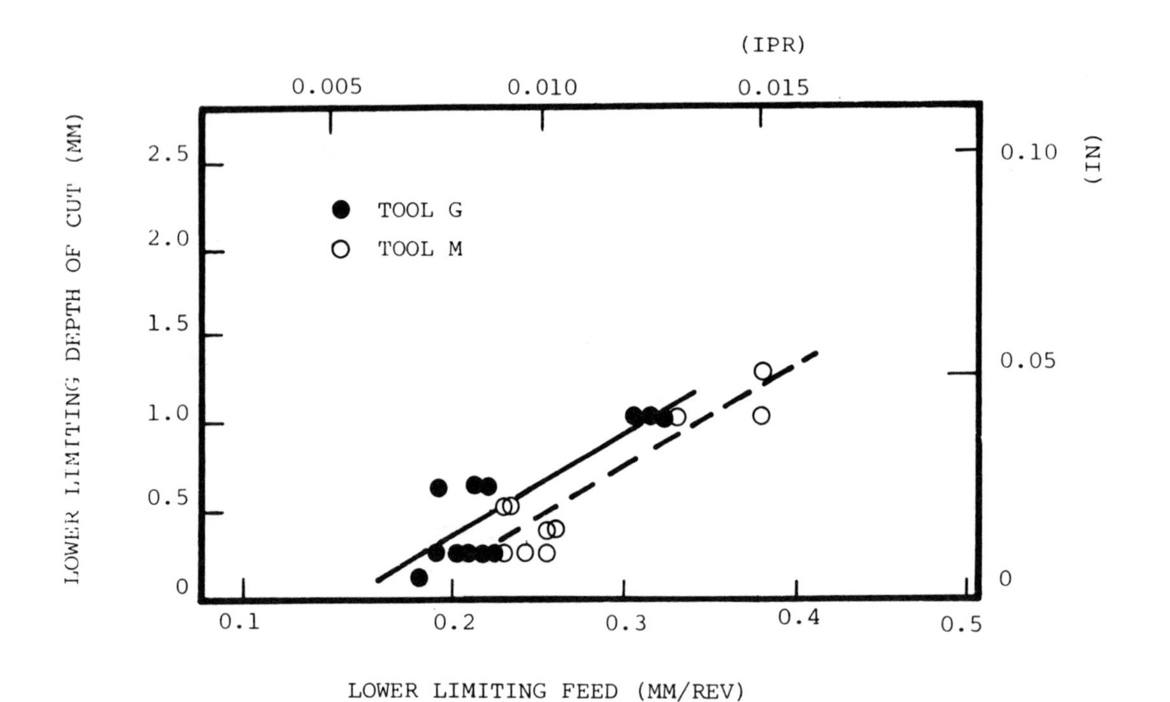

FIGURE 8. RELATIONSHIP BETWEEN LOWER LIMITING FEED AND LOWER LIMITING DEPTH OF CUT

SUMMARY OF CHIP CONTROL TESTS

From the discussion in previous sections, the following phenomena were ascertained:

1. The side cutting edge angle and the cutting speed have only a small effect on chip breakage.

2. Both workpiece material and chip-breaker shape have a significant effect on chip breakage.

3. Lower limiting feed, upper limiting feed and lower limiting depth of cut are interrelated.

In the following sections, lower limiting feed will be quantitatively discussed in terms of chip-breaker geometry and the ways to estimate the lower limiting feed for a combination of work material and cutting tool.

RELATIONSHIP BETWEEN CHIP FLOW RADIUS AND LOWER LIMITING FEED

Reference papers on chip control suggested that the chip breakage was related to both chip thickness and chip flow radius,[8] and that lower limiting feed was a function of the chip flow radius.[2] The chip flow radius can be calculated from the chip breaker configuration. More than one equation of chip flow radius for each type of chip breaker was proposed. The following equations were used in this study:

1. Groove-type chip breaker,

$$R = \frac{H}{2} + \frac{W^2}{8H} \quad (1)$$

   Where R is chip flow radius, H is depth of groove and W is width of groove.

2. Obstruction-type chip breaker,[13]

$$R = W \cdot \cot(\theta/2) \quad (2)$$

   Where W is width of rake face and θ is the angle between rake face and obstacle face.

3. Land-angle type chip breaker

   No equation has been proposed for this type of chip breaker. In this study, the land-angle-type breaker is regarded as identical to the obstruction type which has the width of rake face W + L/2, where W is the rake-face width of the land-angle-type tool, and L is its land length.

While consideration of tool-chip contact length was proposed by Trim and Boothroyd,[3] it was neglected in this study. The geometric values in the shape column used in the calculations and the calculated chip flow radii for the tools are given in Figure 2. The chip flow radius of Tool A cannot be calculated by these equations. There are some equations to estimate the natural chip flow radius for a flat tool,[8] but Tool A will be left out of the discussion for simplicity.

The chip flow radius and the lower limiting feed for some of the present tests are plotted in Figure 9. The data for each material fall on a line which goes through the origin with little scattering, which means the lower limiting feed is described by a linear equation as follows:

$$F = K \cdot R \qquad (3)$$

Where F is lower limiting feed, R is chip flow radius[3] and K is a constant. As easily understood from Figure 9 and the summary in a previous section, K is mainly dependent on the workpiece material and varies only slightly with side cutting edge angle and cutting speed. Notice that the tool chip-breaker geometry is expressed by R and it has no effect on the value of K. In other words, chip breakage or lower limiting feed is affected by tool geometry and workpiece material separately.

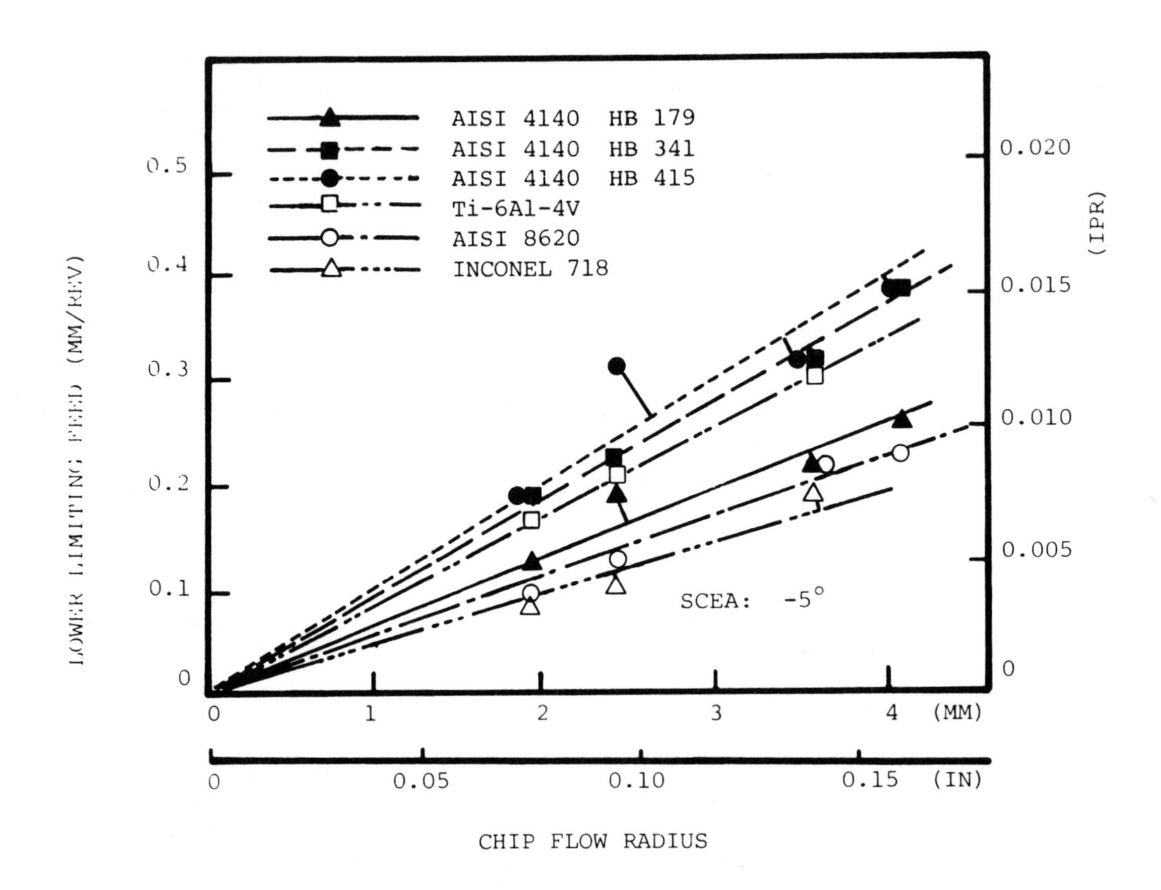

FIGURE 9. RELATIONSHIP BETWEEN CHIP FLOW RADIUS OF TOOLS AND LOWER LIMITING FEED

The values of the constant K for all the tests in this study are summarized in Table 4. Some references suggested the relationship was expressed by a square root function or $F = a \cdot \sqrt{R}$. However, within the range of the tests on this study, Equation 8 is a better fit.

TABLE 4A. CONSTANT K FOR VARIOUS MATERIALS

WORKPIECE MATERIAL:	AISI 4140 HB 179	AISI 4140 HB 341	AISI 4140 HB 415	AISI 8620	INCONEL 718	Ti-6Al-4V
CONSTANT K:	0.065	0.093	0.099	0.057	0.049	0.085

TABLE 4B. CONSTANT K FOR VARIOUS SIDE CUTTING EDGE ANGLES

Material: AISI 4140, HB 179
Cutting Speed: 131 m/min (400 fpm)

SIDE CUTTING EDGE ANGLE:	-5°	0°	15°
CONSTANT K:	0.065	0.064	0.058

TABLE 4C. CONSTANT K FOR VARIOUS CUTTING SPEEDS

Material: AISI 4140, HB 179
Side Cutting Edge Angle: -5°

CUTTING SPEED, M/MIN: (FPM):	66 (200)	98 (300)	131 (400)	164 (500)	197 (600)
CONSTANT K:	0.053	0.059	0.065	0.066	0.068

METHODS OF ESTIMATING LOWER LIMITING FEED

One of the ways to estimate lower limiting feed is to make a calculation using Equation 3 when the tool geometry and the constant K are known. Using the chip flow radius R shown in Figure 2 and the constant K in Table 3, lower limiting feeds were calculated for all the test conditions. A comparison of the values is given in Figure 10. The standard deviation of the difference (experimental value - calculated value) is 0.023 mm/rev (0.009 ipr), which is sufficiently small compared with the normal feed rates used in machine shops.

One of the applications of this method is in building a data base. Once the values of R for each tool and the machinability value of K for each workpiece material are stored in the data file, the lower limiting feed for any combination of the tools and the workpiece materials can be calculated. To determine the value of the constant K for a material, only one chip breakage test with the tool whose chip flow radius is known is necessary.

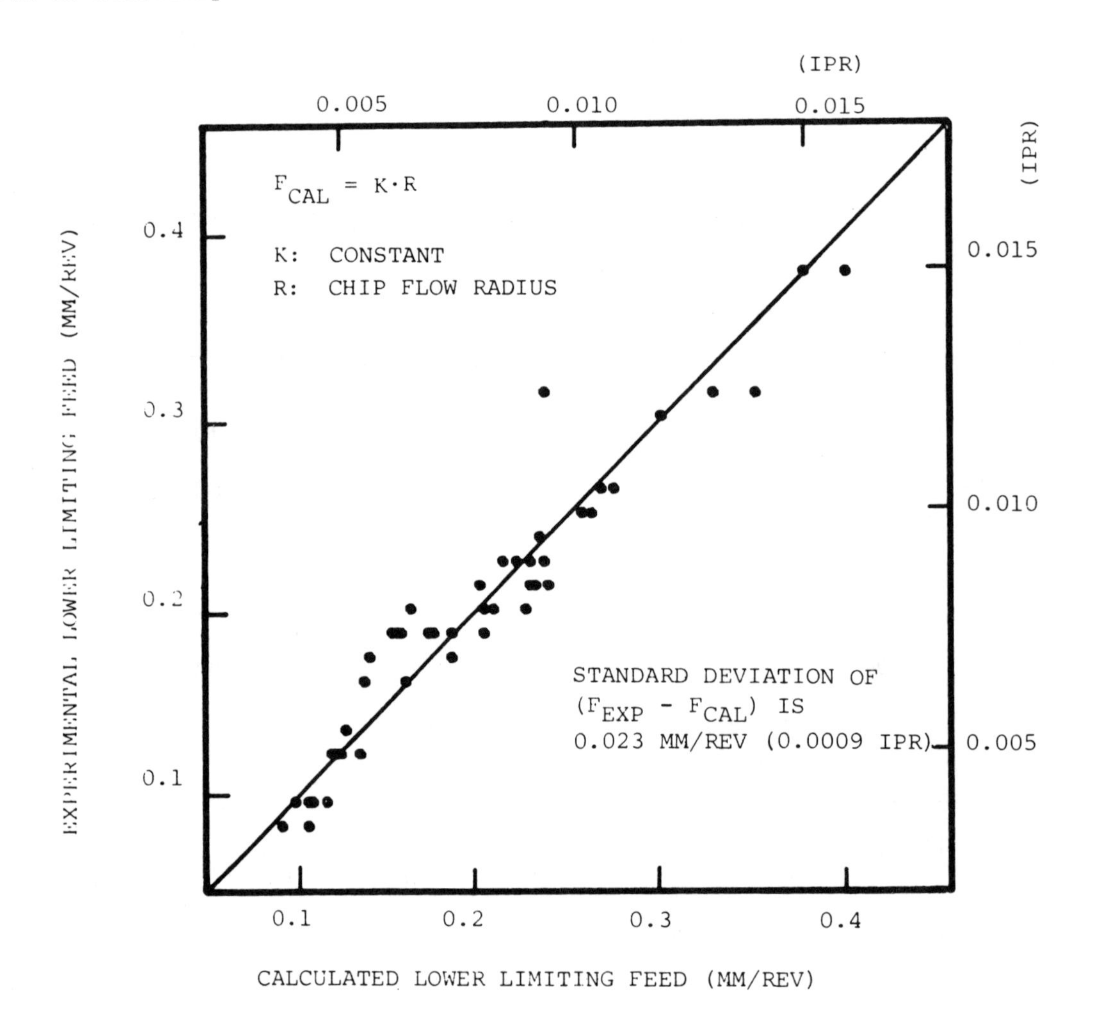

FIGURE 10. COMPARISON BETWEEN CALCULATED AND EXPERIMENTAL VALUES OF LOWER LIMITING FEED

This method cannot handle the case of flat rake face tools, like Tool A in this study, or other tools whose chip-breaker configurations are too complicated to calculate the R value mathematically.

Suppose that the lower limiting feeds for three combinations out of four composed of two types of tools and two types of materials are known. The equations are obtained from Equation 3 as follows:

$$F_{1A} = K_1 \cdot R_A \tag{4}$$

$$F_{2A} = K_2 \cdot R_A \tag{5}$$

$$F_{1B} = K_1 \cdot R_B \tag{6}$$

$$F_{2B} = K_2 \cdot R_B \tag{7}$$

In the above equations, suffixes 1 and 2 refer to material 1 and material 2 and suffixes A and B refer to Tool A and Tool B. F_{1A}, F_{1B} and F_{2A} are the known lower limiting feeds and F_{2B} is the unknown value. Substituting Equation (4) through (6) into (7) and eliminating R_A, R_B, K_1 and K_2, Equation (8) is obtained as follows:

$$F_{2B} = \frac{F_{1B}}{F_{1A}} \cdot F_{2A} \tag{8}$$

Choosing two materials out of six used in this study, and assuming the lower limiting feeds for the tools except Tool G were unknown, the unknown values were calculated by Equation (8). This process was repeated until all the combinations of tools and materials were included. Figure 11 is the comparison of such calculated limiting values to the experimental values. The standard deviation of the difference between calculated and experimental values is larger than that of the order method; however, it is still small enough to accept this method as a first estimation.

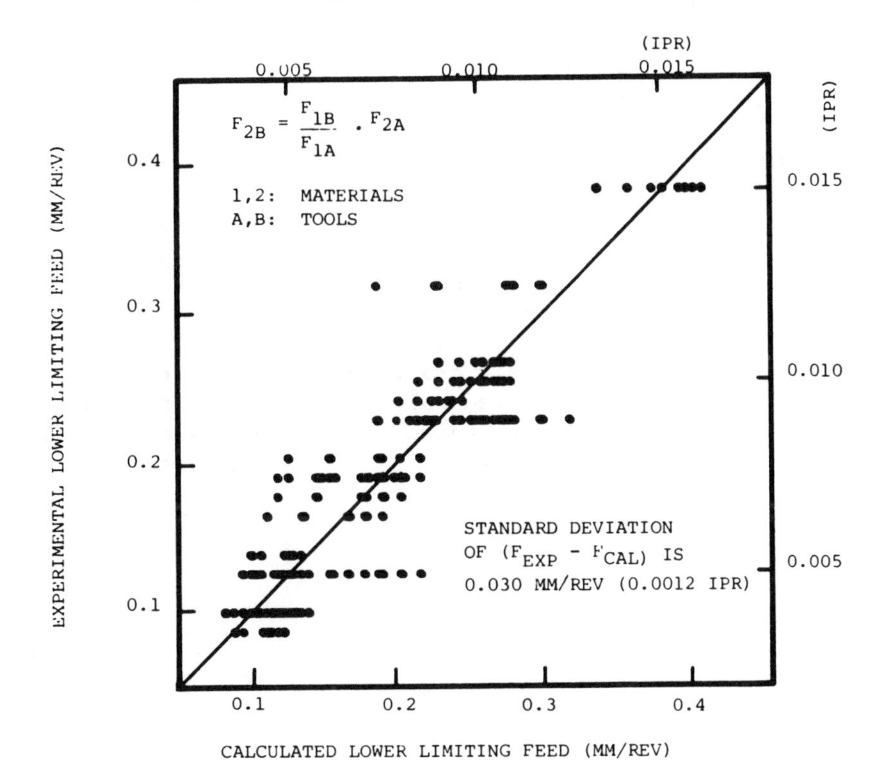

FIGURE 11. COMPARISON BETWEEN CALCULATED AND EXPERIMENTAL VALUES OF LOWER LIMITING FEED

CONCLUSIONS

From this study, the following conclusions were obtained:

1. The side cutting edge angle of the tool and the cutting speed have a small effect on chip breakage phenomena. The increase in side cutting edge angle slightly reduces the lower limiting feed for well controlled chips. The lower limiting feed becomes a little higher when the cutting speed is raised.

2. Both workpiece material and chip-breaker shape have significant effects on the chip breakage phenomena. The effect of these variables can be expressed by the equation:

$$F = K \cdot R \tag{9}$$

Where F is lower limiting speed, R is chip flow radius which is calculated from tool geometry, and K is a constant which is mainly related to the workpiece material.

3. The lower limiting feed, upper limiting feed, and lower limiting depth of cut are interrelated.

4. Lower limiting feeds calculated using Equation 9 coincide with experimental values. This method can be used to estimate lower limiting feeds when the tool geometry and constant K for the material are known.

5. When the chip-breaker geometry and constant K are not known, but the chip control data for three out of four combinations composed of two types of tools and two types of materials are known, unknown values can be estimated from the equation:

$$F_{2B} = \frac{F_{1B}}{F_{1A}} \cdot F_{2A} \tag{10}$$

Where 1 and 2 indicate different materials, and A and B indicate different tools.

ACKNOWLEDGEMENTS

The authors wish to express their appreciation to Kennametal Inc. for their offer of all the tools used in this study, Daido Steel Company, Ltd., who provided the workpiece materials, and to Metcut Research Associates Inc. for their support of this research.

The authors also express their gratitude to Dr. M. Field of Metcut Research Associates Inc. who offered valuable guidance from the beginning of this study and to Susan Harvey, Karen Keiter and Jeanne Brotherton for their assistance in the preparation of the manuscript.

BIBLIOGRAPHY

1. Henriksen, E. K., "Chip Breakers", National Machine Tool Builers' Association, 1953.

2. Okushima, K.; Hoshi, T.; and Fujiwara, T., "On the Behavior of Chip in Steel Cutting, Part II", Bulletin of JSME, Vol. 3, No. 10, 1960, pp. 199-205.

3. Trim, A. R. and Boothroyd, G., "Action of the Obstruction Type Chip Former", International Journal of Production Research, Vol. 6, No. 3, 1968, pp. 227-240.

4. Worthington, B., "The Operation and Performance of a Groove-Type Chip Forming Device", International Journal of Production Research, Vol. 14, No. 3, 1976, pp. 529-558.

5. Spaans, C., "A Systematic Approach of Three-Dimensional Chip Curl, Chip Breaking and Chip Control", SME Technical Paper, MR70-241, 1970.

6. Spaans, C. and Goedemondt, A. A., "The Breakability. An Aspect of the Machinability -- A Computer-Simulation of Chip Formation", SME Technical Paper, MR71-154, 1971.

7. Rasch, F. O. and Tonnesen, K., "Tool Failure and Chip Form as Restrictions when Selecting Cutting Data", Annals of the CIRP, Vol. 26, No. 1, 1977, pp. 45-48.

8. Kluft, W., Konig, W., van Luttervelt, C. A., Nakayama, K., and Pekelharing, A. J., "Present Knowledge of Chip Control", Annals of the CIRP, Vol. 28, No. 2, 1979, pp. 441-455.

9. Elgamayel, J. and Pinto J. G., "Design of a Molded-in Chip Breaker in Throwawa Inserts", New Developments in Tool Materials and Applications, Illinois Institute of Technology, 1977, pp. 73-81.

10. Jones, D. G. and McCreery, J. F., "A Study of Preformed Chip Control Devices in Throwaway Carbide Inserts", SME Technical Paper MR73-215, 1973.

11. "Tool Life Testing with Single Point Turning Tools", ISO 3685-1977(E), International Organization for Standardization, 1977.

12. Machining Data Handbook, Metcut Research Associates Inc., 3rd edition, Machinability Data Center, 1980.

13. Nakayama, K., "A Study on Chip-Breaker", Bulletin of JSME, Vol. 5, No. 11, 1962, pp. 142-150.

CHAPTER 3

BORING

Presented at the SME 1981 International Tool & Manufacturing Engineering Conference, April 1981

Method of Increasing Rigidity of Boring Bars for Boring Small Holes

by Boris Berdichevsky
Sloan Valve Company

Boring of relatively small holes (1 inch diameter and smaller) with a single point boring tool equipped with a throwaway carbide insert is usually rather difficult operation, because the rigidity of the bar itself is insufficient, especially for high feed rates (more than 0.008 inch per revolution) and size of the bar, and hence its moment of inertia has been determined by the size of the bore to be machined.

Besides, the body of a boring bar is weakened by the pocket for a cutting insert, cutting edge of which is to be placed on the neutral axis of the bored hole in order to maintain proper tool geometry (See Fig.1)

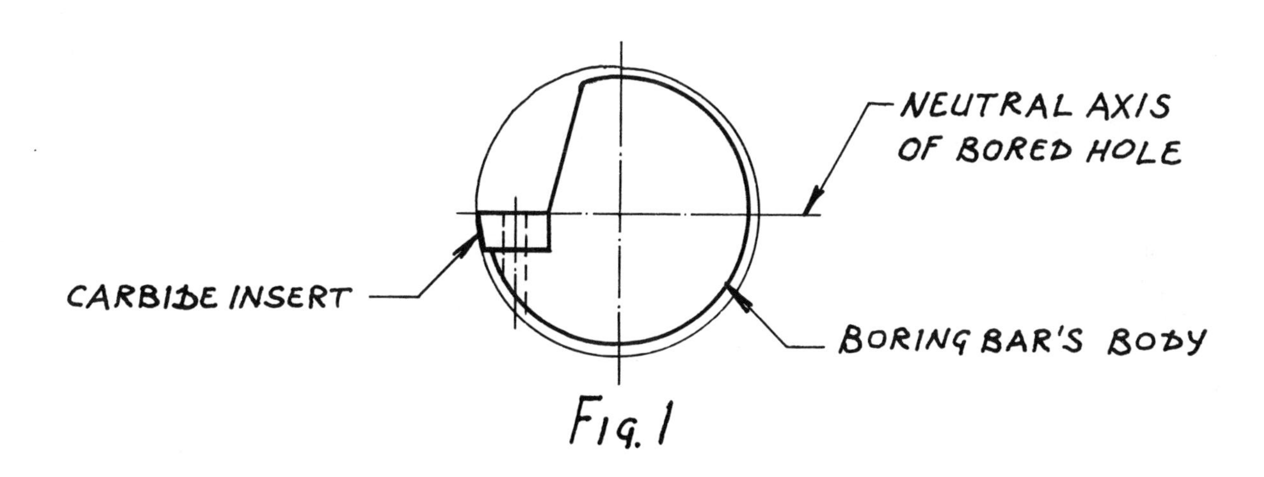

Fig. 1

Changing the tool material to a better grade of steel or heat treatment to a higher hardness does not improve the rigidity of the boring bar since all steels have practically the same modulus of elasticity and the addition of alloys or hardening does not change modulus magnitude and hence does not increase the stiffness of the boring bar.

To increase the moment of inertia of the boring bar and hence its rigidity, which is especially important to provide firm support surface for the carbide insert and diminish as much as possible its deflection, the cutting edge of the carbide insert has been placed above the neutral axis of the bored hole without violating cutting tool geometry (See Fig.2)

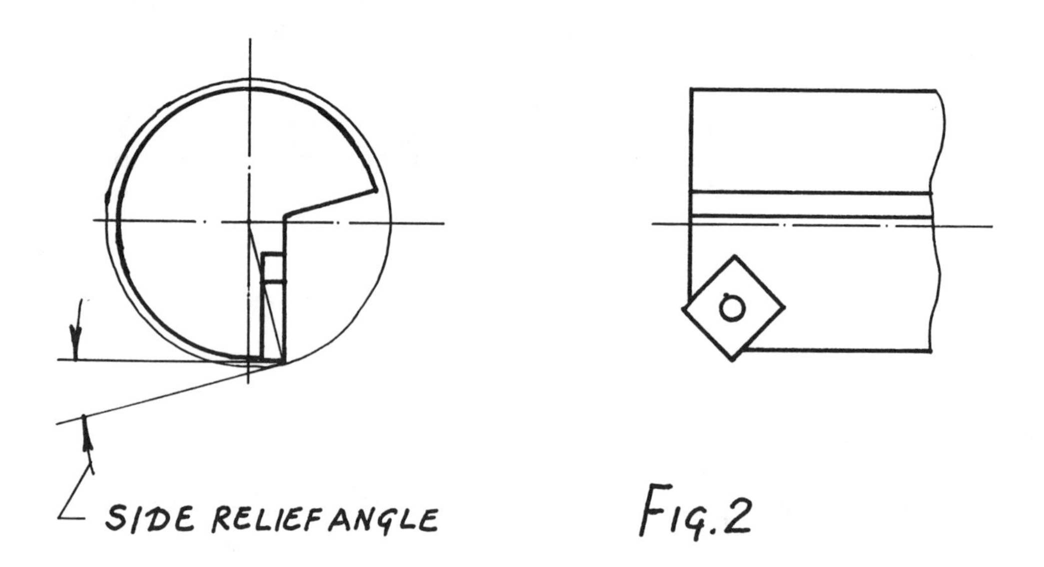

Fig.2

Since the moment of inertia of the boring bar has been increased, the rigidity of the bar also increases and enables to machine parts with the higher feed rates (approximately 20% higher). This method is especially useful when carbide insert with a positive rake angle is utilized and insert is set at angle in the bar.

The following calculations can be applied for determining the height of insert's cutting edge above the neutral axis of the bored hole dependably on required side relief angle (See Fig.3)

Definition of Symbols:

a - Side relief angle normal to bored hole.
a_1 - Side relief angle of cutting insert, normal to its cutting surface for the given hole dia.
H - Height of an insert's cutting edge above the neutral axis of the bored hole.
B - Angle of setting of insert in the boring bar.
R - Radius of the bored hole.

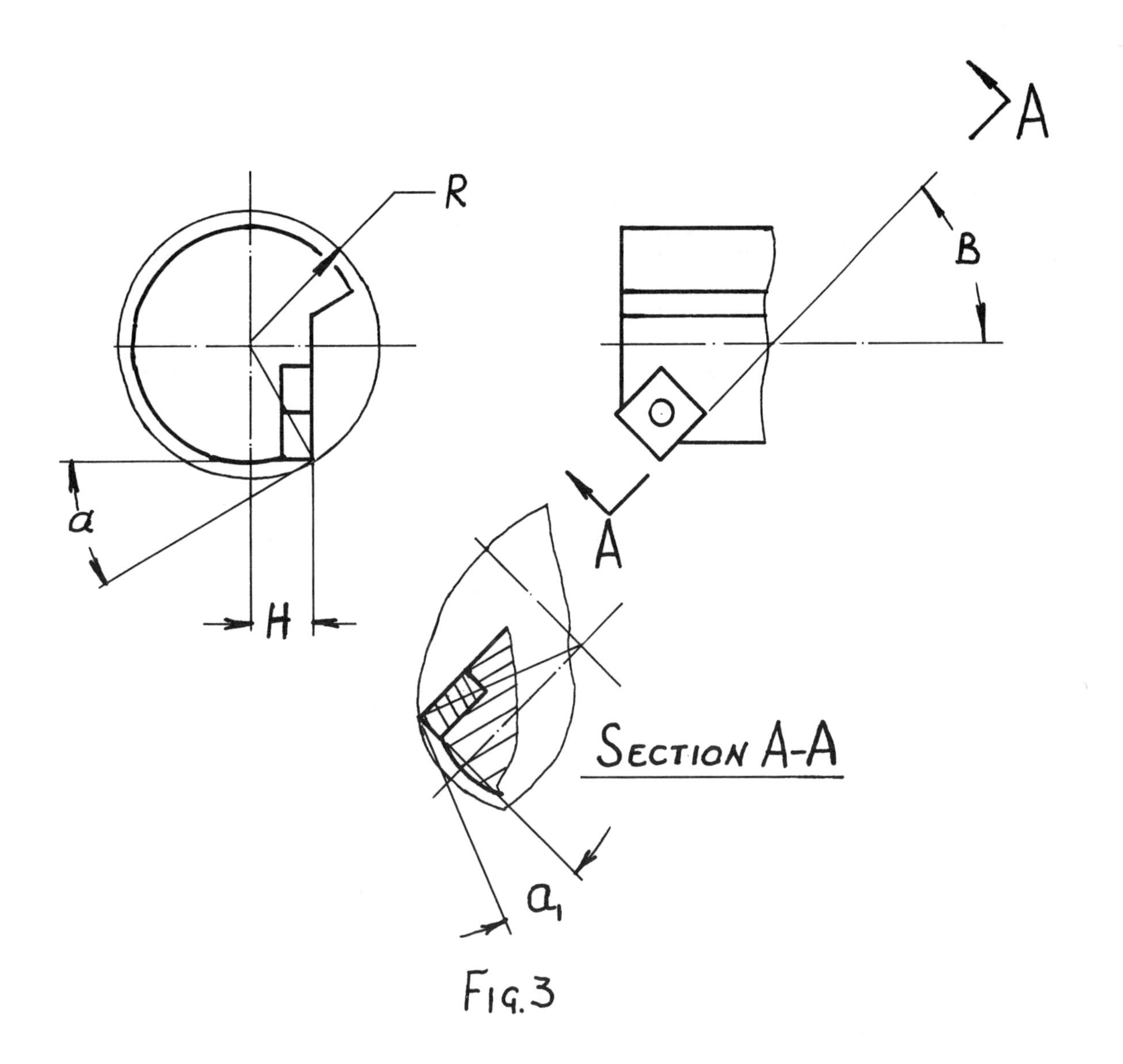

Fig.3

Usually angle a_1 is determined by the kind of material to be machined (steel, cast iron, etc.) and radius R of the bored hole is also known.

The relationship between angles a and a_1 can be expressed by the following equation:

$$\frac{R\sin a \sin B}{R\cos a} = \tan a_1 \text{ or } \tan a \sin B = \tan a$$

Reorganizing the equation with respect to a, we can get

$$\tan a = \frac{\tan a_1}{\sin B}$$

Thus, if angle a is known, it is possible to determine H: $H = R\sin a$.

Applying the above formula it is possible to calculate the height H of an insert's cutting edge above neutral axis of the bored hole dependably on side relief angle that is required for cutting various materials.

Table I shows the magnitude of side relief angle a for various angles a_1 dependably on the angle B of setting cutting insert in a boring bar.

Table I

B Degrees	a_1 Degrees	a Degrees
	7	$25^{o}23^{1}$
15	8	$28^{o}30^{1}$
	9	$31^{o}28^{1}$
	10	$34^{o}16^{1}$
	7	$13^{o}48^{1}$
30	8	$15^{o}42^{1}$
	9	$17^{o}34^{1}$
	10	$19^{o}25^{1}$
	7	$9^{o}51^{1}$
45	8	$11^{o}15^{1}$
	9	$12^{o}38^{1}$
	10	14°
	7	$8^{o}4^{1}$
60	8	$9^{o}13^{1}$
	9	$10^{o}22^{1}$
	10	$11^{o}30^{1}$

Reprinted from *Manufacturing Engineering,* July 1974

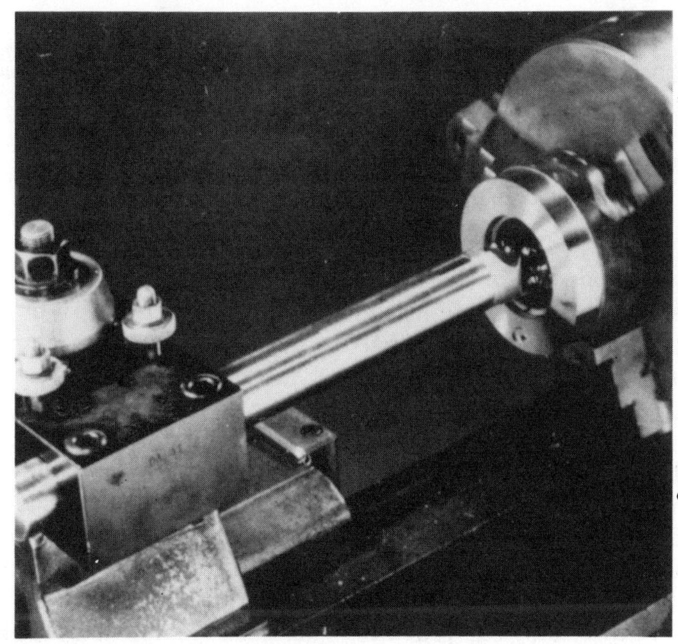

NO-CHATTER BORING—with up to 7:1 overhang ratio—results from tooling system that considers tool material and insert geometry along with a unique inertia-damping concept.

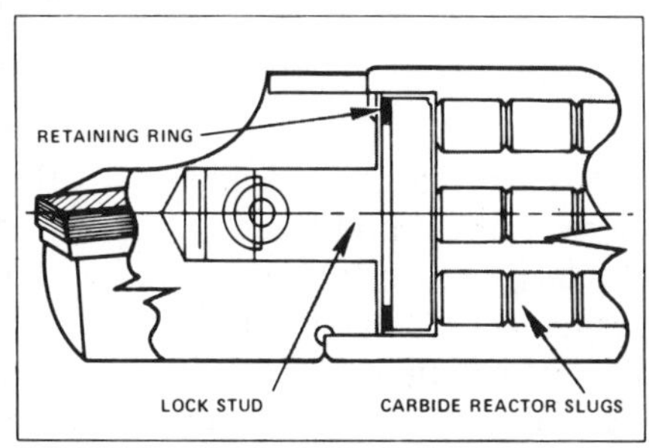

INERTIA-DAMPING: The key is 28 carbide reactor slugs positioned radially. Slug mass, clearance, and tensile strength of bar are all critical.

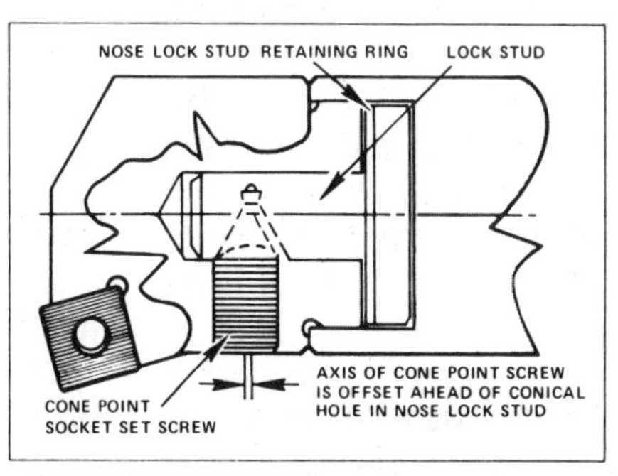

SINGLE SET SCREW speeds nose change. Straight-line holding pressure derives from offset cone-point axis.

Bore Chatter Free —

with up to 7:1 overhang ratio of bar length to diameter. It's done with a tooling system design that considers more than just inertia damping

BERNARD FEINBERG
Executive Editor

CHATTER-FREE FINISH in extended-length boring (up to 7:1 overhang ratio of bar length to diameter) is the performance promised for a tooling system introduced recently by Valeron Corporation's Valenite Div. Called the De-Vi-Bar, the system comprises a unique inertia-damping concept, a boring bar of specific tensile strength, and interchangeable noses designed with a negative-rake pocket to exploit the chip-control capability inherent in Valenite's Positive/Negative inserts. These inserts, which provide positive cutting action for negative-rake holders, eliminate the need for mechanical chip-breakers and bulky clamps and clamp screws. The net effect is a low, clean profile that lowers horsepower requirements while allowing doubled or even tripled feed rates.

How It's Done. Essential ingredients for optimum performance in long-overhang boring include: proper tensile strength of bar material, a cutting edge that's designed to offer the least resistance to cutting action, and a reacting mass positioned to offer maximum resistance to chatter impulses.

In the Valenite approach, carbide reactor slugs, positioned in a circular pattern in the boring bar and/or head, are approximately twice the weight of the steel removed from the holes that contain them. Thus, a reactor mass is positioned up front (a critical consideration) in a pattern that provides for dissimilar reacting forces to a given radial or torsional impulse. A given radial impulse, for example, causes a counteracting force against the outer wall of the slug cavity in the series closest to the impulse, and against the inner wall of the slug cavity in the series on the opposite side of the bar. In effect, dissimilar forces react against the initial impulse.

In general, chatter patterns take an elliptical shape in a direction approximately 15 to 20 deg relative to the tool's cutting edge. Thus, damping efficiency is a function of the reacting mass' weight, pattern, and its ability to react in any direction. For this reason, slug clearance too is critical and ranges from 0.002 to 0.003 and 0.005 to 0.006 inch, depending on bar size.

Quick-Change Capability. At the present time, a full range of bore sizes is accomplished with two sizes of noses in six bar diameters. When changing sizes, the bar remains in the chuck. Noses are retained by a lockscrew design in which the centerline of the screw hole is slightly forward of the conical seat centerline. Downward travel of the lockscrew causes the conical end of the pin to contact the forward side of the conical seat in the nose lock stud, pulling the nose tightly into the base.

Nose styles available include standard triangles, squares, and diamonds plus a vernier-adjustable nose, a threading and grooving nose, and one to accept high-speed or brazed carbide tool bits. Where large bores require light cuts, an adapter allows combining small noses with large diameter bars.

Significantly, operating conditions and carbide grade selection are similar to those normally used in turning operations — surface speeds are in the order of 350 to 400 fpm for most steels. Speeds drop to 250-300 fpm on harder or tougher materials, and rise to 700 fpm on softer, nonferrous materials. ■

Reprinted from *Cutting Tool Engineering,* March/April 1979

Block-Type Boring Bars Have Important Advantages . . . Even on N/C Applications

Norman H. Lovendahl, President
NL Tool Company

BORING to close tolerances at high metal-removal rates can be accomplished by using block type boring tools. Primary reasons for their superior performance on heavy metal-removal cuts are the generous chip clearances and the larger carbide inserts in the cutter blocks.

Block type boring bars are so named because of their design. The bars are slotted in the location where boring is to be done. Cutter blocks that hold standard indexable inserts are held in the slot by a taper pin which forces the block to the back of the slot and centrally in the bar.

This makes it possible for one size bar to accommodate any number of interchangeable cutter blocks within a wide bore size range, providing maximum versatility for different size rough boring, finish boring, counterboring, chamfering, and facing operations.

The cutting geometry in a standard tool is 5° negative. This can be changed by using a different industry standard insert. For example: a tool using a SNMG 432 would be 5° negative, using a SNMP 432 it would be 5° positive, and using a SNMS 432 it would be 15° positive.

Roughing passes are done with two-insert blocks with both inserts cutting the same diameter. This balances the cutting load, minimizes spindle deflection, and attains maximum tool life. Bars may be arranged with additional slots for performing multiple operations in one pass. For example, a two-slot bar would permit rough boring followed by finish boring or counterboring. A three-slot bar provides three-operation capability such as boring, counterboring, and chamfering. At NL Tool we have produced special line bars to accommodate up to sixteen operations.

Solid made-to-size blocks are used for high volume runs. Adjustable blocks are used for low volume runs because of maximum versatility. They are available with 1/2″, 3/4″, and 1″ adjustment. They are also recommended for close tolerance finish boring. Adjustable two-cutter blocks are made of two identical pieces mounted on a center section. Size is adjusted from either end by inserting a hex wrench through the access hole. As the size-adjust screw is rotated, each end of the tool advances equally from center. They can be preset in a fixture or at the machine with micrometers. For extremely close tolerance bores of .0005″ or less, final adjustment on the machine may be necessary.

Feed-out cutter blocks are made for special operations such as step boring, grooving, large diameter counterboring, etc. The block is made up of two end pieces, feed screw, dial and locking screw. The block rotates with the bar. The operator or fixture holds the handle to keep the actuator stationary. In each revolution the pin engages with the dial which rotates the feed-out screw increment, moving the cutter into the workpiece. The tool is arranged with a rapid return for long cuts.

Block-type boring bars, being preset indexable tools, are ideal for N/C programming. A few boring bars can be adapted to many parts without any presetting fixtures. A boring bar with several tool slots can be used to reduce the number of operations such as a rough bore followed by a finish bore and chamfer. More operations can be accommodated by using a multi-cutter block arranged with several tool positions. When an adjustable cutter block is used, the axial dimension remains constant when the

Single Slot Bar

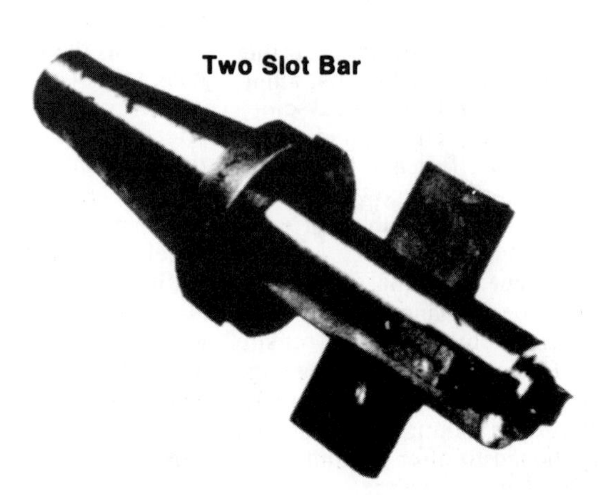

Two Slot Bar

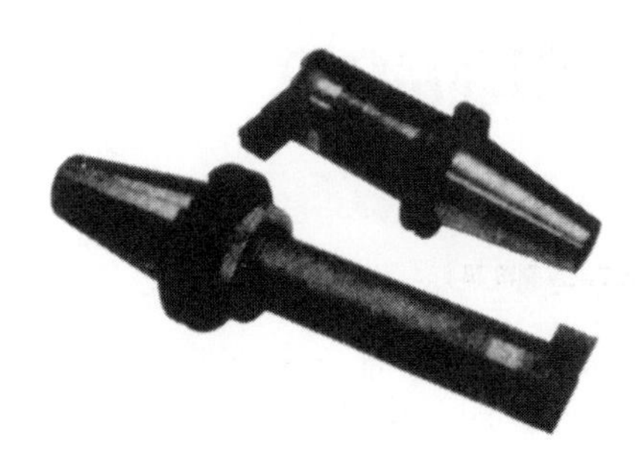

Single Point Tools

bore size is adjusted.

Four NL Tool cutter block series cover four size ranges for a total range of bore sizes from 1-13/32″ to 17-1/2″. The smallest size accommodates a range of bore sizes from 1-13/16″ to 4-1/2″, and the largest in the series accommodates a range from 5-1/2″ to 17-1/2″. Each series has a specific slot size for interchangeability of cutter blocks.

In contrast, cartridge type boring tools are designed for a specific narrow range because the cartridge adjustment is less than 1/16″. Bore sizes beyond the 1/16″ range require a complete new boring tool assembly. This compares to a block type boring tool with a 1-3/4″ bar that can accommodate bores from 2″ to 7″.

The cost of a new two-cutter block to accommodate a different bore is about the same as a replacement cartridge for a cartridge-type boring tool. Thus, while the cost of a block-type boring tool assembly is greater than a cartridge-type tool, the tooling costs for accommodating a number of different bores are much less.

For single point boring a modified block design also accommodates a wide range of bores plus fine adjustment in .0001″ increments. This is accomplished with an eccentric head that holds the single-cutter block. The eccentric head also accommodates a toolholder for straight shank boring bars. Four NL Tool bars accommodate a bore size range from 3/16″ to 7-1/4″. The smallest bar has an adjustment range of .015″ and the largest an adjustment range of .040″. Eight single-point blocks, two for each bar, accommodate the complete bore size range. Standard square shank brazed tip or HSS tools can be substituted for the single-insert cutter block.

Two slot bar machining a 1-3/4″ cast hole to 2.187± .0005 in one operation, the first tool roughing and the second tool finishing.

The versatility of block-type boring bars plus the economic factor when accommodating a number of different size bores are important advantages for both N/C and conventional operations. • • •

Presented at the SME 1975 International Tool and Engineering Conference, April 1975

Parameters in Selecting Boring Bar Designs

by Leon G. Kosker
Kennametal, Incorporated

This paper will explain various boring bar designs and give the parameters in selecting the following boring bars for non-rotating applications: solid steel boring bars; solid carbide shank boring bars with cone brazed steel heads; steel DeVibrator bars; carbide DeVibrator bars; and composite DeVibrator boring bars. It will also state some general rules to use in selecting a bar for a particular machining job and a "rule of thumb" for the minimum holding length. The paper will touch on what causes a bar to "chatter" and how the damping device works to counteract it. It will end up by giving some specific examples of successful application.

INTRODUCTION

Chatter: Webster's definition of chatter is, "...to vibrate rapidly in cutting". For production engineers, tool designers and tool manufacturers who are involved in machining deep holes with a boring bar, chatter can be an "Excedrin headache". One way to fight against chatter is to increase the dynamic stiffness of the boring bar. This paper will explain the reasoning behind different designs and also the theories for the general rules used in selecting a boring bar.

PROBLEMS

Boring bar problems have always had a certain mystery about them. All of us in the metalcutting field have had deep hole machining jobs to do and have wondered, "what boring bar size and design will I need to successfully machine this part?" "What diameter and length of bar is required?" "What lead angle would work best?" "Should I use a boring bar with positive or negative inserts?" "How much chip clearance is required--not counting the feeds, speeds and depths of cut?" Any one of these things, if not applied correctly in a machining set-up for deep hole boring, can cause the bar to vibrate. The violent and detrimental form of forced and self-excited vibration is known as chatter. Chatter has always been a prime concern in deep boring, but it is even more so when the workpieces are machined on transfer lines, automatic equipment or numerically controlled machines.

Theory of metalcutting assumes a steady-state cutting process free of vibration with constant values of feed and cutting speed. It is obvious that the cutting force can never be exactly constant. A small disburbance influencing the steady-state motion, such as a hard spot in the workpiece, can initiate a free vibration of the tool. Once vibration is started, the variable speed of the tool point relative to the workpiece can change the cutting conditions and generate a force which is the function of the displacement as well as of the velocity. If this force is larger than the total damping of the system, then a self-excited vibration comes into being, and the system undergoes dynamic instability, which we call chatter.

Modern machine tools chatter only when several cutting conditions occur simultaneously. But, a boring bar with its relatively long overhang is more susceptible to chatter than other tooling or other members of the machine. If the small influence of the machine tool at the tool point is neglected, then the vibration in a cutting process may be reduced to the vibration of the boring bar.

NATURAL FREQUENCY

Like all bodies possessing mass and elasticity, a boring bar is capable of vibration and represents an elastic cantilever beam with infinite number of degrees of freedom. The natural frequency of a boring bar is a function of the static stiffness-to-weight ratio.

An example of how weight affects the natural frequency of a bar would be to picture a young swimmer out on the end of a diving board just bending his knees and making the board spring up and down--this would be low natural frequency. Then, when he jumps off, the board vibrates at a high natural frequency because of less weight.

Static stiffness is given by the following formula:

$$K = \frac{3\ EI}{L\ 3}$$

Where:

K = static stiffness (pounds per inch)

E = the modulus of elasticity of the boring bar material

I = the amount of inertia $\left(\frac{\pi D^4}{64}\right)$ where D = bar diameter

L = the unsupported length of the boring bar

In cutting operations, high natural frequencies are required for two reasons:

(1) They make it possible to avoid a vibration condition at a near resonance, which is created within the machine by imbalances, gears, bearings, pumps and the like.

(2) They reduce the amplitude of boring bars in response to impact disturbances created by interrupted cutting, entrance of the tool point into the workpiece, irregular surface conditions or hard spots in the work material.

Since the natural frequency of a boring bar is a function of the static stiffness-to-weight ratio, a high degree of this ratio can be attained by (1) using a material with a high modulus of elasticity and (2) reducing the effective mass at the cutting end of the bar while maintaining a high stiffness at the clamped end. This can be accomplished by tapering the bar to a smaller diameter at the cutting end or by forming a cylindrical cavity at the cutting end.

A boring bar of "composite" design (see Figure #1) uses the advantages stated above for attaining a high natural frequency. It has a carbide shank at the clamped end which is

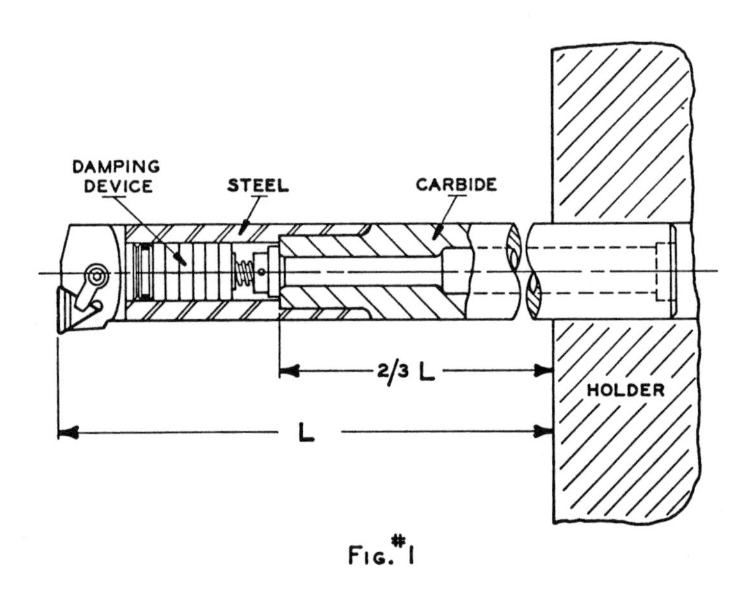

FIG. #1

a material with a high modulus of elasticity and a steel tube at the cutting end which weighs approximately one-half as much as carbide. The steel tubing with its cylindrical cavity reduces the effective mass at the cutting end, and is easily machined to accept various cutting heads.

Figure #2 shows the actual natural frequencies of two carbide bars with and without a cavity as a function of overhang. The test results indicate that the natural frequency of a boring bar with relatively long overhang can be increased about 35-40 percent by forming a cylindrical cavity at the cutting end. This cavity helps not only to increase the natural frequency, but it also serves as a housing for the damping device now used in modern boring bars.

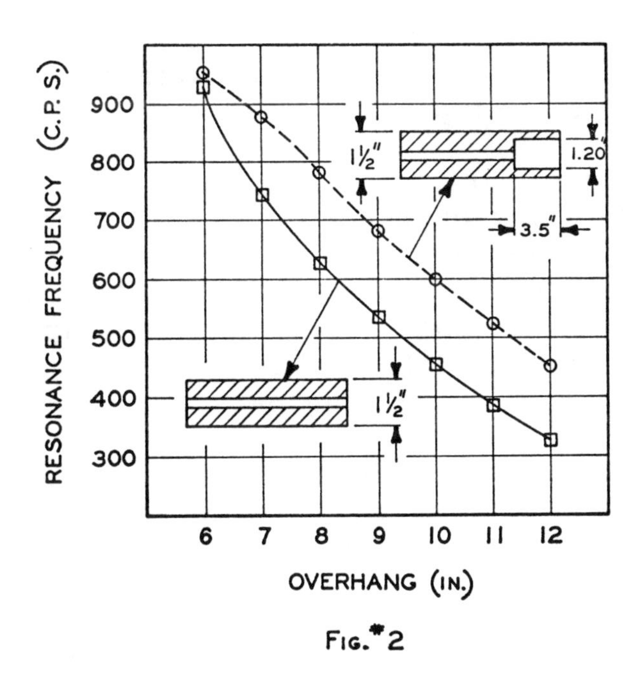

FIG. #2

DYNAMIC STIFFNESS OF BORING BARS

Recent studies indicate that in any self-excited chatter, the criterion for stability is dynamic stiffness. The ability of a boring bar to suppress chatter is measured by its dynamic stiffness which is defined as the ratio of exciting force to amplitude at resonance, and it is expressed by:

$$K_d = 2 \zeta k$$

Where: K_d = Dynamic stiffness

δ = Damping ratio (dimensionless)

k = Static stiffness of the boring bar (pounds per inch)

This expression clearly shows that the dynamic stiffness is directly proportional to the static stiffness (k) and the damping ratio δ (delta). This means that if two boring bars have the same static stiffness, the one with the higher damping ratio would achieve better chatter-free cutting performance than the other. Or, if two boring bars have the same damping ratio, the one with the higher static stiffness would offer more resistance to chatter than the other.

Dynamic stiffness is the most important parameter for the dynamic stability of a boring bar.

The actual damping involved in the vibration of a boring bar consists generally of four different types of dampings: (1) structural damping in the shank, (2) frictional damping due to relative movement between bar and tool holder, (3) damping due to kinetic friction at tool point and (4) damping of vibration absorbers or DeVibrators which are now used as a tool element in the design of boring bars.

Structural damping constitutes only a very small percentage of the overall damping. It is the internal friction of the molecules in the material used for the boring bar, so this damping is very small.

Damping between bar and holding device varies with the degree of fixation of the boring bar or cutting insert. This damping is not desired because looseness in holding reduces the static stiffness and natural frequency. This could be movement of the insert in the pocket or the bar in the holder, which is certainly undesirable.

Damping at the tool point is the most efficient in limiting the amplitude. The value of the damping force depends on the axial component of the cutting force and the coefficient of friction between the workpiece and tool point. This would be the rubbing of the cutting edge on the workpiece.

DEVIBRATORS

DeVibrators are successfully used to suppress chatter and their damping capacity depends upon the design.

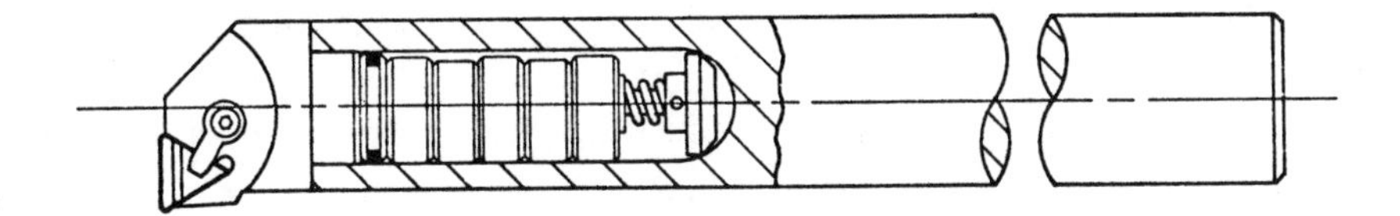

FIG. #3

The DeVibrator shown in Figure #3 consists of several spring loaded high inertia discs, loosely housed within a cylindrical cavity near the cutting element of the bar.

This device utilizes viscous, friction and random impact damping. Viscous damping occurs as the atmosphere within the cavity is compressed by the inertia action of the discs. Vibration of the boring bar tends to cause relative motion between the discs and the bar. When this occurs, the air surrounding the discs dissipates vibration energy in the form of viscous damping. Friction damping is produced by the relative lateral movement of the interfaces of the discs under spring load. The principle of random impact utilizes repeated collisions of the discs against the cavity wall to arrest the transverse movement of the bar.

Some of the important factors that influence the DeVibrator's damping capacity are the location and size of the discs, the spring load and the critical clearance between the O.D. of the slugs and cavity of the boring bar. The greatest displacement during vibration naturally occurs at the cutting end of the bar. It is important, therefore, that the discs be placed as close to this end as possible and be as large in diameter as practical. The discs in this design (Figure #3) are made of heavy tungsten alloy (0.61 lbs./cu. in.) to attain an optimum inertia necessary for effective damping. The number and thickness of the discs will vary depending on bar diameter and unsupported length.

The faces of the discs are ground smooth to control friction and prevent interlocking. Spring load is relatively light and dependent upon the weight of the discs. The spring controls the amount of friction between the discs. The DeVibrator cavity and all its components must be absolutely clean and dry at assembly and sealed from external contamination.

The required clearance between the discs and the inside diameter of the boring bar range from 0.002" upward depending on the diameter and the unsupported length of the boring bar. The test results for the dynamic stiffness of a 2" diameter carbide DeVibrator boring bar, with and without discs, and the effect of disc clearance to unsupported length is shown in Figure #4.

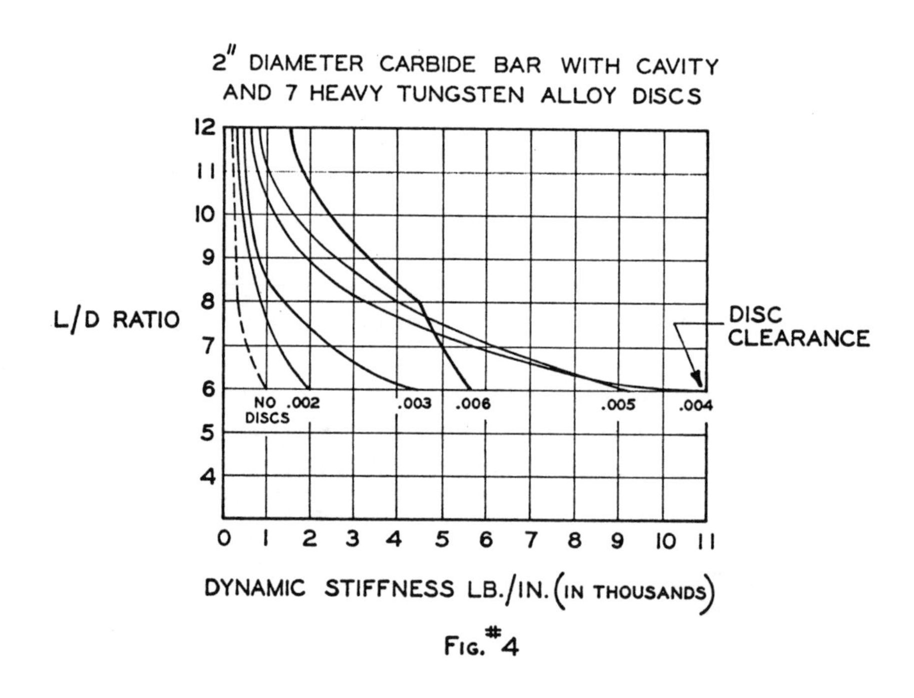

FIG. #4

For example, if you look at the carbide bar without discs at a 6:1 ratio, which would be a 12" unsupported length, it has a dynamic stiffness of 1,000 pounds per inch. Then at 8:1 ratio, the dynamic stiffness drops to about 300 pounds per inch. But, the same bar with heavy tungsten alloy discs in the cavity with 0.002" clearance shows that the dynamic stiffness is doubled at a 6:1 ratio.

This test shows that the boring bar can be tuned for an 8:1 or larger ratio by using 0.006" for the disc clearance. But, if that same boring bar was used at lower ratios, such as 6:1, it would not work as well as a boring bar with disc clearance of 0.004" or 0.005".

The test results charted in Figure #4 show that the dynamic stiffness of this boring bar at an 8:1 ratio, without discs, is 300 pounds per inch and has been improved to 4,500 pounds per inch with discs at a 0.006" clearance. In summary, standard Kennametal carbide and composite bars are designed using the above parameters.

To visualize the action of the DeVibrator, consider the bar at an idealized point of vibration; at the bottom of its cycle, starting to move upward and with all discs resting on the bottom of the cavity. (Refer back to Figure #3.) When the bar reaches the top of its cycle and starts down, the inertia of the discs will carry them upward until they bump into the top of the cavity.

The first impact will come from the largest disc because it has the least clearance or shortest distance to travel. Then, successive impacts will tend to occur with random timing because of the different sizes of the discs and because of their different positions.

Further assurance of random timing is given by the light spring at the end of the row of discs. This causes a slight amount of friction between the faces of the discs and helps delay the impacts.

Because of friction, the smaller discs will be delayed in their upward path just as soon as the larger ones start downward. By the same token, the larger ones will be restrained from "bouncing ahead" and catching up with the bar.

Thus, the impacts are almost invariably opposed to the vibrational movement. In essence, the bar imparts energy to the discs, and then the discs return the energy at the wrong time, so far as the continuation of vibration is concerned; but at the right time as far as the machine operator is concerned.

BORING BAR DESIGN

The boring bar is normally the most elastic component in a modern machining set-up and is, therefore, most susceptible to vibrations which may intensify into chatter. Today's technology has afforded us the means to suppress these vibrations and, thus, has expanded the design horizons. The following boring bar designs will be put into their proper perspective, and parameters will be given later for selecting each design: solid steel bars; solid carbide shank boring bars with cone brazed steel heads; steel DeVibrator; carbide DeVibrator bars and composite DeVibrator boring bars.

Remember that, dynamic stiffness is of prime importance in the design of a chatter-free boring bar. As previously mentioned, it is a function of the static stiffness and damping ratio.

As mentioned before, it is important that a high modulus material be used in boring bar shanks to increase static stiffness and thus attain a high degree of dimensional accuracy. Steel has a modulus of elasticity of approximately 30,000,000 psi. Tungsten carbide has a modulus of elasticity of approximately 90,000,000 psi, which gives a 3:1 improvement in rigidity over steel. Solid tungsten carbide bars with cone brazed steel heads are commercially available in sizes from 3/8" diameter through 1" in diameter. (See Figure #5.)

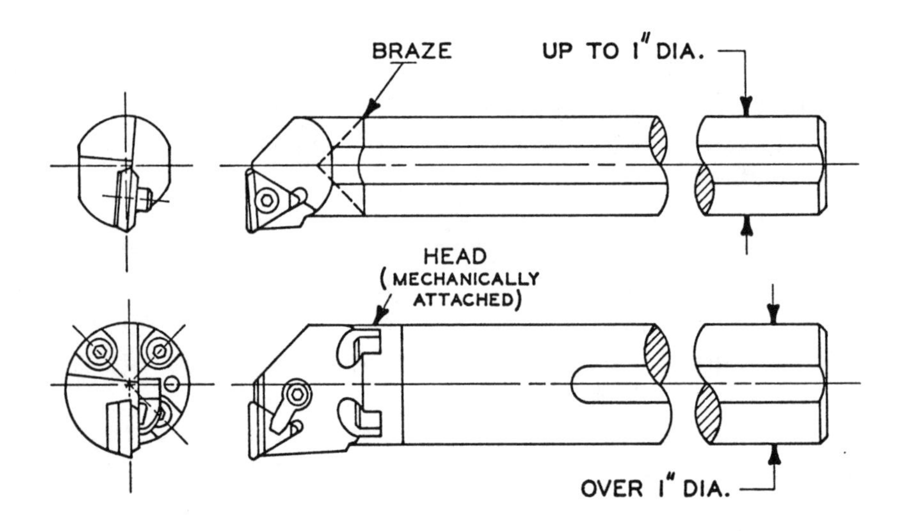

FIG.* 5

The reason for 1" being the maximum diameter that we recommend for brazing steel to carbide is because of the difference in the coefficient of thermal expansion at brazing temperature Steel has a mean coefficient of thermal expansion of 7.6 x 10^{-6} in./in./° F. from room temperature to 1200° F. Tungsten carbide has a mean coefficient of thermal expansion of 3.3 x 10^{-6} in./in./° F. from room temperature to 1200° F., which means that the steel expands over twice the amount that the carbide expands, which causes very high braze strains when the parts cool to room temperature. For boring bars that are larger than 1" in diameter we recommend using heavy tungsten alloy collars brazed to the carbide bar. This can be accomplished because the heavy tungsten alloy has approximately the same coefficient of thermal expansion as carbide and also has good wetting properties for brazing. The heavy tungsten alloy can then be machined, drilled and tapped for mechanically holding the head to the bar. Carbide DeVibrator bars 1-1/4" diameter through 2-1/2" diameter are commercially available with bolt-on head design.

SELECTING THE PROPER BORING BAR

In selecting the proper boring bar for a job, we have some general rules that we start with.

The first rule is to select a boring bar of the largest possible diameter. This will always be a compromise in which you are sacrificing bar stiffness for chip clearance and vice versa. The optimum balance is a matter of judgment and depends on the material and cutting conditions.

The reason for selecting a bar with the largest diameter is because the static stiffness of the bar increases with the fourth power of the diameter. For example, a 1" bar to the fourth power equals one. But, a 1-1/4" diameter boring bar to the fourth power equals 2.441". Therefore, a 1-1/4" diameter boring bar is almost 2-1/2 times as stiff as a 1" diameter bar.

Another general rule is to keep the unsupported length of the boring bar as short as possible because the static stiffness of the bar increases when the unsupported length of the bar decreases. This is inversely proportional to the cube of the length. What this means is, if you have a boring bar with an overhang of 10", it would be stated as one over ten to the third power which equals one over one thousand. Now, if you took that same bar and had an 8" overhang, it would equal one over 512. Therefore, a bar at 8" of unsupported length is almost two times as stiff as the same bar at 10" of unsupported length.

Rule number three is the minimum holding length of the bar in a holder should be 2-1/2 times its diameter for 1" diameter bars or larger, and to allow 2-1/2" for holding length on boring bars that are under an inch in diameter.

Rule four, keep the tool pressure on the boring bar to the minimum. Increased tool pressure normally makes the cutting system more susceptible to chatter, which the relative lack of rigidity in a boring bar magnifies. Therefore, it is advisable to decrease tool pressure through the use of positive rake tool geometry, sufficient relief, minimum lead angle and the smallest possible nose radius.

The following parameters can be used as a guide in selecting standard, commercially available boring bars for non-rotating applications. (See Figure #6.)

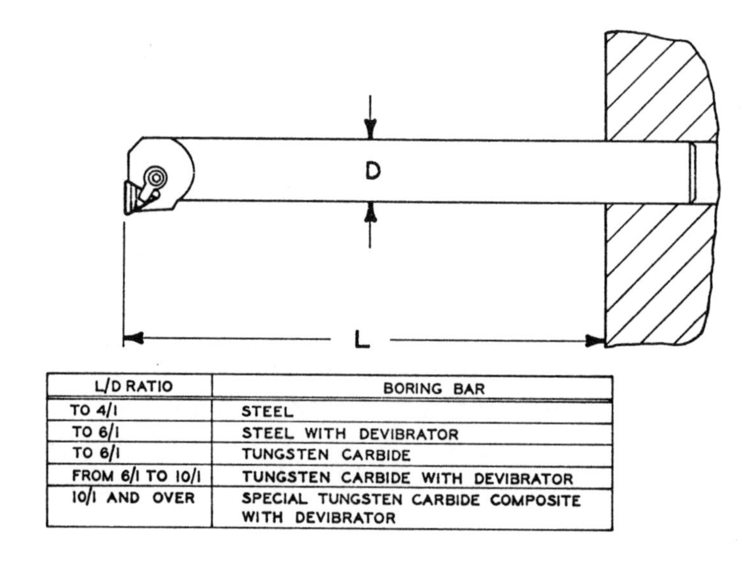

L/D RATIO	BORING BAR
TO 4/1	STEEL
TO 6/1	STEEL WITH DEVIBRATOR
TO 6/1	TUNGSTEN CARBIDE
FROM 6/1 TO 10/1	TUNGSTEN CARBIDE WITH DEVIBRATOR
10/1 AND OVER	SPECIAL TUNGSTEN CARBIDE COMPOSITE WITH DEVIBRATOR

FIG. #6

1. For unsupported length to diameter ratios under 4:1, a solid steel boring bar would be adequate; but a steel DeVibrator bar would offer some improvement.

2. For unsupported length to diameter ratios between 4:1 and 6:1, use a steel DeVibrator. If the hole to be bored is small in diameter, a solid carbide shank with cone brazed steel head can be used at this ratio.

3. For unsupported length to diameter ratios between 6:1 and 10:1, a carbide DeVibrator boring bar is required. Composite design boring bars are recommended for bar diameters between 2-3/4" and 6" diameter.

If is important to keep in mind that carbide DeVibrator bars offer more than just chatter control. Because of their high modulus of elasticity, they improve dimensional accuracy and finish and reduce the possibility of "out of roundness" since deflection is minimized.

4. Unsupported length to diameter ratio in excess of 10:1 requires special considerations with carefully selected cutting conditions. A composite design bar can be designed, but the minimum bar diameter would be 1-1/2" diameter and the DeVibrator package would have to be tuned for a given unsupported length.

SPECIFIC EXAMPLES

When considering the boring bar geometry, the accepted cylindrical shape with one or more longitudinal flats for chip clearance and clamping is by far the most economical and versatile. However, the optimum boring bar requires maximum beam strength at the point of greatest bending moment, plus high dynamic stiffness without regard to shape, ease of manufacturing or versatility.

For high production runs where versatility is not a factor, a special steel bar may sometimes be more economical. As an example, see Figure #7. To machine this part with a standard of one diameter may require a tungsten carbide DeVibrator bar due to the length to diameter ratio of 6:1. However, by increasing the root diameter as permitted by the part geometry, an effective length to diameter ratio of 3:1 is obtained. The special steel bar would then be satisfactory although limited to that specific job.

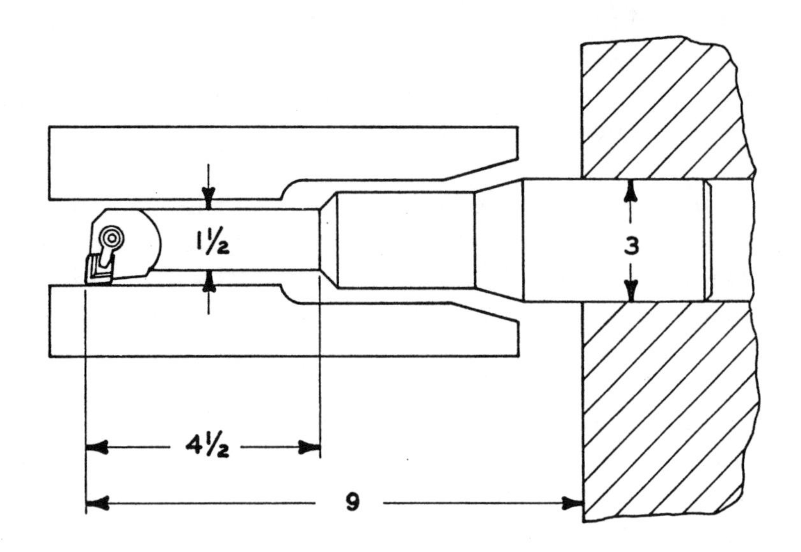

FIG. #7

General design requisites mentioned, plus additional consideration for part and chip clearance, prompted the design of this bar (Figure #8). This concentrically tapered tungsten carbide DeVibrator boring bar was originally designed to machine a familiar shape (beer bottles) for the glass bottle mould industry. The shape and construction of this bar lends itself well to the static and dynamic parameters previously discussed, but its versatility is limited. This bar also brought into focus the need for a replaceable head on small diameter boring bars.

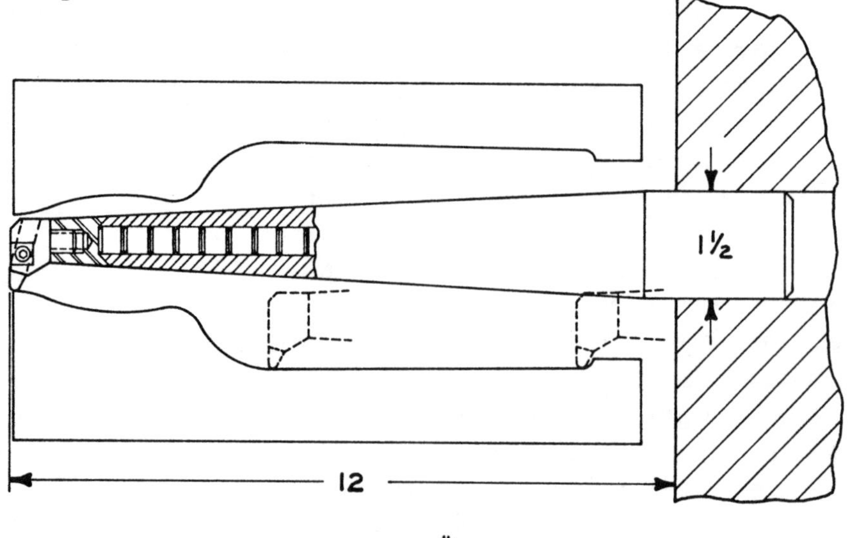

FIG. #8

Figure #9 shows a later innovation of this bar that is now a Kennametal standard. This straight-sided eccentrically tapered bar can do straight contouring in addition to tapered work. With this bar, moulds are now contoured at up to four times greater speeds with fewer cuts and nearly 100 percent improvement in surface finish. It has replaced a multitude of steel bars which were time consuming to use and a great deal less versatile. Its versatility is broadened through its ambidextrous construction--right and left hand heads may be used without bar modification.

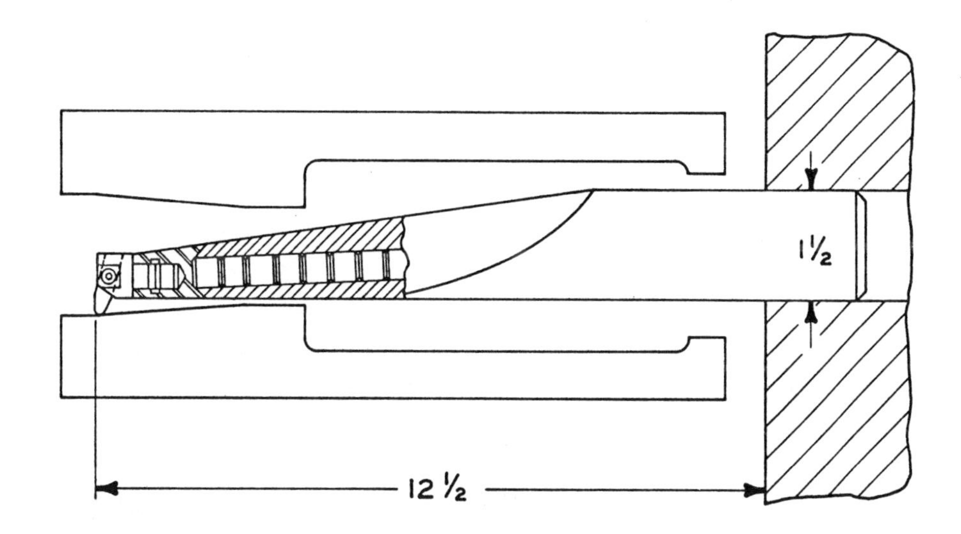

FIG. #9

Figure #10 shows a 5" diameter by 73" long tungsten carbide composite bar with DeVibrator which was required to bore a 12" diameter hole 50" deep. Contouring to the center of the closed end of the bore ruled out the use of a larger diameter bar. A diametral tolerance of $\pm$ 0.0003" with 45 rms surface finish was maintained with this bar.

The elliptically-shaped bar (Figure #11) is also a tungsten carbide bar with DeVibrator unit. It was designed for a special workpiece where the part's geometry dictates its shape. The addition of a pull bore head permits complete machining with one bar.

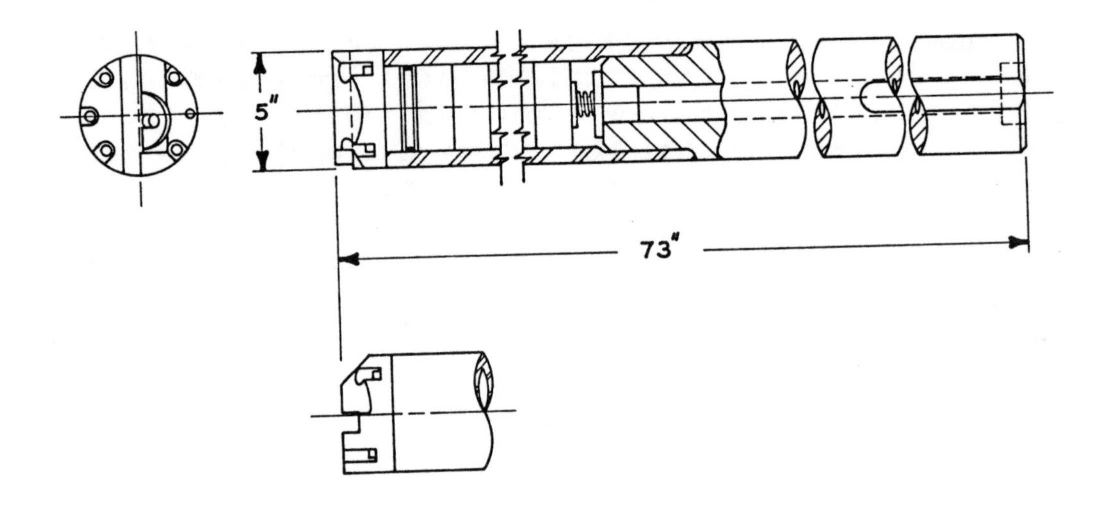

FIG. #10

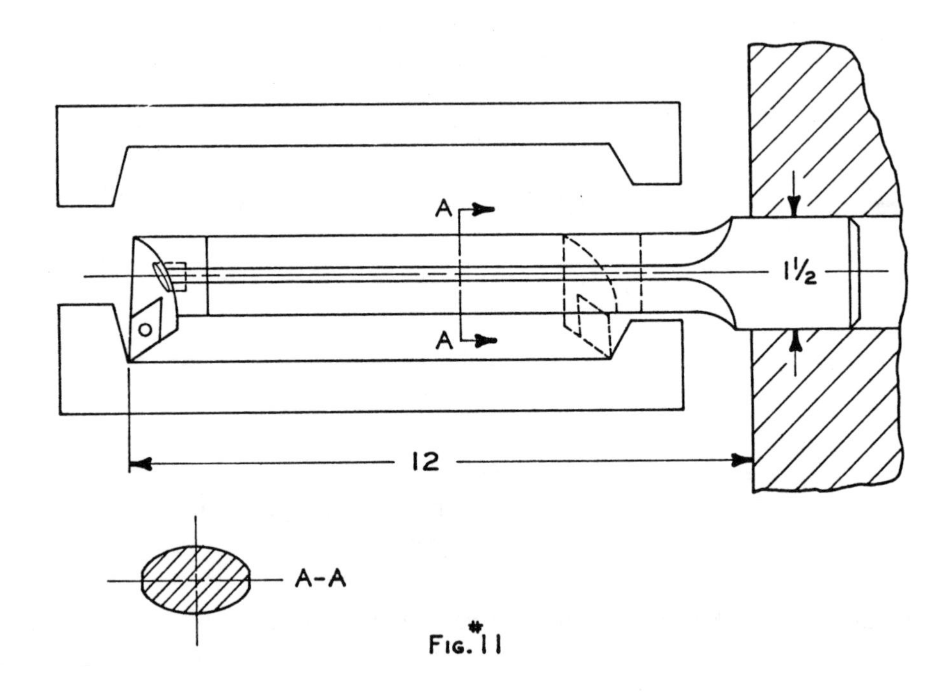

FIG. #11

CONCLUSIONS

Keep in mind that the proper selection of the boring bar for a particular job is very important. The bar diameter should be as large as possible, but this will always be a compromise in which you will be sacrificing bar stiffness for chip clearance and vice versa. The optimum balance is a matter of judgment and depends on the material being machined and the cutting conditions. The overhang of the boring bar should be as short as possible. The minimum holding length should be 2-1/2 times the bar diameter for boring bars 1" diameter and larger and 2-1/2" for boring bars under 1" in diameter. The bar must be held rigidly at the supporting end or the advantages of the best design in the world will be nullified. Poor support or gripping could be like trying to write with a pencil by holding onto its eraser.

When selecting the proper boring bar design, don't forget to use the following unsupported length to diameter ratios as a guide:

solid steel boring bar -- up to 4:1 ratio
steel DeVibrator boring bar -- up to 6:1 ratio
carbide DeVibrator boring bar -- up to 10:1 ratio

Other practical considerations include boring bar configuration and tool pressure. The length to diameter ratio of unsupported overhang may sometimes be reduced by tailoring the bar configuration as the part permits. Tool pressure which deflects the boring bar can be reduced through the use of positive rake tooling with adequate relief and the smallest possible nose radius. Severe cutting conditions, lack of chip control, and chip interference which increase tool pressure should, of course, be avoided.

There are many factors in a given metalcutting system which can create chatter. The more common ones are sometimes overlooked due to the relative difficulty of internal machining. Even the best boring bar will not suppress chatter which originates from another component in the system.

ACKNOWLEDGMENTS

I wish to express my sincere appreciation and acknowledgment to Mr. W. L. Kennicott, Vice President of Engineering; to Mr. J. W. Heaton, Engineering Manager; to Mr. E. L. Sorice, Design Supervisor; and to Kennametal Inc. for the opportunity to author and present this technical paper. Also, to J. Dicesere; T. Lockard; D. Burick and L. Smithley for their assistance in conducting tests, preparing the illustrations and typing the manuscript.

REFERENCES

Machine Tool Vibrations, S. A. Tobias, John Wiley and Sons, Inc., New York

Mechanical Vibrations, W. T. Thomson, Prentice-Hall, Inc., Englewood Cliffs, New Jersey

Design and Devices for Chatter-Free Boring Bars, A. R. Alev and W. C. Eversole, Kennametal Inc., Latrobe, Pennsylvania

Reprinted from *Production*, January 1985

Shankless Boring Unit Replaces Boring Bar

The change from conventional to shankless cutting tools is doubling tool life in machining axle shaft tube bore diameters in nodular iron rear axle differential carriers for use in Thunderbird, Cougar, LTD and Mercury models at Ford Motor Co.'s Transmission & Chassis Div., Van Dyke Plant in Sterling Heights, MI. The boring units, similar to an 8-tooth multi-flute end mill, are providing savings because of longer tool life, less scrap and its precision is providing a higher quality product.

Reasons that led to utilizing the shankless boring units on a trial basis are explained by Peter J. Mousseau, manufacturing engineer: "On occasion we got back from repair conventional V-shape and taper-shank tools that were either improperly heat treated or repaired. When these tools were put back on the job we experienced .015-in. to .030-in. runout and I've even seen the shanks crack and break off. This results in a $500 loss because both the tool and workpiece are scrap.

"Shankless tools, on the other hand," he continues, "are so massive and sturdy compared to conventional types that it seemed to me that it would have no runout problem. I became convinced of system rigidity because a spiroid gear form with 36 teeth, similar to a curvex coupling, locates the tool in the spindle adapter."

The mounting surface of the body and adapter consists of two gear-like halves, one the mirror image of the other, seating accurately to provide precise indexing. Held together by a conventional draw bolt, the curved gear teeth provide self centering and high torque transmission.

Fifty Percent Tool Life Increase. Initially, Mousseau purchased two 2.735-in.-dia. shankless boring units and one adapter from Eclipse/Detroit, a division of Illinois Tool Works, Detroit. This unit was installed to finish bore axle tube holes in the fourteenth machining station on one of two identical 20-station Cross (Fraser, MI) Transfer-Matic transfer lines.

At a feed of .0497 in. per revolution and speed of 264 sfm at 368 rpm, the tool bores .036 in. off the diameter 3¼ in. deep. Cycle time is 13.6 seconds.

In its first test, the shankless boring unit with eight inserted C-2 carbide tipped blades was running strong after 4000 parts—the point at which the conventional boring tool was replaced. Continued testing demonstrated consistently that 6000 parts can be machined before requiring regrinding and one test operation even successfully machined 8900 parts. In addition, five or six regrinds are obtained from the Eclipse/Detroit tools whereas conventional tools only provided four.

Second Purchase. After realizing what benefits were gained during four months of operation, Mousseau purchased and installed one of two 2.699-in.-dia. boring units to rough bore the axle tube diameters. This was installed in the twelfth station of the same transfer line. This station has been a trouble spot because of the high number of broken shanks occurring here. "I've seen up to five boring bars break in a three day period," he says. "The shankless unit with eight C-2 carbide indexable inserts, however, has not broken after six months of operation. Additionally, the inserts machine 2500 parts before indexing whereas the conventional tooling was set up to machine only 1500 pieces."

With a feed of .0318 in. per revolution and a speed of 261 sfm at 368 rpm, the shankless cutting tools bore .120 in. off the diameter 2 in. deep machining 1100 pieces per day. Cycle time is 14.2 seconds.

A 2¾-in.-dia. shankless boring unit shown in the tube bore diameter of the rear axle differential carrier it machines

Mousseau estimates a $25,000-$30,000 annual savings will be realized by utilizing shankless boring tools for roughing and finishing operations on the two differential carrier transfer lines. And if that is successful, he plans conversion of an 8.8-in. rear axle differential carrier used for Mercury Marquis, Crown Victoria and Lincoln Continental models. *MJW*

Reprinted from *Machine and Tool BLUE BOOK*, February 1985

Rod Boring Rate Triples at Trane

When The Trane Co. switched to die cast con rods for the compressors for its commercial air conditioners, its old rod boring machine couldn't keep up with production needs. A new production boring machine equipped with semiautomatic tool compensation not only provided the extra throughput capacity but halved the scrap rate as well.

By switching from permanent mold castings to die castings for the one-piece aluminum connecting rods that go into the reciprocating compressors of its commercial air conditioning units, The Trane Co. was able to reduce manufacturing costs at its La Crosse, Wis. plant. The permanent mold casting had required a roughing operation, milling of the rod faces, and chamfering of the edges of both bores, in addition to finishing of the crank pin and wrist pin bores. By switching to a die casting of near-net shape, only boring was needed. Even the chamfer was cast in. After boring, the die casting is washed and then is ready for assembly.

Trane personnel made a fixture for the die-cast rod and put the job on the 17-year-old boring machine that had been used to machine the permanent mold castings. Both wrist-pin and crank bores were done in one setup. Tolerances were held and production rate approximated 50 parts an hour.

The scrap rate of approximately 1 percent definitely was acceptable, thanks to the good fixture. The chief reasons for the scrap were tool-wear adjustments, required every 150 to 200 parts, and set-up adjustments for change-over from one rod size to the next.

Three different lengths of die-cast connecting rods are run, for Trane's 6, 7.5 and 8-hp compressors. Center-to-center distances between the bores for the three sizes are approximately 2.30, 2.35 and 2.42 inches. Change-over from one size to the next took place every three to four days, with each new setup likely to add to scrap.

A capacity problem was forecast for the old machine—it represented a potential bottleneck. Rebuilding the old boring machine was out of the question—it was barely keeping up with production and no hedge of material could be accumulated.

Trane personnel decided to replace the old boring machine with a new production-boring machine. "You don't need a CNC machine for this kind of part," said the manufacturing engineer responsible for the operation.

Trane is working to a tolerance of 0.0003 inch on both bores. Because of draft in the aluminum die casting, the amount of material removed goes up to 0.03-inch of stock on each side of the bore, which is the total for both semifinishing and finishing passes. The finished diameters for the crank bore and wrist-pin bore are 1.562 and 0.687 inch respectively. Surface finish needs to be 16 microinches, AA, or better.

A DeVlieg horizontal, three-spindle, single-ended boring machine (Fig. 1) was the choice. The machine, which has S60 precision spindles, is equipped with a two-station fixture and chute-fed automatic loader.

Tool compensation was the next matter of concern. Trane opted for pneumatic-hydraulic tool positioners built by Samsomatic Ltd., Detroit.

Semiautomatic Operation

After nine months of experience with using the new fixture on the old boring machine, production was switched to the DeVlieg machine. The process now requires the operator to load approximately 15 parts into each of two chute-fed loaders on the DeVlieg. The upper chute receives the parts crank-pin end first; the lower chute, wrist-pin end first.

Two parts are clamped simultaneously and automatically into a fixture on the machine table (Fig. 2). Spindle-mounted tool-comp positioners, one on each of three spindles, are energized to radially position the cutting tools at the correct point in their 0.008-inch travel range. The coolant comes on, and then the spindles start.

After a rapid traverse left of the table, the table feeds the fixtured parts to the two rear boring spindles to do the crank bore in the rear part and the pin bore in the front part (Position 1 in Fig. 3).

Upon completion of the feed at 0.003 ipr, the tool-comp positioners are de-energized and automatically retract the tools before the table rapids to the right. This retraction, which prevents tool drag marks during withdrawal, is a characteristic of the pneumatic-hydraulic positioners.

In Position 2 of the machining cycle, the cross slide with fixture rapids forward and the three positioners are energized. Next, the table rapids left and then feeds the fixtured parts onto the front two boring spindles. The machine cycle ends with the parts positioned over an unload conveyor, ready for unclamping, and two new parts in position for clamping and the start of a new automatic cycle.

Finish-bored parts are conveyed to the operator's position. The operator makes visual checks for porosity and surface defects. The operator also has an air-gage fixture and checks dimensions of every twentieth part: bore diameters, bore centerline parallelism and center distance between bores. One piece out of every 100 is checked for bore surface finish. A roving inspector also checks surface finish and machined dimensions on random parts.

With two parts produced in every machine cycle, production rate averages 140 parts per hour, which includes rest allowances for the operator. The machine is capable of 150 parts per hour—three times the rate of the old boring machine.

Scrap rate has been very low—less than 0.5 percent, a definite improvement over the 1 percent before. Bores are well within the allowed tolerance of 0.0003 inch.

Surface finish has been as good as 10 microinches. When finish is worse than 16, it is time for a tool change. Life of the inserts is extremely good, according to the Trane engineer. "This is probably because we don't make the many manual tool adjustments we would make without tool comp, using a gage on the tool," he said. That is, an inadvertent overcompensation during some manual adjustments can mean that a very heavy cut may be taken on a trial part, causing excessive tool wear. Tool compensation reduces this likelihood.

Tool life is in excess of 10,000 parts for each of the two types of tools on the machine. Both types of tools are of polycrystalline diamond. Those for the wrist-pin boring bar, on the center spindle, are cartridge-mounted. Those on the front and rear crank boring bars are inserts. All three boring bars are of twin-tool design, one tool semifinishes and the other finishes.

With the original boring method, also using polycrystalline diamond, tool life was only 4000 to 5000 parts. Coolant, in both cases, was a fully synthetic type.

Tool Compensation Needed

Correction for tool wear is important for several reasons. First, when starting up in the morning, there is a slight size difference because the machine is cold. The bores would be a little small. In about an hour, the machine is warmer and the operator has to change the size. To do this, he uses the tool compensation. Second, there is some tool wear, and the operator may comp the tool once a day to correct for it. Finally, the tool comp reduces lost time following tool changes.

On the control panel for the Samsomatic system, each of the three tool positioners has its own readout showing how far the positioner has extended the tools on that spindle. Plus and minus pushbuttons next to that readout let the operator step the tool without opening the machine guard.

The routine in changing tools is to: install the new cartridge or insert, watch the readout while stepping the tool to the assumed position, and cut a trial part. If the size of the trial bore isn't correct, the operator knows exactly how many times to press the pushbutton to bring the size into the tolerance range. The process is fast, easy and cuts down on scrap.

Step size is set to 0.00005 inch for this workpiece. Therefore, each time a pushbutton is pressed, that bore diameter is changed by twice the

Fig. 1. Top: Trane Co.'s new DeVlieg production boring machine equipped with Samsomatic semi-automatic tool comp. Bottom: Each of the machine's three spindles is individually compensated. Each boring bar is a roughing/finishing tool; inserts are polycrystalline diamond.

amount, 0.0001 inch, making it easy for an operator to compensate a tool accurately. Step size can be set from 0.00002 to 0.00035 inch.

According to Samsomatic engineers, a slip-stick frictional effect is a characteristic of the conventional drawbar positioner. When close-tolerance machining is being done, that effect impairs the attainable repeatability of tool positioning. Slip-stick is caused by friction as a pin is forced up a wedge-like ramp or taper to move the tool point.

In contrast, friction in the pneumatic-hydraulic positioner is lower because the only two sources are a hydraulic shaft seal and a rolling-diaphragm seal for a piston. The result is a positioning repeatability of ±0.00002 inch or better.

At Trane, return on investment for the new production boring equipment has not yet been calculated. However, the anticipated cost reduction is in the vicinity of $30,000 annually for the three sizes of connecting rods.

In the near future, a second family of larger connecting rods will be bored on the machine. Center distances and bore sizes will be greater. The second family will be for 20, 25 and 30-hp compressors in Trane commercial air conditioning units with 20 to 30 tons of cooling capacity. The rods will be loaded manually and machined one at a time, rather than two at once, because of their size.

When the second family is run on the machine, any cost savings will be in addition to the anticipated $30,000 annual cost reduction for the family of smaller rods. Purchase of the machine was initially justified on the basis of running both families of parts. A return on investment in excess of 30 percent is the expectation. **BB**

Fig. 2. Two connecting rods are clamped automatically into the fixture of the boring machine. The two chute-fed loaders are above the fixture and to the right.

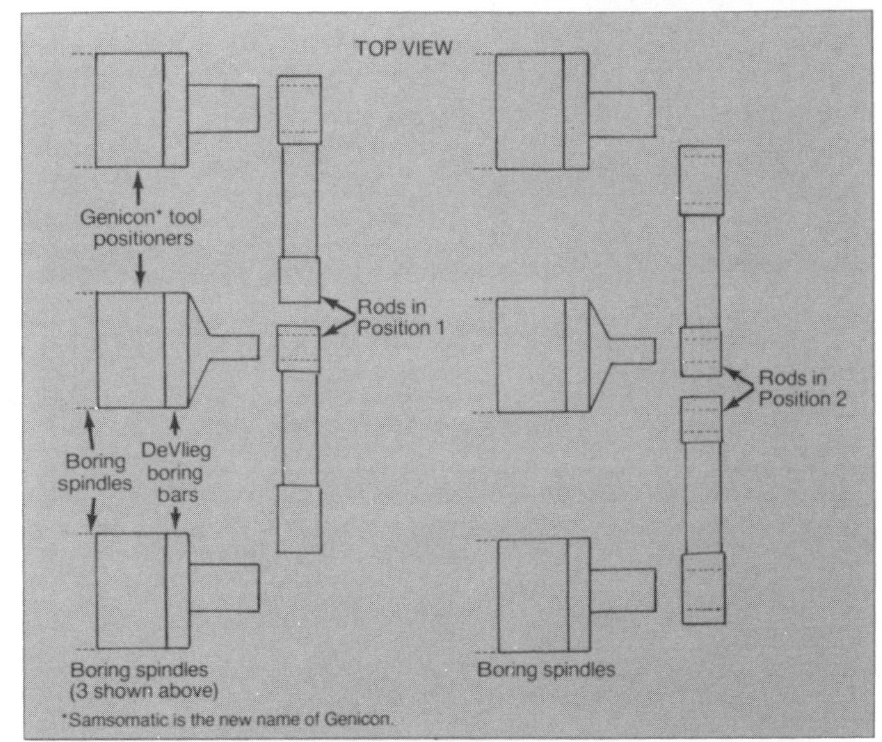

Fig. 3. Two rods are machined per cycle. In the first stage (left) the crank-pin bore of the trailing rod and the wrist-pin bore of the leading rod are bored. In the second stage (right), the table shifts to present the crank-pin bore of the lead rod to the "front" spindle while the center spindle machines the wrist-pin bore of the trailing rod.

CHAPTER 4

THREADING

Reprinted courtesy of Sandvik, Incorporated, Coromant Division
Fair Lawn, New Jersey

Thread Turning

Turning is one of the most common methods for producing screw threads. At the same time, thread turning is regarded as one of the most complicated machining operations in the engineering industry. A number of difficult problems are often encountered with regard to chip control, accuracy, abnormal wear etc. — problems that need not arise if tools are used that have been designed to meet the demands of modern machining.

T-MAX threading tools have been developed to meet demands for high quality and good production economy. The threading inserts of the full-profile type form the entire thread, including the crest, in a single operation, which entails a considerable simplification of the thread-cutting operation. With the conventional V-profile insert, threading is carried out in two operations: diameter turning and profile turning. The tolerance problems that can then arise can be avoided by machining in a single set-up with full-profile inserts. The major and pitch diameters will always be concentric. And not only will the costs of rejection due to incorrect thread profile be eliminated, but operations such as deburring and, in some cases, preturning can be omitted as well.

T-Max thread turning tools are indexable insert type and eliminate costly regrindings with subsequent thread profile inspection. A rational grinding procedure in production gives low cutting edge costs, motivating high cutting data and thereby increased production. The interchangeable chip formers included in the range provide wide opportunities to adjust chip form to the application at hand.

Sandvik Coromant cemented carbide grades for threading tools cover most applications. The coated grade, GC 225, provides exceptionally long tool life owing to its high resistance to flank wear, plastic deformation and built-up edge formation. With GC 225, it is possible to thread 2—3 times longer than with conventional threading grades, while the high wear resistance of the grade permits higher cutting data.

This manual serves as a guide to the use of T-Max threading tools. It also provides a general description of the construction of the screw thread, common thread profiles, thread tolerances and standards.

The screw thread

The screw thread is one of our most common machine elements and can be used for many different purposes. Two main groups can be distinguished: fastening threads for fixed joints and translation threads.

Fastening threads, e.g. bolts and nuts, are used for fastening parts that are mounted together and remain fixed.

Translation threads are used for translating rotational movement into linear movement. Besides providing linear movement, such threads may be intended for transmitting forces. An example of this is the screw jack, which is built up around a screw thread and is used for heavy lifts. Translation threads are also used in various instruments for precision measurement.

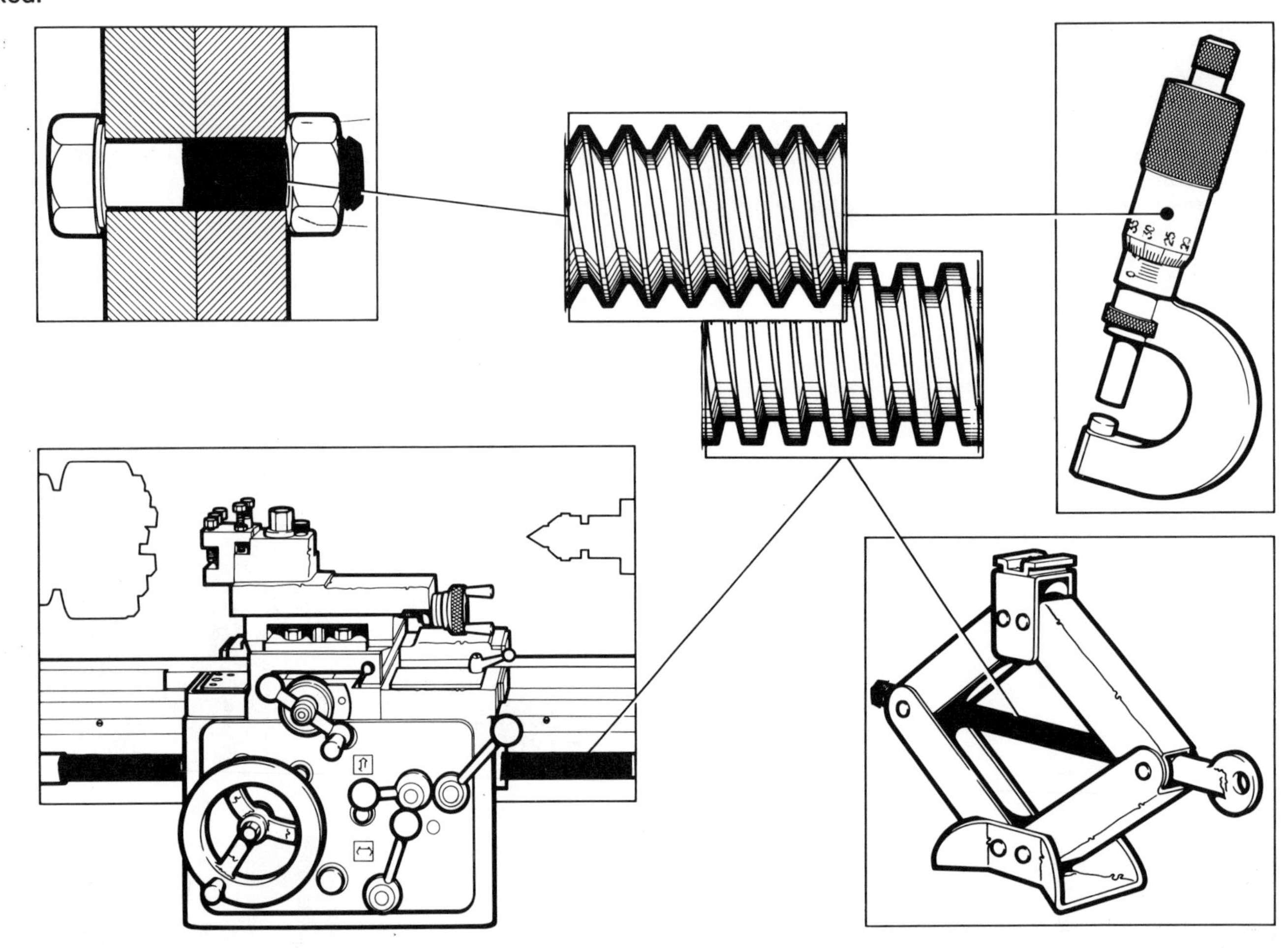

Construction

Imagine that a right-angled triangle is wound around a straight cylinder in such a manner that the base of the triangle coincides with the circumference of the cylinder. The hypotenuse of the triangle will then describe a helix around the cylinder. A thread is obtained in the same manner, the profile of the thread following a helix. The height of the right triangle is equal to the pitch, p, of the thread. The pitch is also defined as the distance between corresponding points on adjacent threads and is therefore expressed in terms of a unit of length (mm) or in terms of the number of threads per unit of length (threads per inch). The lead L of the thread is defined as the axial distance through which a point on the thread advances during one turn of the thread. The angle ρ is called the lead angle.

Normally, threads have only one start, in which case the pitch and the lead will be the same. In the case of multiple-start threads, the lead is a multiple of the pitch. This is explained in greater detail under "Multiple-start threads".

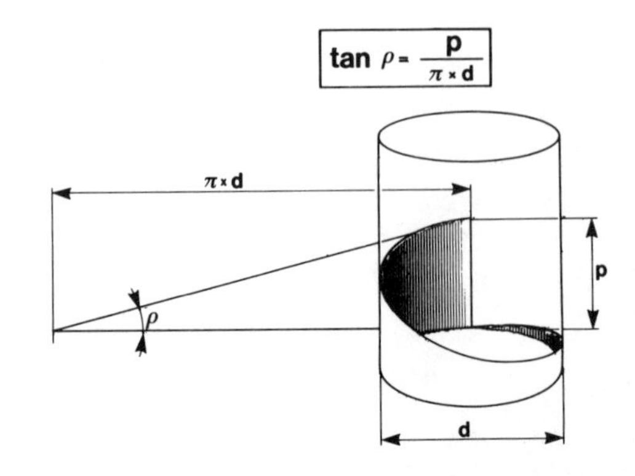

At a given pitch, the lead angle will vary depending upon the diameter. The lead angle is calculated from the following equation: tan $\rho = \frac{p}{\pi \times d}$

The thread profile is the shape enclosed by the contours of the thread in an axial plane section. The thread profile is characterized by the thread angle, which is measured between the flanks. If we start with a fundamental triangle, the sides of the triangle will constitute the flanks of the thread profile. By truncating the triangle (i.e. slicing off its crest) as shown in the figure, we can give the thread profile flat or rounded roots and crests. The root is the rounded or flat part of the thread groove between two flanks and will be on the minor diameter of external threads (bolt threads) and on the major diameter of internal threads (nut threads). Similarly, the crest, which is the flat or rounded top of the thread ridge will be on the major diameter of external threads and the minor diameter of internal threads. The depth of the thread is measured between the crest and the root at an angle perpendicular to the centreline.

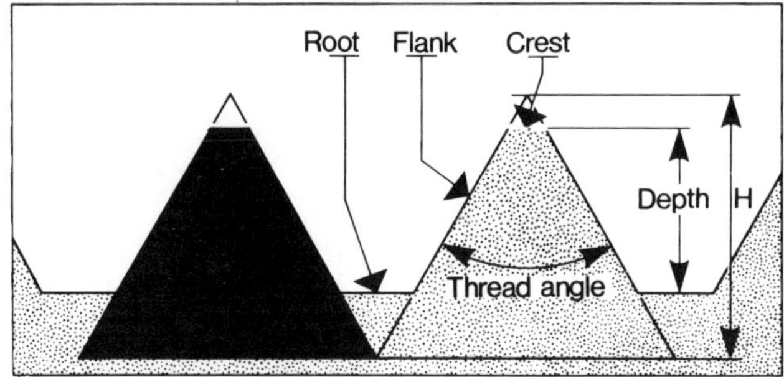

The major diameter is the largest diameter on an external or internal thread. The minor diameter is the smallest diameter on an external or internal thread. The pitch diameter is the diameter that cuts through the thread profile at a level where the width of the thread ridge is equal to the width of the thread groove. If the position of the lead angle is not defined, it is assumed to be at the pitch diameter.

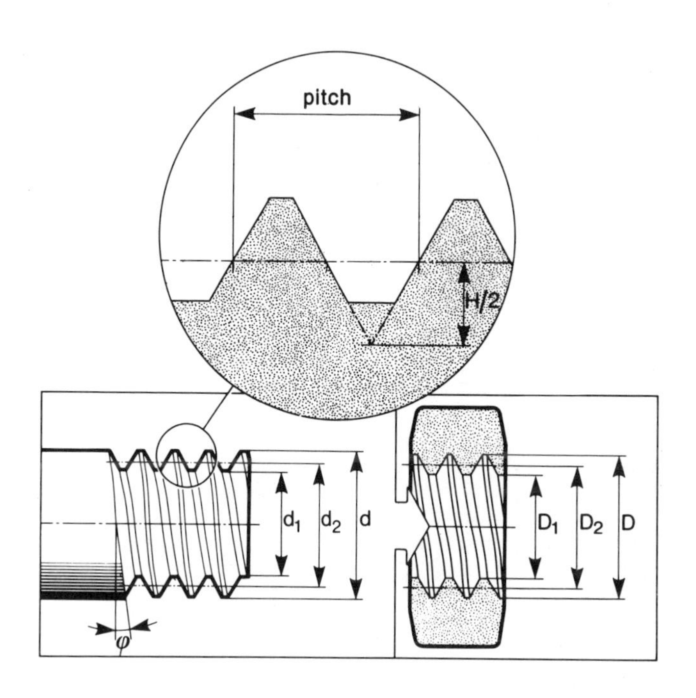

D, d = Major diameter
D_2,d_2 = Pitch diameter
D_1,d_1 = Minor diameter

Thread systems

The thread was one of the first machine elements to be subject to standardization. This standardization covers entire thread systems and includes thread profile, symbols, diameters, pitches, tolerances and gauging systems. If each manufacturer were to have his own thread system, customers would have problems getting hold of spare parts. This is avoided through standardization. In order to increase interchangeability between different countries, an attempt has been made to create a world wide standard for screw threads. The ISO thread is adopted as an international thread and international tolerance standards have been established. The SI and the Whitworth threads are examples of other widespread thread systems. They are, however, intended to be replaced by the ISO thread in the future.

The **ISO thread** is available in a metric version called the M thread and in an inch version called the UN (unified) thread. The basic profile, i.e. the theoretical profile to which the tolerances are related, is obtained by truncating the fundamental triangle H/8 for the major diameter and H/4 for the minor diameter, where H is the height of the fundamental triangle. The thread angle is 60°.

The **SI thread** has the same thread angle as the ISO thread, i.e. 60°. The basic profile is obtained by truncating the fundamental triangle H/8 for both the major and minor diameters.

The thread angle of the **Whitworth thread** is 55°. The basic profile is obtained by truncating the fundamental triangle H/6 for both the major and minor diameters. The basic profile has a rounded crest and root with a radius of 0.1373 × the pitch.

Tolerance standards for these three thread profile are dealt with on page 20—25.

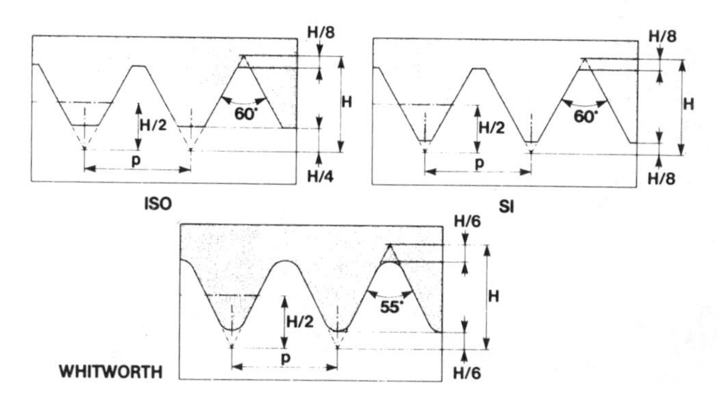

Coarse and fine threads

One might be misled to think that the terms "coarse" and "fine" refer to the quality of the thread, but this is not the case at all.

"Coarse" and "fine" refer to the combination of pitch and diameter. The ISO thread is made with a standard pitch (coarse thread) and with a finer pitch (fine thread) for each diameter.

For metric threads, the pitch is given in mm and for unified threads in number of threads per inch. Metric coarse threads are designated Md (d = diameter) e.g. M30 for a diameter of 30 mm. According to the table, this thread has a pitch of 3.5 mm. For a metric fine thread, the pitch must be indicated in the designation, Mdxp, since these threads are available with a number of different pitches for the same diameter when the diameter is greater than 7 mm. For example, for the diameter 30 mm, there are fine threads with pitches of 1, 1.5, 2 and 3 mm. They are then designated as follows: M30 × 1, M30 × 1.5, M30 × 2 and M30 × 3. At diameters above 68 mm, there are only fine threads.

The unified thread is designated by the diameter in inches followed by UNC for coarse threads and UNF for fine threads. The pitch can be omitted, since only one coarse and one fine pitch are made for each diameter.

Thread designation	Pitch P	Major diameter d = D	Pitch diameter $d_2 = D_2$	Minor diameter $d_1 = D_1$
M28 × 2	2	28	26,701	25,835
M28 × 1,5	1,5	28	27,026	26,376
M28 × 1	1	28	27,350	26,917
M30	3,5	30	27,727	26,211
M30 × 3	3	30	28,051	26,752
M30 × 2	2	30	28,701	27,835
M30 × 1,5	1,5	30	29,026	28,376
M30 × 1	1	30	29,350	28,917
M32 × 2	2	32	30,701	29,835
M32 × 1,5	1,5	32	31,026	30,376
M33	3,5	33	30,727	29,211
M33 × 3	3	33	31,051	29,752
M33 × 2	2	33	31,701	30,835
M33 × 1,5	1,5	33	32,026	31,376
M35 × 1,5	1,5	35	34,026	33,376
M36	4	36	33,402	31,670
M36 × 3	3	36	34,051	32,752
M36 × 2	2	36	34,701	33,835
M36 × 1,5	1,5	36	35,026	34,376
M38 × 1,5	1,5	38	37,026	36,376
M39	4	39	36,402	34,670
M39 × 3	3	39	37,051	35,752
M39 × 2	2	39	37,701	36,835
M39 × 1,5	1,5	39	38,026	37,376

(Taken from SMS 1701)

Thread designation D—n UNC d—n UNC	Major diameter d = D		Pitch		Pitch diameter $d_2 = D_2$	Minor diameter $d_1 = D_1$
	inch	mm	No of threads per inch n	P		
1/4—20 UNC	1/4	6,35	20	1,27	5,524	4,976
5/16—18 UNC	5/16	7,938	18	1,411	7,021	6,411
3/8—16 UNC	3/8	9,525	16	1,588	8,494	7,805
7/16—14 UNC	7/16	11,112	14	1,814	9,934	9,149
1/2—13 UNC	1/2	12,7	13	1,954	11,430	10,584
9/16—12 UNC	9/16	14,288	12	2,117	12,913	11,996
5/8—11 UNC	5/8	15,875	11	2,309	14,376	13,376
3/4—10 UNC	3/4	19,05	10	2,54	17,399	16,299
7/8—9 UNC	7/8	22,225	9	2,822	20,391	19,169
1—8 UNC	1	25,4	8	3,175	23,338	21,963
1 1/8—7 UNC	1 1/8	28,575	7	3,629	26,218	24,648
1 1/4—7 UNC	1 1/4	31,75	7	3,629	29,393	27,823
1 3/8—6 UNC	1 3/8	34,925	6	4,233	32,174	30,343
1 1/2—6 UNC	1 1/2	38,1	6	4,233	35,349	33,518
1 3/4—5 UNC	1 3/4	44,45	5	5,08	41,151	38,951
2—4 1/2 UNC	2	50,8	4 1/2	5,644	47,135	44,689
2 1/4—4 1/2 UNC	2 1/4	57,15	4 1/2	5,644	53,485	51,039
2 1/2—4 UNC	2 1/4	63,5	4	6,35	59,375	56,627
2 3/4—4 UNC	2 3/4	69,85	4	6,35	65,725	62,977
3—4 UNC	3	76,2	4	6,35	72,075	69,327
3 1/4—4 UNC	3 1/4	82,55	4	6,35	78,425	75,677
3 1/2—4 UNC	3 1/2	88,9	4	6,35	84,775	82,027
3 3/4—4 UNC	3 3/4	95,25	4	6,35	91,125	88,377
4—4 UNC	4	101,6	4	6,35	97,475	94,727

(Taken from SMS 1716)

Thread designation D—n UNF d—n UNF	Major diameter d = D		Pitch		Pitch diameter $d_2 = D_2$	Minor diameter $d_1 = D_1$
	inch	mm	No of threads per inch n	P		
1/4—28 UNF	1/4	6,35	28	0,907	5,761	5,367
5/16—24 UNF	5/16	7,938	24	1,058	7,249	6,792
3/8—24 UNF	3/8	9,525	24	1,058	8,837	8,379
7/16—20 UNF	7/16	11,112	20	1,27	10,287	9,738
1/2—20 UNF	1/2	12,7	20	1,27	11,874	11,326
9/16—18 UNF	9/16	14,288	18	1,411	13,371	12,761
5/8—18 UNF	5/8	15,875	18	1,411	14,958	14,348
3/4—16 UNF	3/4	19,05	16	1,588	18,019	17,330
7/8—14 UNF	7/8	22,225	14	1,814	21,046	20,262
1—12 UNF	1	25,4	12	2,117	24,026	23,109
1 1/8—12 UNF	1 1/8	28,575	12	2,117	27,201	26,284
1 1/4—12 UNF	1 1/4	31,75	12	2,117	30,376	29,459
1 3/8—12 UNF	1 3/8	34,925	12	2,117	33,551	32,634
1 1/2—12 UNF	1 1/2	38,1	12	2,117	36,726	35,809

(Taken from SMS 1717)

Thread Turning

The operation

In thread turning, the tool is normally fed along the rotating workpiece just as in an ordinary longitudinal turning operation. The feed rate is, however, much higher.

In a previous chapter, we have already defined the lead as the axial distance through which a point on a screw helix moves during one turn of the thread. In a turning operation, the feed is defined as the axial advance of the tool as the workpiece rotates one turn. This means that the feed coincides with the lead of the thread (and the pitch in the case of single start threads). The shape of the groove, i.e. the thread profile, is determined by the shape of the insert. Normally, the thread is formed in a number of passes. This is because the depth of cut will be too great if the entire thread is machined in a single pass. Instead, the depth of the thread is divided into a number of passes, each with a smaller depth of cut, a new infeed being made for each pass until a fully formed thread has been obtained.

Coordinated feed — rotation

In thread turning, careful coordination is required between the longitudinal feed of the tool and the rotation of the work piece in order to obtain the desired pitch of the thread. This coordination can be obtained by means of a lead screw, cams or numerical control.

The fundamental movements are the same no matter what kind of machine is used. With radial infeed, the movements of the tool will follow a rectangular path. The tool is traversed longitudinally on the machining pass. It is then withdrawn in the transverse direction so that it runs clear of the workpiece as it returns to the starting position. When the tool has been returned to its starting position, a new infeed is made in the same thread groove as on the preceding pass.

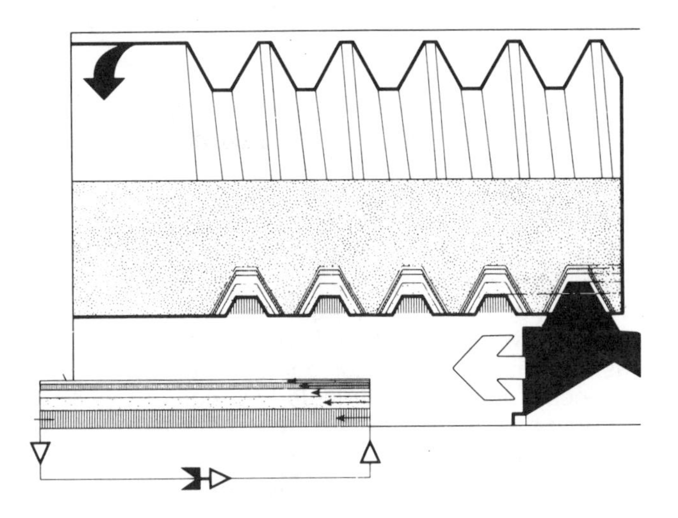

The threading of short production runs or individual pieces is often done in a centre lathe. A lead screw is then used to drive the carriage with the tool. The principle of coordination between the rotation of the workpiece and the feed movement of the tool is as follows: The workpiece is connected to the machine spindle. The rotation of the machine spindle is transmitted to the lead screw via a feed gearbox. A fixed nut, which is connected to the carriage, surrounds the lead screw. When the lead screw rotates, the nut is threaded onto the lead screw. In this manner, a linear movement is transmitted to the carriage. The purpose of the feed gearbox is to alter the ratio between the speed of rotation of the machine spindle and the speed of rotation of the lead screw so that the desired feed (and thereby pitch) is obtained.

The nut consists of two halves and can be opened so that it runs free when normal longitudinal feed is to be used. The feed rate, expressed in mm/min, increases with increasing cutting speed v in accordance with the following formula:

$$s' = \frac{1000 \times v \times p}{\pi \times d} \quad \text{(mm/min)}$$

where v is expressed in m/min, p is the pitch of the thread in mm and d is the diameter of the workpiece in mm.

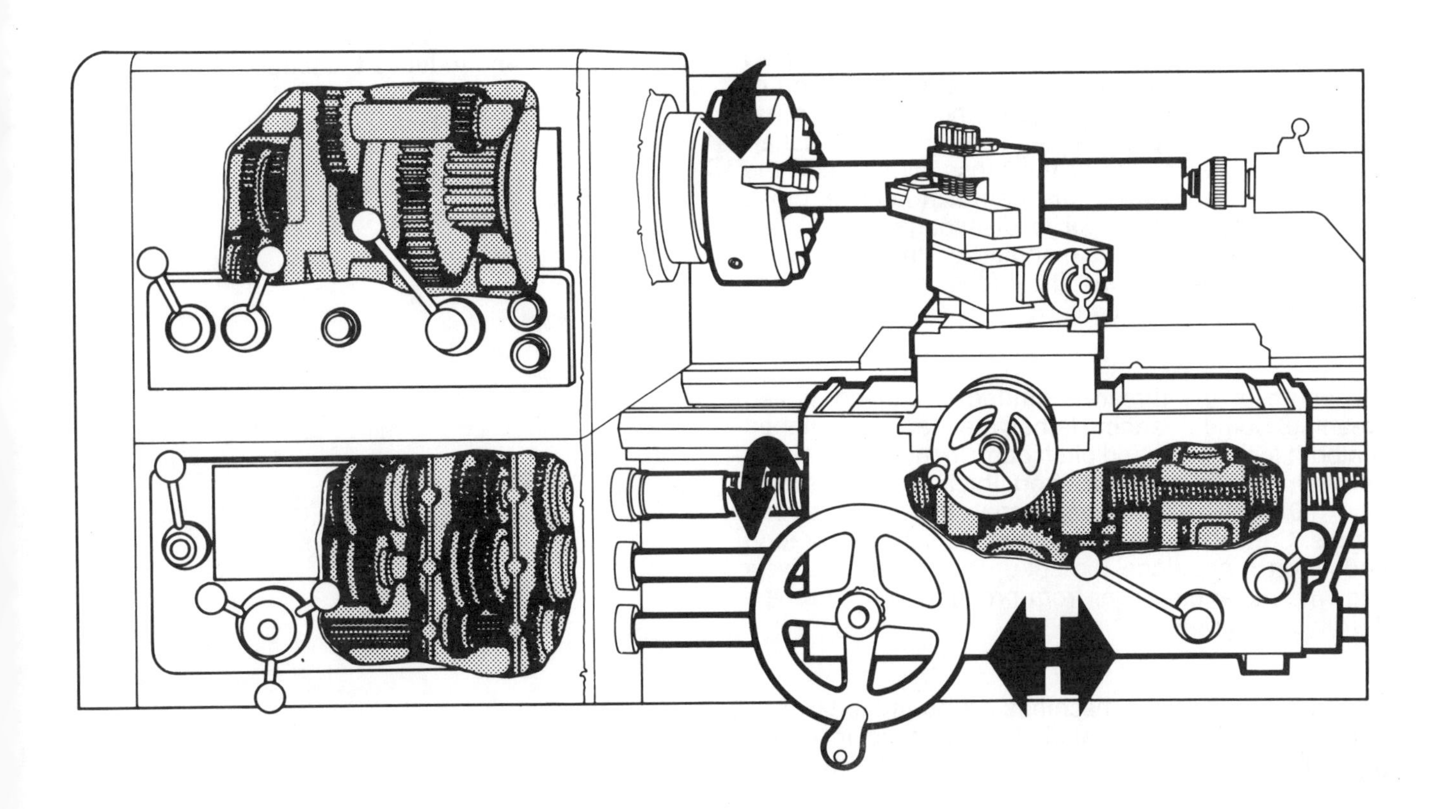

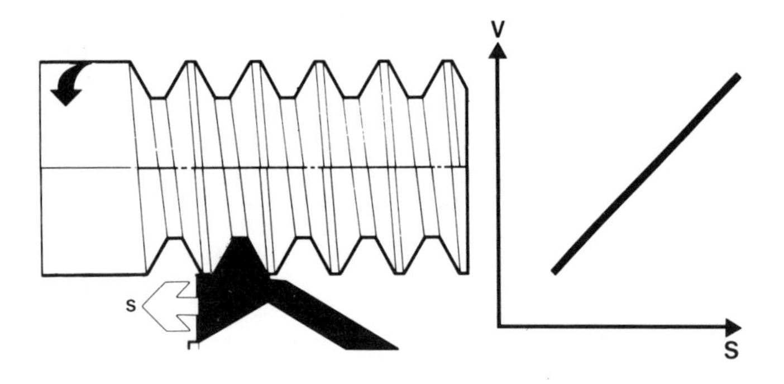

At high cutting speeds, the feed rate can be so great that it is impossible to interrupt the longitudinal movement at a given point. The difficulties increase with increasing pitch and decreasing diameter of the workpiece. This can be particularly dangerous when approaching a shoulder or the chuck. When a thread turning operation is being performed with a cemented carbide insert, which is suited for high cutting speeds, the lathe must sometimes be equipped with a threading attachment that automatically interrupts the feed movement and disengages the threading insert. In addition, the lathes are equipped with a threading dial that indicates where the tool should stop on its return so that it can start at the right point on the next pass.

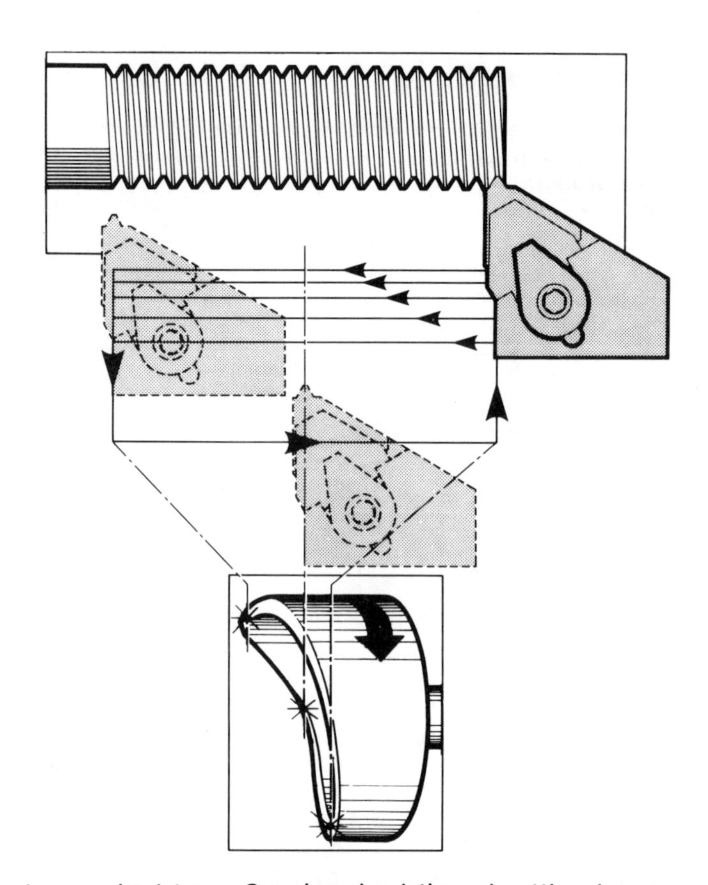

In threading with a lead screw, the direction of rotation must be changed at the start and end of each cut. This is avoided when the feed drive is controlled by a cam. Most lathes that are made specially for threading are cam-controlled. The entire operation is carried out automatically after the number of passes and the infeed depth (penetration) have been preset. The rotation of the machine spindle is transmitted to a camshaft via a gearbox. A series of cams are mounted on the camshaft. In cam control, a follower is connected to the machine slide to be controlled. The cam that controls the longitudinal slide is called the pitch cam. The feed during the operation is obtained through the pitch of the cam. As shown in the figure, the cam has a flat where the feed stops. Another cam on the camshaft controls the movements of the cross slide. When the longitudinal feed stops, this cam actuates a cross movement so that the tool is disengaged. The pitch cam is designed in such a manner that the direction of feed reverses after the flat. The infeed on the next pass is controlled by a feed cam.

The length of the thread is limited in cam-controlled threading. Some threading lathes are therfore equipped with both lead screw and pitch cam. In these machines, the pitch cam is used for short thread lengths (up to about 80 mm) and the lead screw for longer thread lengths. Cam-controlled machines are suitable for long production runs, since the set-up times are relatively long when changing cver from production of one type of article to another.

Numerically controlled machines are controlled by programs where the information is expressed in digital form. The programs are normally fed into the machine via punched tape. Synchronized thread-cutting is normally used for thread turning in NC machines. A sensor is then connected directly to the spindle. On each signal from the sensor, the machine slide will travel a given distance. In this manner, the feed is synchronized with the rotation of the workpiece. An infinite number of pitches can be chosen within a given range. Different control units have varying performance with regard to programmable pitch, permissible feed rate and permissible spindle speed. Some control units are made with fixed thread-cutting cycles. This simplifies the programming work and reduces the length of the punched tape.

Positioning and infeed of the tool

In threading, as in other turning operations, the tool should be applied at the centreline of the workpiece in order to ensure correct machining performance.

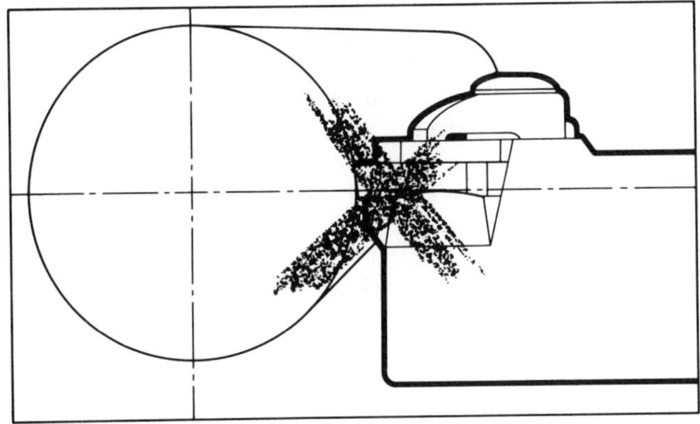

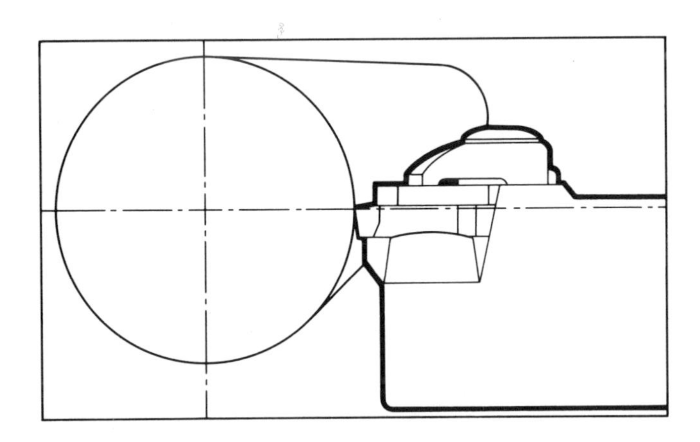

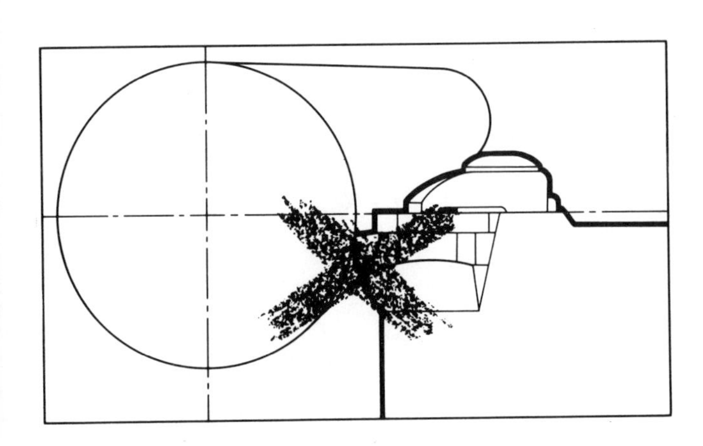

Since the thread profile is perpendicular to the centreline of the workpiece, the insert must be adjusted to a perpendicular position as shown in the figure below. Otherwise, a crooked thread profile will be obtained.

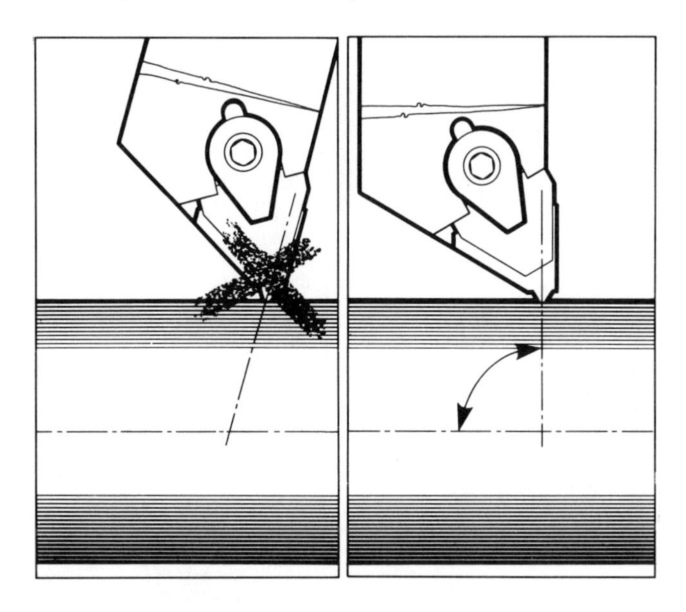

There are three different types of infeed.

In general, radial infeed should be chosen. The tool is then fed perpendicular to the centreline of the workpiece and metal is removed on both sides of the insert simultaneously, providing smooth chip formation.

At larger pitches, the length of engagement of the insert is often too large with radial infeed, causing vibration. It is then possible to use flank infeed, where the tool is fed along one flank. Although wear is then greater on the non-cutting side, flank infeed is used at large pitches in order to obtain better chip control.

For very large threads, incremental infeed is usually employed. An example of this type is shown in the figure, where one side of the profile is first machined in small increments to a given depth. The tool is then advanced and the next part of the profile is machined, and so on until full width has been obtained. This process is then repeated in a number of passes to full depth.

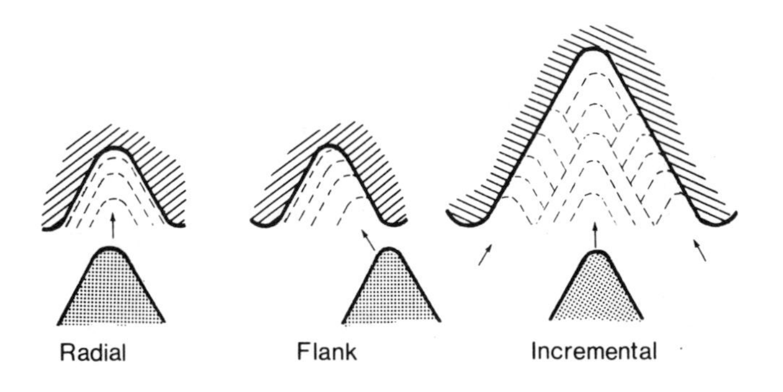

Thread outlet — thread relief groove

In some cases, threads are made with thread outlets. The thread outlet is the part of the thread that, owing to the design of the tool, is not given a full profile.

In the case of an external thread with thread outlet, as shown in the figure below, the matching internal thread must have a thread relief groove, i.e. a recess at the end of the thread (thread clearance) in order to permit it to be threaded against the shoulder. Another alternative is that the external thread be made with a thread relief groove. Then the nut does not require any thread clearance to be threaded against the shoulder.

A thread relief groove is sometimes used as a safety precaution when thread-cutting in machines where there is a risk that the tool will not be withdrawn at exactly the same point after each pass. If the tool is withdrawn too late, the depth of cut will be equal to the total depth of cut for several passes, which can damage the tool. This is avoided if the thread is provided with a thread relief groove, since the tool will run clear of the workpiece if it continues too far.

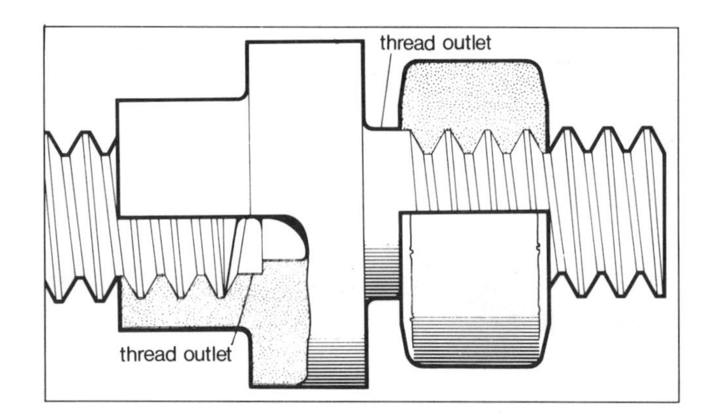

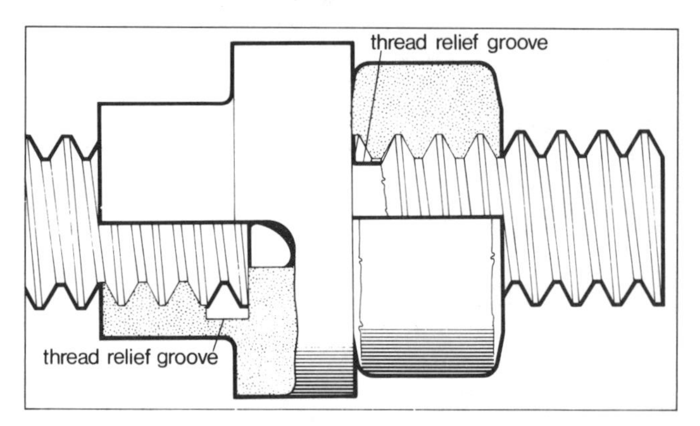

T-MAX threading tools

Insert types

The V-profile insert is the simplest tool for thread turning. Threading is carried out in two operations: diameter turning and profile turning. There is then a risk that the major diameter will not be concentric with the pitch diameter. These problems can be avoided by using T-MAX full-profile inserts. The thread profile including the crests, is then formed in a single operation. After the right pitch diameter has been set, the tool generates the other dimensions. Provided there is sufficient material for a full thread, good concentricity, the correct form and the right dimensions are obtained. Deburring of the crests is eliminated and the preceding turning operation can be carried out with less dimensional accuracy or, in some cases, omitted entirely.

T-MAX threading tools cover pitches from 0.75 to 5.0 mm or 20—5 threads/inch. The cemented carbide indexible inserts are available in three full-profile types: A, B and C.

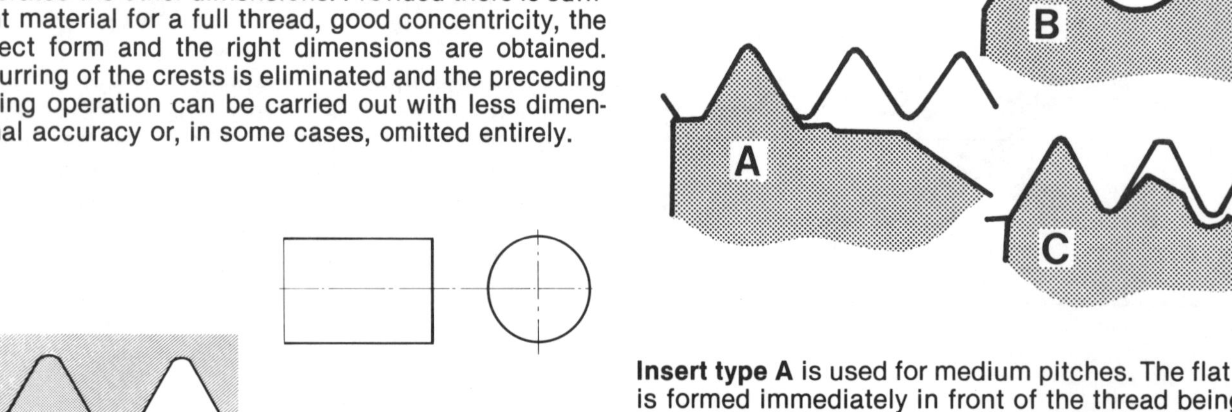

Insert type A is used for medium pitches. The flat crest is formed immediately in front of the thread being cut. The thread outlet can be narrow or omitted entirely. A slight rounding of the crest is obtained owing to the fact that there is a radius (≈ 0.1 mm) between the insert's flank and the straight crest-forming part. However, this is of no practical importance at the pitches for which this type of insert is intended.

Insert type B is used for small pitches (p < 1.5 mm). Crestforming takes place behind the thread being cut, thereby avoiding rounding of the crest. The last thread will be uncrested if there is no thread outlet.

Insert type C is used for coarse pitches in order to obtain adequate support for reliable positioning. Crestforming takes place immediately behind the thread being cut.

The indexable inserts are made with ISO (mm), ISO (inch), SI and Whitworth profiles. Insert type B is not used for Whitworth profile, however.

The indexable inserts are available for external and internal threading. The inserts for internal threading are a mirror image of the corresponding external inserts. Both external and internal inserts are available in right- and left-hand versions. Since tolerances and cutting geometry differ between external and internal inserts, it is important that they should not be confused. Inserts for internal threading are marked with a circle around the pitch number. On external inserts, the pitch number is underlined. The threading inserts are available in following grades: GC 225, S10T, S30T and H1P. S10T is a modification of the previous P10 grade S1P. Compared to S1P, which it fully covers or can substitute, S10T give longer tool life and less plastic deformation.

These grades have the following characteristics:

GC 225 (ISO P15—P30)	High wear resistance permits a very high cutting speed. Very high resistance to plastic deformation
S10T (ISO P01—P15)	Very good edge sharpness. High resistance to flank wear. Covers up for high surface finish demands and accuracy. Also for stainless steel.
S30T (ISO P15—P40)	Good edge sharpness. Covers up for great toughness or stainless steel and other material that gives unfavourable conditions.
H1P (K01—K20)	For threadcutting in cast iron, alloyed cast iron, bronze and brass. Good wear resistance.

With T-MAX threading tools, the same holder can be used for many different inserts. This is made possible by the fact that different profiles and pitches are ground in the same insert size.

Toolholders

There are two types of toolholders for external threading. On the one type, 166F, the shim is fixed by a pin. The insert with loose chipformer is clamped with a T-MAX S clamp. On the other type of toolholder, 166N, the insert seat is milled directly in the shank in order to permit work to closer tolerances.

There is one type of toolholder (166K) for internal threading. The insert is fixed and located in the same way as on the 166N. This type has no chipformer. The insert seat is milled directly in the shank, since cutting a recess for the shim would weaken the seat. In internal threading, it must be possible to adjust the overhang of the tool according to the depth of the bore and the length of the thread. A sleeve type holding tool is available for T-MAX internal toolholders. Its design permits the overhang of the tool to be varied within wide limits.

In thread turning, the axial cutting force changes direction. The threading insert should therefore be supported both on its rear face and on the side face parallel to the holder. In T-MAX toolholders for threading, the insert is supported on its side face by a shoulder on the shim. The inserts have a positive basic form, which means that the shoulder only has to extend slightly beyond the body of the insert. The tool can therefore be used for threading close to a shoulder.

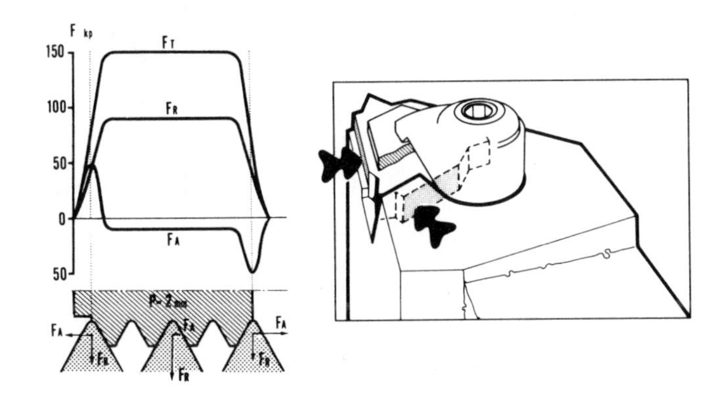

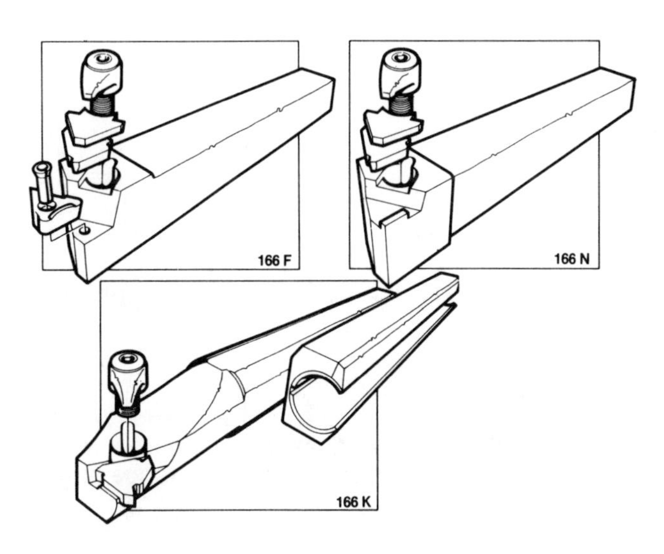

Insert geometries

T-MAX threading inserts are triangular with a positive basic form.

Mounted in the holder, external inserts have a $+2^{\circ}$ rake angle. If the rake angle is too large, the cutting edge will be weakened, reducing tool life. Moreover, the shape of the thread profile will vary in the machining of small and large diameters when the rake angle is greater then 0°. At a $+2^{\circ}$ rake angle, however, the thread will remain within the tolerance zone, regardless of the diameter. This positive rake angle has been chosen to give lower radial forces, which often improves surface finish.

When external inserts are mounted in the holder, the clearance angle on the point of the insert is 8° and the clearance angles on the flanks perpendicular to the cutting edge are 5°. The insert has an angle of inclination of 1°.

For internal inserts, a 0° rake angle, 5° clearance angle on the flanks and 1° angle of inclination are obtained when the insert is mounted in the holder. The clearance angle α on the point of the insert is 10° or 12°, depending upon the size of the insert (see figure).

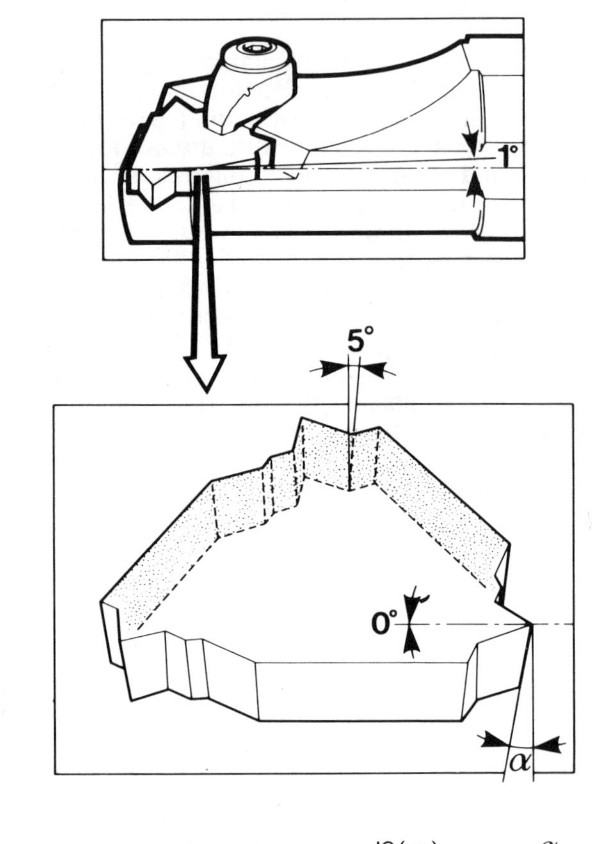

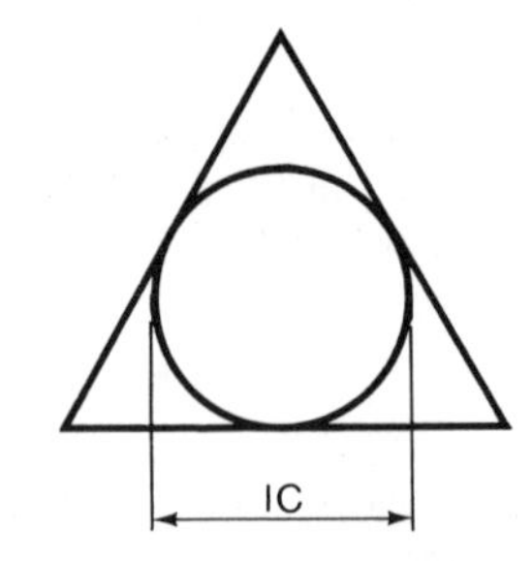

IC (mm)	α
6.35	12°
9.52 — 12.7	10°

Compensation for lead angle of the thread

The clearance angles on the flanks of the tool are dependent on the lead angle of the thread. When the thread's lead angle ρ does not coincide with the insert's angle of inclination λ, the clearance angle on the one flank will be smaller than on the other. This means that the flank wear land will grow more rapidly on the one flank and shorten the life of the insert.

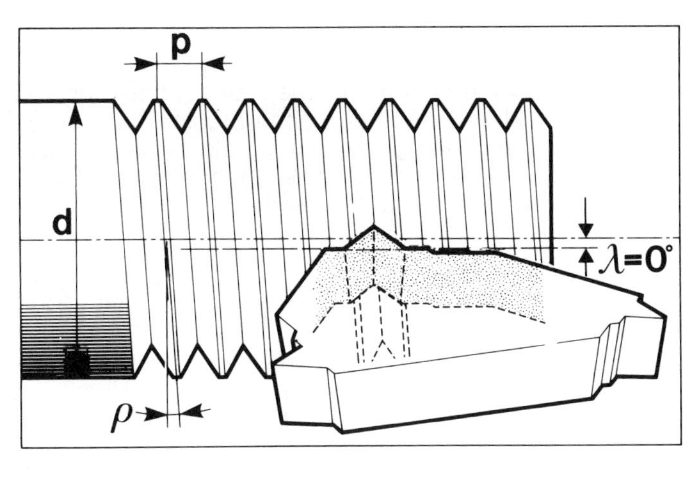

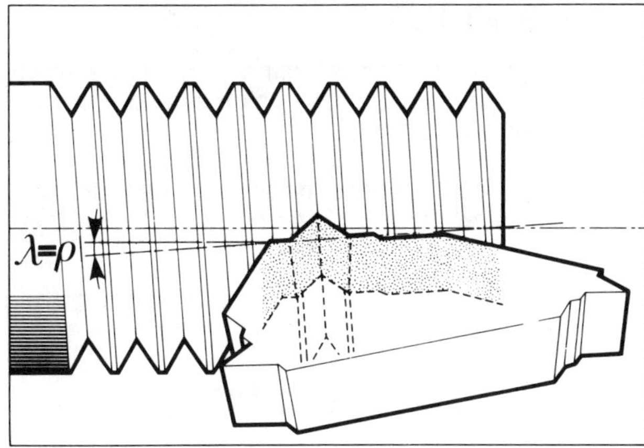

The lead angle is obtained from the formula:

$$\tan \rho = \frac{p}{\pi \times d}$$

and is therefore dependent on the combination of diameter and pitch. T-MAX threading inserts have a 1° lateral inclination when they are mounted in the toolholder. This provides sufficiently large clearance angles for the most frequent combinations of diameter and pitch. In order to limit the maximum difference between the lead angle of the thread and the lateral inclination of the insert to 0° 30' a wedge-type packing shim is used.

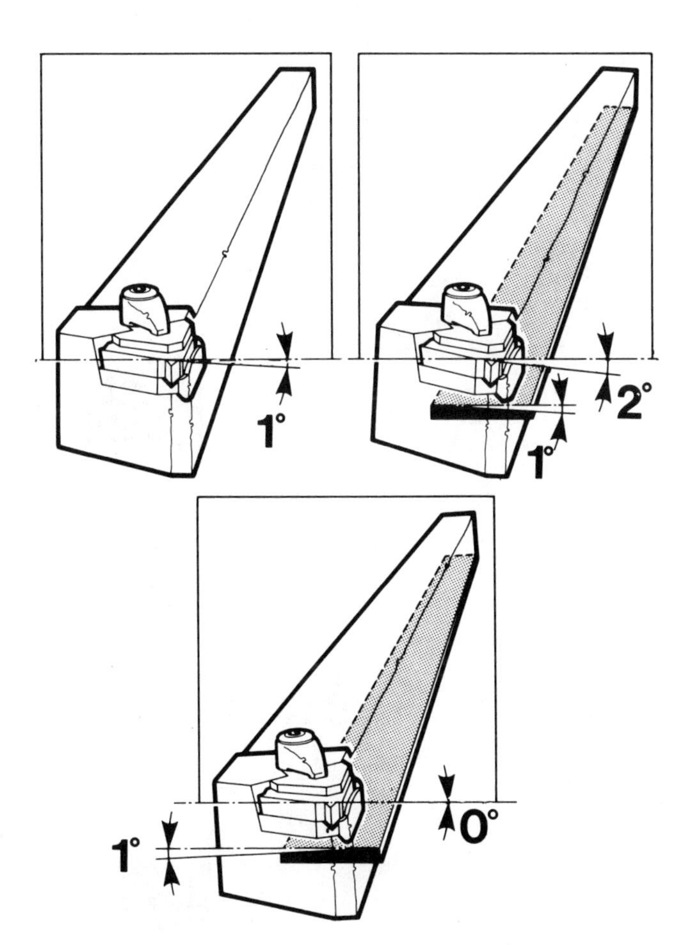

The wedge has a 1° inclination. When it is placed under the holder, a lateral inclination of 0 or 2° is obtained, depending upon how it is turned. Since the cutting tip is raised when the wedge is placed under the toolholder, the centre height must be adjusted. Wedges can therefore not be used in machines where such adjustment is not possible (e.g. NC machines). Compensation for the lead angle of the thread can then be obtained by machining the top and bottom sides of the toolholder to the desired angle.

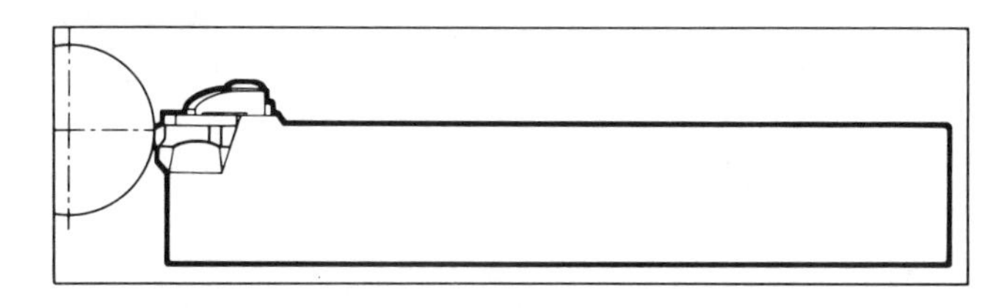

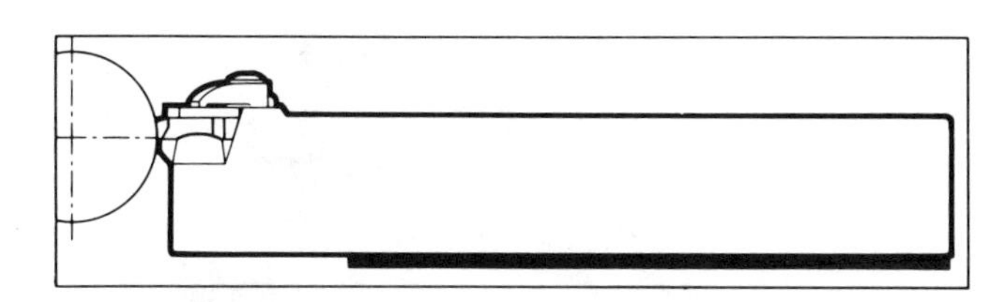

A suitable angle of inclination can be chosen with regard to thread diameter and pitch with the aid of the diagram below.

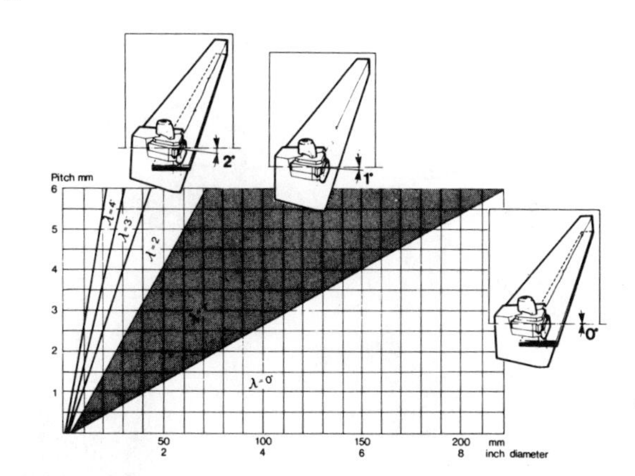

Right- and left-hand threads

There are right-hand and left-hand threads. Ordinarily, right-hand threads are used, in which case the thread is screwed in or on clockwise.

The figures below illustrate the possibilities for changing the feed direction, the direction of rotation of the spindle and the position of the tool in connection with the external and internal machining of right- and left-hand threads. The figures also illustrate the possibility of using right-hand tools for left-hand threads and left-hand tools for right-hand threads.

EXTERNAL THREADS

Right-hand threads

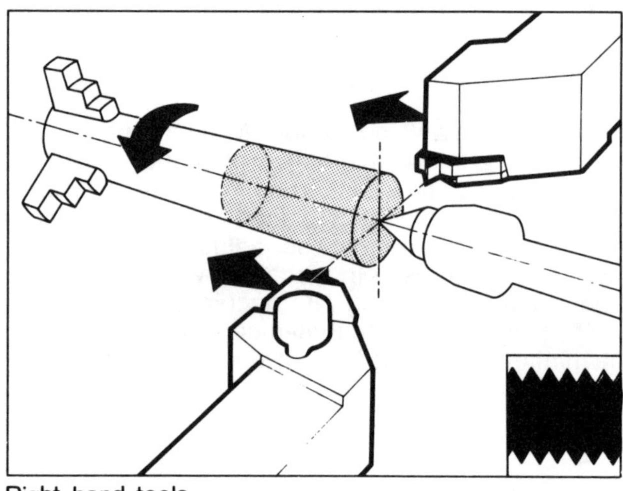

Right-hand tools

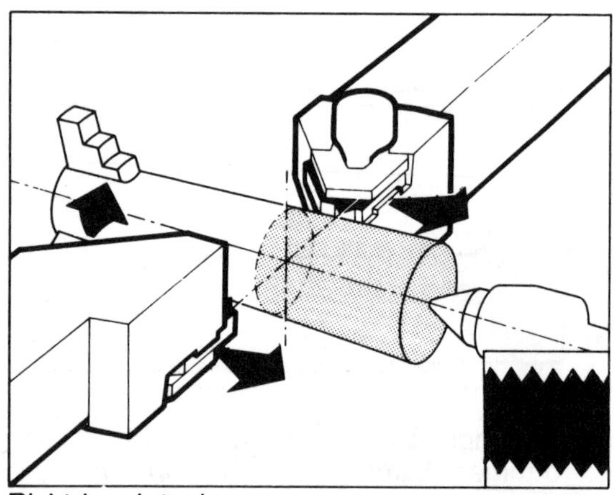

Right-hand tools

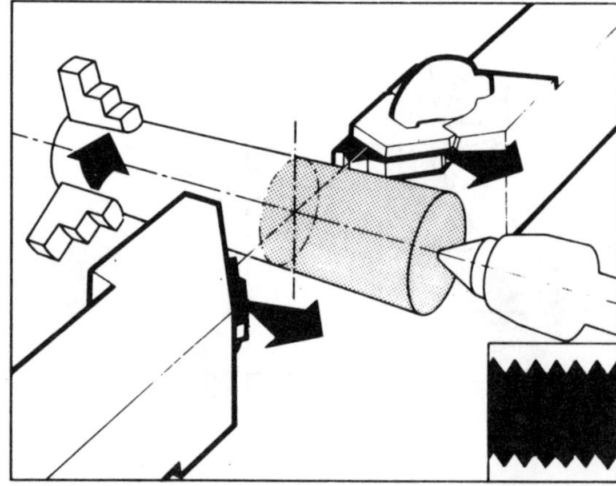

Left-hand tools

Left-hand threads

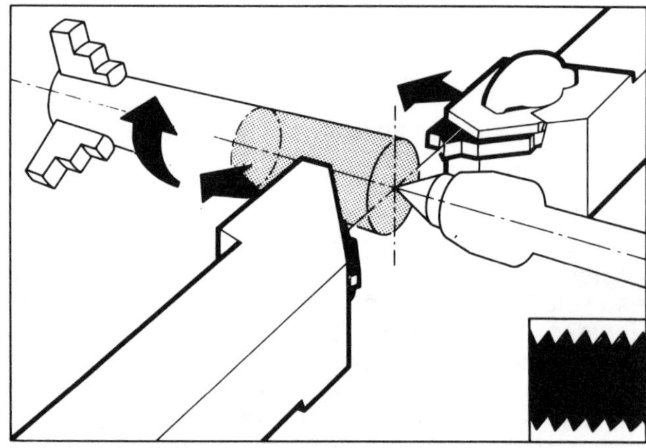

Left-hand tools

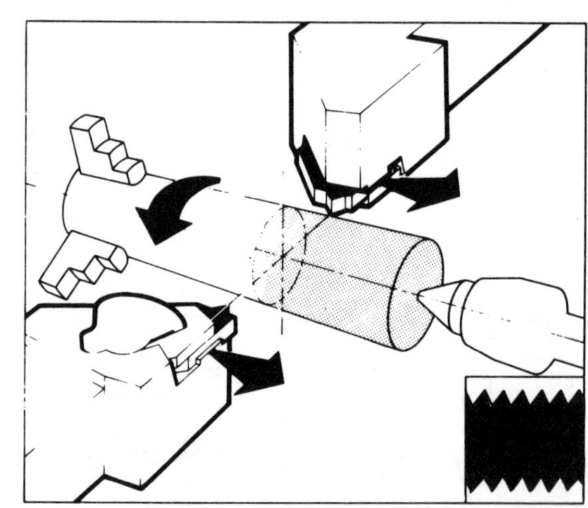

Left-hand tools

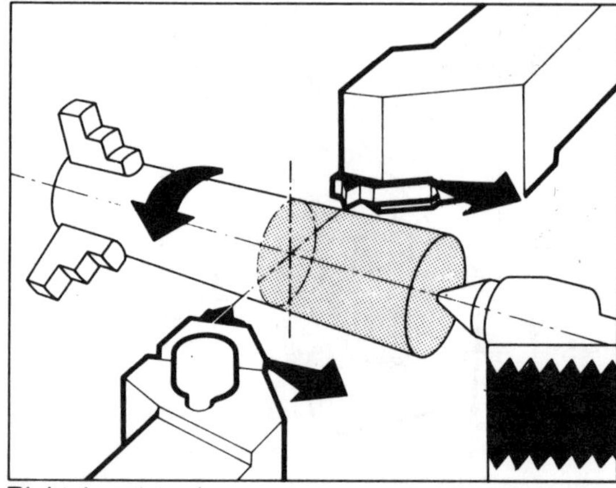

Right-hand tools

INTERNAL THREADS

Right-hand threads

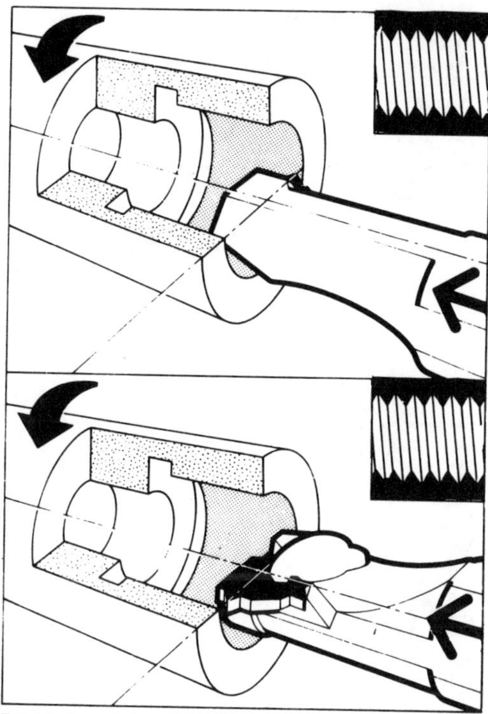

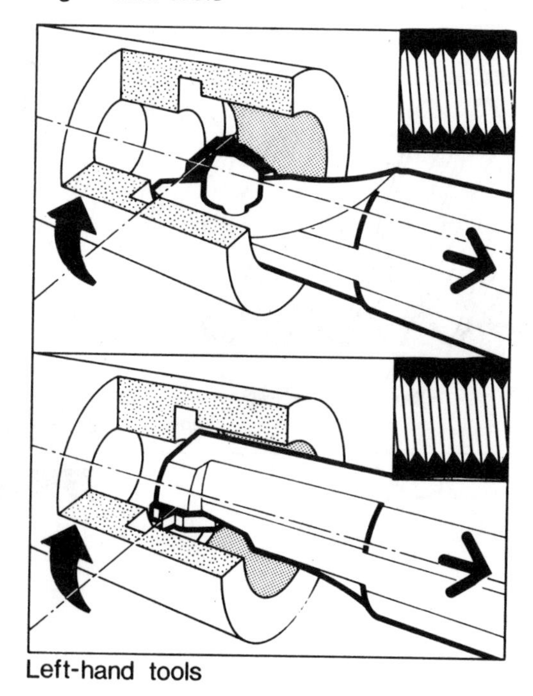

Right-hand tools

Left-hand tools

Left-hand threads

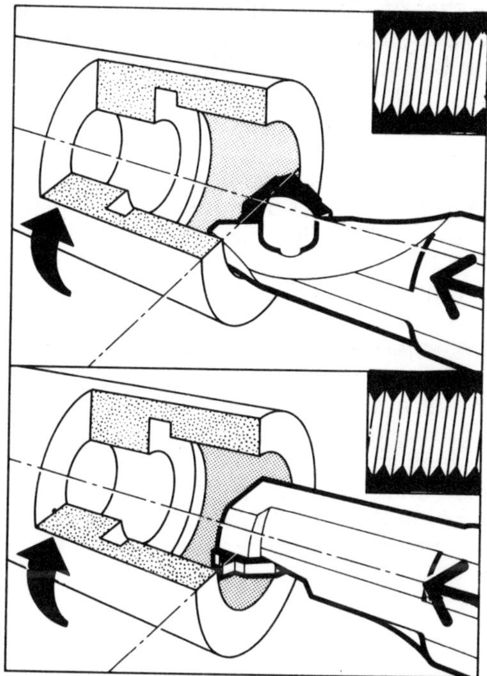

Left-hand tools

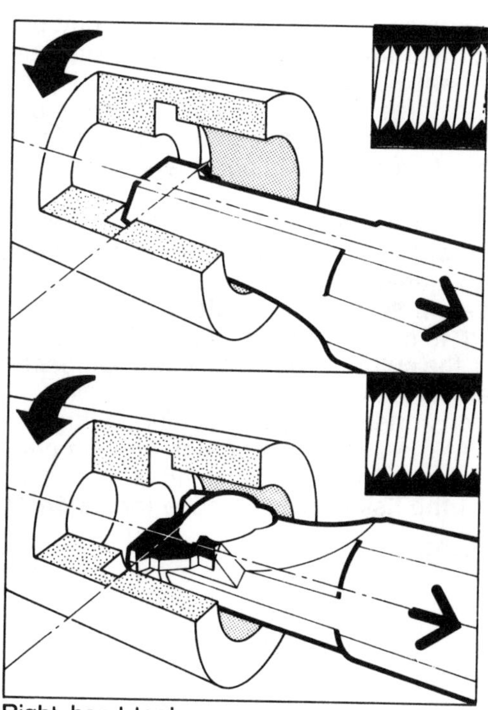

Right-hand tools

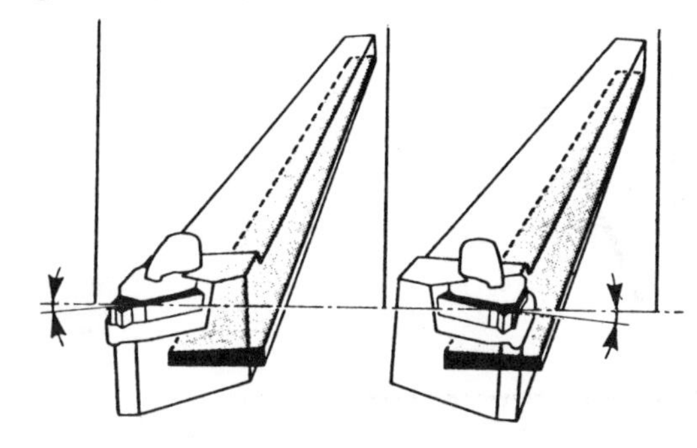

If compensation is to be provided for the lead angle of the thread in the machining of left-hand threads, the insert is inclined in the opposite direction compared to right-hand threading.

Multiple-start threads

Some threads, especially translation threads, have two or more parallel thread grooves, i.e. two or more starts. The lead of a thread with, for example, two starts will be twice that of a single-start screw. A large lead permits the thread to be screwed in or on faster. The axial load on the thread increases, however, with increasing lead angle. This load is distributed among more thread grooves on a multiple-start thread.

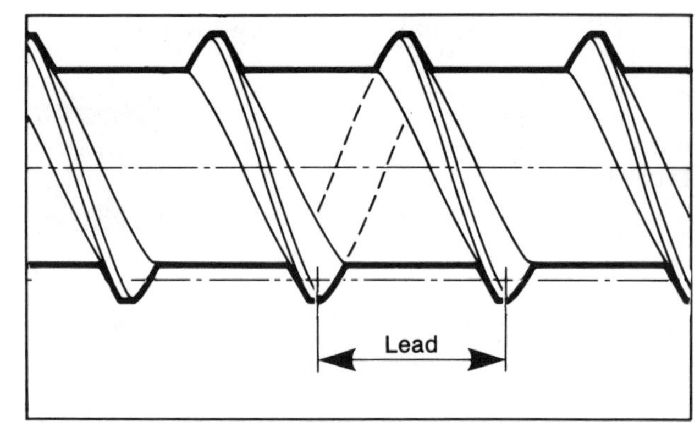

One start is completed

The lead increases in relation to the pitch by a multiple equal to the number of starts. On a single-start thread, the pitch and the lead are equal; with two starts, the lead is twice the pitch; with three starts, the lead is three times the pitch etc.

Multiple-start threads can be manufactured in two ways. The first way is to complete a single thread groove in a number of passes and then start with the next groove (start). Another alternative is to adjust the first infeed and machine one pass on a groove, and then go back and machine the first pass on the next groove, after which the second infeed is set and the next pass is machined on each groove, and so on until all the grooves are finished.

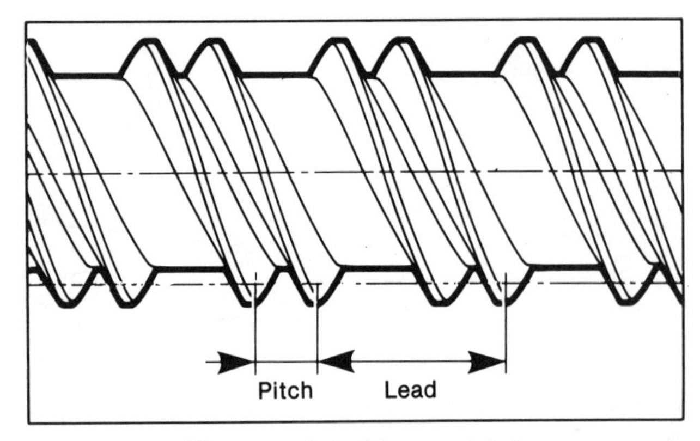

The second start is completed

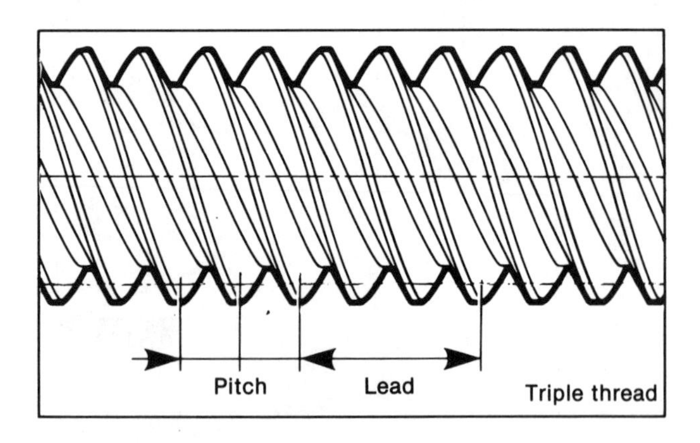

Machining data

In an ordinary longitudinal turning operation, accuracy and surface finish can be balanced with optimum tool performance by varying cutting speed, depth of cut and feed. In thread turning, these parameters cannot be varied independently of each other; certain limiting factors must be taken into consideration.

The **cutting speed** is normally 25% lower in threading than in ordinary turning. This is partly due to the fact that the shape of the threading insert limits heat dissipation so that high cutting speeds result in high cutting temperatures. At high chip load and cutting speed, the cutting temperature can approach the sintering temperature of the cemented carbide. The binder can then soften, resulting in deformation of the cutting edge. At the same time, a high enough cutting speed must be used when machining with carbide tools to prevent the risk of built-up edge formation.

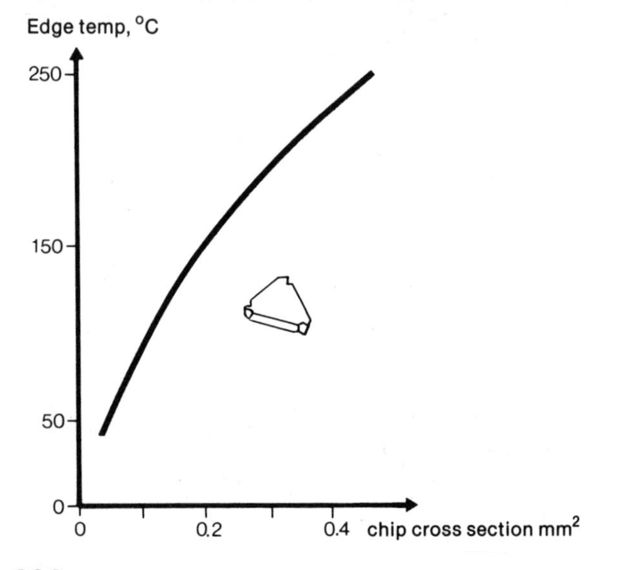

The ability of heat-dissipation for a threading-insert is represented of the edge temperature as a function of the chip cross-section.

Built-up edge formation, also known as loading or pickup of metal, is caused by the chips being welded together with the cutting edge and later being broken off, taking along a piece of the edge. If the machine does not permit a high cutting speed, the toughest possible cemented carbide grade should be used. However, the cutting speed should not be less than 40 m/min when machining with cemented carbide. The table below gives recommended values for the choice of cutting speed for thread turning in different materials.

CMC No.	Material		Hardness HB	GC 225	S10T	S30T	H1P
				Cutting speed			
01.1	**Unalloyed steel**	C max 0.25%	90—130	260—220	210—170	130—100	—
01.23		C max 0.25—0.8%	125—180	240—180	190—140	120— 80	—
01.4		C max 0.8—1.4%	180—250	210—150	170—120	100— 70	—
02.1	**Alloyed steel**	Annealed-normalized	125—225	210—150	170—120	100— 70	—
03.11.12		High-alloyed- annealed	150—250	180—140	140—100	80— 60	—
05.1.2	**Stainless steel**	Ferritic and austenitic	150—270	190—100	150— 80	100— 70	—
06	**Cast steel**		100—250	180—140	140—100	80— 60	—
08.1	**Cast iron**	Ferritic (pearlitic)	160—210	240—180	190—140	120— 80	190—140
08.2		Pearlitic	190—260	210—150	170—120	100— 70	210—150

The **feed s** (mm/rev) must coincide with the desired thread pitch (or lead, in the case of multiple-start threads). This means that the feed rate s' (mm/min) must increase with increasing cutting speed in order for the feed per revolution to be kept constant. We have already discussed how the cutting speed affects the feed rate. We then noted that problems may be encountered in certain machines in stopping the longitudinal movement of the tool at the right place at high cutting speeds, since the feed rate s' may be too great. In the diagram below, the feed rate s' can be read off for v = 120 m/min at varying thread diameters and pitches p.

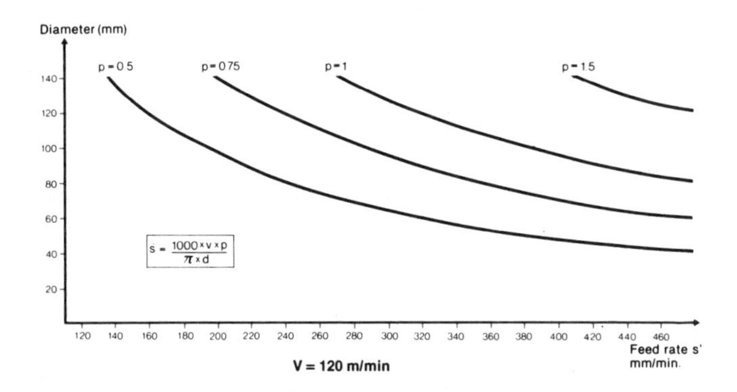

Depth of cut is a critical factor in threading. With each pass, an increasingly large portion of the cutting edge is engaged in the work and the load on the tool increases. If the depth of cut is kept constant during several passes, the chip removal rate can increase by up to three times with each infeed. In order to keep the stress on the cutting edge as uniform as possible, the depth of cut should be reduced with each pass.

In order to obtain a fine surface or close tolerance, the threading operation can be concluded with one or two finishing passes. The light tool pressure eliminates elastic deflections in the tool-machine-workpiece system, which improves accuracy. Such "spring passes" cannot be made in stainless steel, since this material is subject to work hardening, which leads to high tool wear. A nominal chip thickness of less than 0.03 mm should be avoided, since the material will deform elastically instead of being cut. For austenitic steels, the infeed should not be less than 0.08 mm.

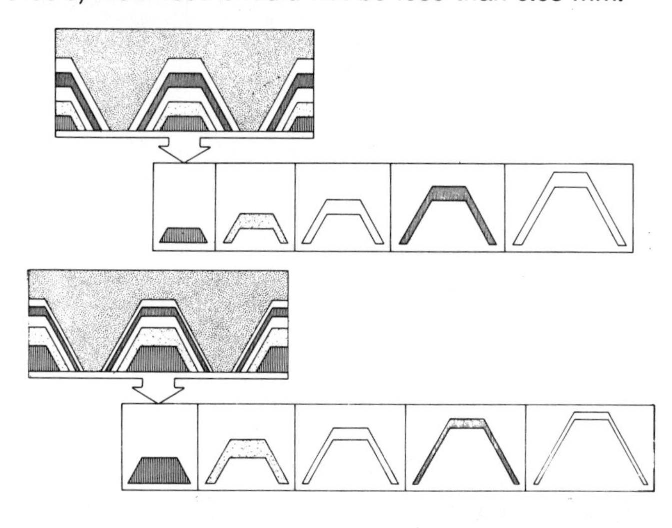

In general, radial infeed should be used for the pitches covered by the T-MAX thread turning range. The tables below provide recommendations for depth of cut for the different passes.

ISO-Metric EXTERNAL

No of infeeds	Infeed mm for different passes and pitches											
1	0.19	0.21	0.22	0.26	0.24	0.27	0.28	0.29	0.34	0.34	0.39	0.43
2	0.15	0.17	0.19	0.23	0.21	0.24	0.26	0.27	0.31	0.32	0.36	0.40
3	0.11	0.13	0.14	0.17	0.16	0.18	0.20	0.21	0.25	0.25	0.29	0.32
4	0.05	0.11	0.12	0.14	0.14	0.16	0.17	0.18	0.21	0.22	0.24	0.27
5		0.06	0.10	0.12	0.12	0.14	0.15	0.16	0.18	0.19	0.22	0.24
6			0.06	0.06	0.11	0.12	0.13	0.14	0.17	0.17	0.20	0.22
7					0.10	0.11	0.12	0.13	0.15	0.16	0.18	0.20
8					0.06	0.06	0.11	0.12	0.14	0.15	0.17	0.19
9							0.11	0.12	0.14	0.14	0.16	0.18
10							0.06	0.11	0.13	0.13	0.15	0.17
11								0.10	0.12	0.13	0.14	0.16
12								0.06	0.08	0.12	0.14	0.15
13										0.12	0.13	0.15
14										0.08	0.10	0.10
Pitch mm	0.75	1.0	1.25	1.5	1.75	2.0	2.5	3.0	3.5	4.0	4.5	5.0

ISO-Metric INTERNAL

No of infeeds	Infeed mm for different passes and pitches											
1	0.19	0.21	0.23	0.27	0.25	0.29	0.30	0.32	0.37	0.36	0.42	0.47
2	0.14	0.16	0.17	0.21	0.20	0.23	0.25	0.26	0.31	0.31	0.36	0.40
3	0.10	0.11	0.13	0.15	0.15	0.17	0.18	0.20	0.23	0.23	0.27	0.30
4	0.05	0.09	0.10	0.12	0.12	0.14	0.15	0.16	0.19	0.20	0.22	0.25
5		0.06	0.09	0.11	0.11	0.12	0.13	0.14	0.17	0.17	0.20	0.22
6			0.06	0.06	0.10	0.11	0.12	0.13	0.15	0.16	0.18	0.20
7					0.09	0.10	0.11	0.12	0.14	0.14	0.16	0.18
8					0.06	0.06	0.10	0.11	0.13	0.13	0.15	0.17
9							0.10	0.10	0.12	0.12	0.14	0.16
10							0.06	0.10	0.11	0.12	0.13	0.15
11								0.09	0.11	0.11	0.13	0.14
12								0.06	0.08	0.11	0.12	0.13
13										0.10	0.12	0.13
14										0.08	0.10	0.10
Pitch mm	0.75	1.0	1.25	1.5	1.75	2.0	2.5	3.0	3.5	4.0	4.5	5.0

ISO-Inch EXTERNAL

No of infeeds	Infeed mm for different passes and pitches												
1	0.22	0.25	0.24	0.25	0.27	0.29	0.28	0.29	0.29	0.31	0.35	0.36	0.44
2	0.19	0.21	0.21	0.22	0.24	0.26	0.26	0.26	0.27	0.29	0.33	0.34	0.41
3	0.14	0.16	0.16	0.17	0.18	0.20	0.20	0.20	0.21	0.22	0.26	0.27	0.32
4	0.12	0.13	0.13	0.14	0.15	0.16	0.17	0.17	0.18	0.19	0.22	0.23	0.28
5	0.10	0.12	0.12	0.12	0.13	0.14	0.15	0.15	0.16	0.17	0.19	0.20	0.24
6	0.06	0.06	0.11	0.11	0.12	0.13	0.13	0.14	0.14	0.15	0.17	0.18	0.22
7			0.06	0.10	0.11	0.12	0.12	0.13	0.13	0.14	0.16	0.17	0.20
8				0.06	0.06	0.06	0.11	0.12	0.12	0.13	0.15	0.16	0.19
9							0.06	0.11	0.12	0.12	0.14	0.15	0.18
10								0.06	0.11	0.12	0.13	0.14	0.17
11									0.06	0.11	0.13	0.13	0.16
12										0.06	0.08	0.13	0.15
13												0.12	0.15
14												0.10	0.10
Pitch threads/inch	20	18	16	14	13	12	11	10	9	8	7	6	5

ISO-Inch INTERNAL

No of infeeds	Infeed mm for different passes and pitches												
1	0.23	0.25	0.25	0.26	0.28	0.30	0.30	0.30	0.31	0.33	0.38	0.39	0.47
2	0.17	0.19	0.20	0.21	0.22	0.24	0.24	0.25	0.26	0.28	0.32	0.33	0.40
3	0.13	0.14	0.14	0.15	0.16	0.18	0.18	0.18	0.19	0.21	0.23	0.25	0.30
4	0.11	0.12	0.12	0.13	0.13	0.15	0.15	0.15	0.16	0.17	0.20	0.21	0.25
5	0.09	0.10	0.10	0.11	0.12	0.13	0.13	0.13	0.14	0.15	0.17	0.18	0.22
6	0.06	0.06	0.09	0.10	0.11	0.11	0.12	0.12	0.13	0.13	0.15	0.16	0.20
7			0.06	0.09	0.10	0.10	0.11	0.11	0.12	0.12	0.14	0.15	0.18
8				0.06	0.06	0.06	0.10	0.10	0.11	0.11	0.13	0.14	0.17
9							0.06	0.10	0.10	0.11	0.12	0.13	0.16
10								0.06	0.09	0.10	0.12	0.12	0.15
11									0.06	0.10	0.11	0.12	0.14
12										0.06	0.08	0.11	0.14
13												0.11	0.13
14												0.10	0.10
Pitch threads/inch	20	18	16	14	13	12	11	10	9	8	7	6	5

SI Metric EXTERNAL

No of infeeds	Infeed mm for different passes and pitches											
1	0.21	0.23	0.24	0.29	0.27	0.30	0.32	0.33	0.39	0.44	0.49	
2	0.16	0.18	0.20	0.24	0.22	0.25	0.27	0.29	0.34	0.35	0.39	0.44
3	0.12	0.13	0.15	0.17	0.17	0.19	0.21	0.22	0.26	0.27	0.30	0.33
4	0.05	0.11	0.12	0.14	0.14	0.16	0.17	0.19	0.22	0.22	0.25	0.28
5		0.06	0.11	0.13	0.12	0.14	0.15	0.16	0.19	0.20	0.22	0.25
6			0.06	0.06	0.11	0.12	0.14	0.15	0.17	0.18	0.20	0.22
7					0.10	0.11	0.13	0.14	0.16	0.16	0.18	0.21
8					0.06	0.06	0.12	0.13	0.15	0.15	0.17	0.19
9							0.11	0.12	0.14	0.14	0.16	0.18
10							0.06	0.11	0.13	0.14	0.15	0.17
11								0.11	0.12	0.13	0.15	0.16
12								0.06	0.08	0.12	0.14	0.15
13										0.12	0.13	0.15
14										0.08	0.10	0.10
Pitch mm	0.75	1.0	1.25	1.5	1.75	2.0	2.5	3.0	3.5	4.0	4.5	5.0

SI Metric INTERNAL

No of infeeds	Infeed mm for different passes and pitches				
1	0.24	0.32	0.34	0.36	0.37
2	0.18	0.24	0.26	0.28	0.30
3	0.13	0.17	0.19	0.21	0.22
4	0.11	0.14	0.16	0.17	0.18
5	0.06	0.12	0.14	0.15	0.16
6		0.06	0.12	0.14	0.14
7			0.11	0.12	0.13
8			0.06	0.11	0.12
9				0.11	0.11
10				0.06	0.11
11					0.10
12					0.06
Pitch mm	1.0	1.5	2.0	2.5	3.0

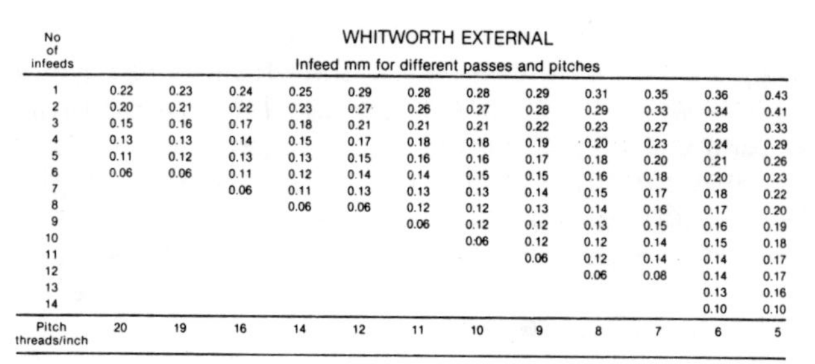

WHITWORTH EXTERNAL

No of infeeds	Infeed mm for different passes and pitches											
1	0.22	0.23	0.24	0.25	0.29	0.28	0.28	0.29	0.31	0.35	0.36	0.43
2	0.20	0.21	0.22	0.23	0.27	0.26	0.27	0.28	0.29	0.33	0.34	0.41
3	0.15	0.16	0.17	0.18	0.21	0.21	0.21	0.22	0.23	0.27	0.28	0.33
4	0.13	0.13	0.14	0.15	0.17	0.18	0.18	0.19	0.20	0.23	0.24	0.29
5	0.11	0.12	0.13	0.13	0.15	0.16	0.16	0.17	0.18	0.20	0.21	0.26
6	0.06	0.06	0.11	0.12	0.14	0.14	0.15	0.15	0.16	0.18	0.20	0.23
7			0.06	0.11	0.13	0.13	0.13	0.14	0.15	0.17	0.18	0.22
8				0.06	0.06	0.12	0.12	0.13	0.14	0.16	0.17	0.20
9						0.06	0.12	0.12	0.13	0.15	0.16	0.19
10							0.06	0.12	0.12	0.14	0.15	0.18
11								0.06	0.12	0.14	0.14	0.17
12									0.06	0.08	0.14	0.17
13											0.13	0.16
14											0.10	0.10
Pitch threads/inch	20	19	16	14	12	11	10	9	8	7	6	5

WHITWORTH INTERNAL

No of infeeds	Infeed mm for different passes and pitches											
1	0.21	0.22	0.23	0.24	0.28	0.27	[illegible]	.28	0.30	0.34	0.35	0.42
2	0.19	0.20	0.21	0.22	0.26	0.26	0.26	.27	0.28	0.32	0.33	0.40
3	0.15	0.15	0.17	0.17	0.20	0.20	0.20	0.21	0.23	0.26	0.27	0.33
4	0.12	0.13	0.14	0.15	0.17	0.17	0.17	0.18	0.19	0.22	0.23	0.28
5	0.11	0.11	0.12	0.13	0.15	0.15	0.16	0.16	0.17	0.20	0.21	0.25
6	0.06	0.06	0.11	0.12	0.14	0.14	0.14	0.15	0.16	0.18	0.19	0.23
7			0.06	0.11	0.12	0.13	0.13	0.14	0.15	0.17	0.18	0.21
8				0.06	0.06	0.12	0.12	0.13	0.14	0.16	0.17	0.20
9						0.06	0.11	0.12	0.13	0.15	0.16	0.19
10							0.06	0.11	0.12	0.14	0.15	0.18
11								0.06	0.12	0.13	0.14	0.17
12									0.06	0.08	0.13	0.16
13											0.13	0.16
14											0.10	0.10
Pitch threads/inch	20	19	16	14	12	11	10	9	8	7	6	5

These recommendations are intended as starting values for machining in steel. The suitable **number of infeeds** must be determined by trial and error. If insert breakage occurs, the number of infeeds should be increased. In general, the number of infeeds should be increased when machining cast iron. The greater the number of passes the operation is divided into, the smaller the depth of cut and the smaller the stresses on the cutting tip will be. If too many infeeds are used, the tool will not cut due to insufficient depth of cut and the material will deform elastically. This leads to higher insert wear. In other words, if insert wear is high, the number of infeeds should be reduced.

Comparison of cutting forces with threading and turning

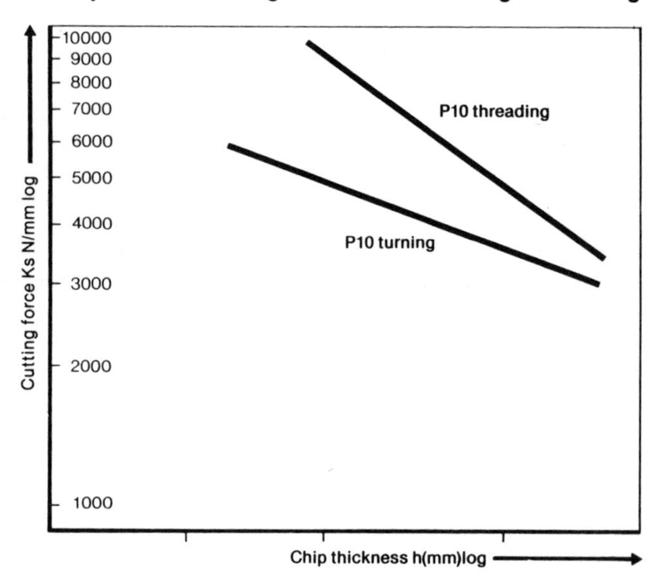

The **cutting forces** and thereby the power requirement are considerably higher in thread turning than in normal turning operations, especially when the chip thickness is small. However, the cutting forces approach the values for plain turning when chip thickness increases.

Chipforming

In thread turning, the tool is fed in radially in stages. The result is that the chip takes on an increasingly stiff V shape, which can be very troublesome. Tests show that chipbreaking is normally too soft in the first cuts and too hard in the last ones. Hard chipbreaking increases the cutting forces and heat input, which reduces tool life and increases the risk of failure. Chipbreaking should be avoided in threading operations, since the cutting forces will break the little cutting tip. Instead, chipformers have been developed.

The chipformer is placed on top of the insert and is provided with a vertical edge that gives the chip an easy-to-handle spiral (curled) shape.

T-MAX threading tools for external machining are based on indexable and interchangeable chipformers. The rear edge of the chipformer has the same angle as the rear edge of the insert seat. If the rear edge of the chipformer is flush up against the edge of the insert seat, it will be pushed forward 0.35 mm (0.014 in) when it is turned.

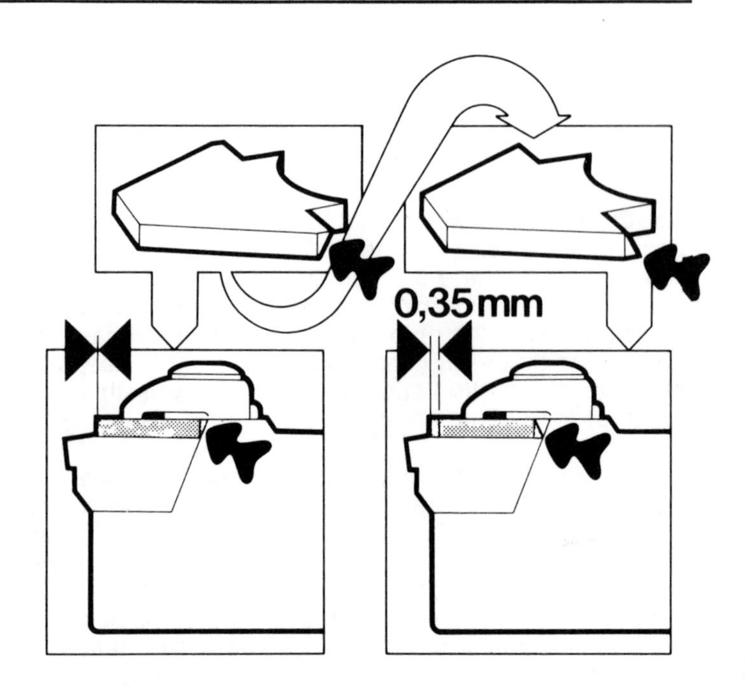

IC 9.52 mm (3/8") chipformers, which are available in three versions, provide six possibilities for adjusting chipforming. IC 12.7 mm (1/2") chipformers, which are available in two versions, afford four possibilities for forming the chips to the desired shape. It is also possible to machine without a chipformer.

Material, diameter, external or internal machining, chip thickness and cutting data affect chipforming. Adjustable chipformers offer an advantage for avoiding production disturbances, since they can be adjusted to the conditions at hand. Trial and error must be employed to find the best chipformer for each application. In thread turning, chip formation varies between different passes. The aim then is to choose a chipformer that gives the best possible results for the operation as a whole. If the chipformer is selected soley with a view towards the best result in the first pass, breakage of the chip may be obtained in subsequent passes, which can damage the tool.

A natural chip formation is obtained during internal threading, since a concave surface is being machined. These tools are therefore not equipped with chipformers. The use of chipformers in internal threading would, moreover, reduce the available space and impede the flow of chips. Under favourable conditions, the chip will automatically be transported out of the hole between the tool and the workpiece. In other cases, cutting fluid, coolant or compressed air can be used to facilitate the flow of chips.

Wear

Tool wear in connection with thread turning is concentrated mainly on the flanks of the insert. Because the threading operation is divided into a number of passes, the part of the flank nearest the tip will have a longer engagement time than the part of the flank that cuts on the last pass. This leads to a change in the profile height (see photo).

The change in shape undergone by the insert profile due to wear have been taken into account in the manufacture of T-MAX threading inserts. The inserts therefore have a long life with respect to profile error.

We have touched upon factors that can affect tool wear in various connections in previous chapters. Here is a brief summary:

- Tool life increases with adequate clearance on the flanks. This can be obtained by compensating for the lead angle of the thread.
- Too many passes can lead to abnormal wear owing to insufficient depth of cut and elastic deformation in the material. (Too few passes leads to insert breakage).
- If the cutting speed is too high, the working temperature can approach the sintering temperature of the cemented carbide and the insert may deform plastically.

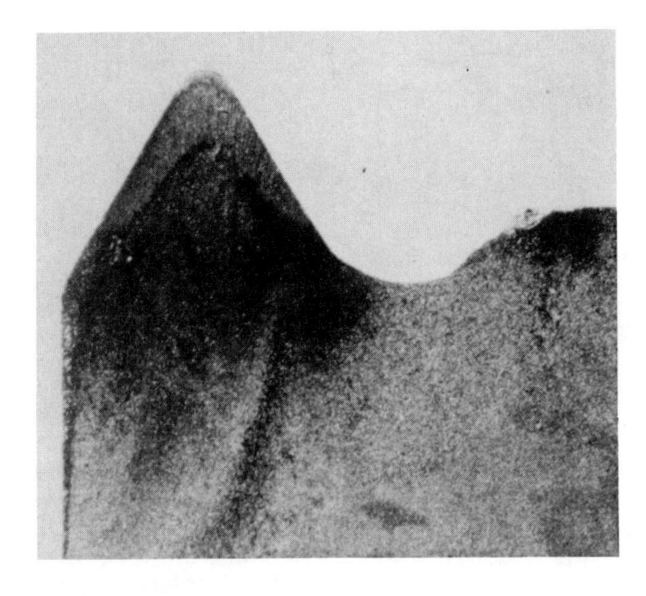

Thread tolerances and standards

Tolerances and fits

Before we get into thread tolerances, we should clarify certain concepts by taking a look at shafts and holes.

The dimension of a part is defined by a nominal size. In reality, the actual size deviates from the nominal size, and the permissible deviation is called tolerance. The tolerances on the diameters of a shaft and a hole that are to be mounted together affect the type of fit that is obtained. A differentiation is made here between clearance and interference. In clearance, the difference between the sizes of the hole and the shaft is always positive. In interference, the difference between the sizes of the hole and the shaft is negative prior to assembly.

When the tolerance zones for the diameters of the hole and the shaft overlap each other, either clearance or interference can be obtained, depending upon where in the tolerance zones the actual sizes lie. This type of fit is called a transition fit.

When the tolerances are positioned in such a manner that the fit always has clearance, it is known as a clearance fit. If the tolerances on the shaft and the hole are positioned in such a manner that interference is always obtained, the fit is called an interference fit. The positions of the tolerances are always given in relation to a basic size. The basic size constitutes a zero line. As far as shafts and holes are concerned, the nominal diameter is the basic size. The permissible tolerance on a shaft can, for example, be indicated as 400 mm +0.02, -0.01. This means that the maximum diameter the shaft may have is 400.02 mm and the minimum 399.99 mm. In other words, the tolerance of the shaft is (400.02 — 399.99 =) 0.03 mm.

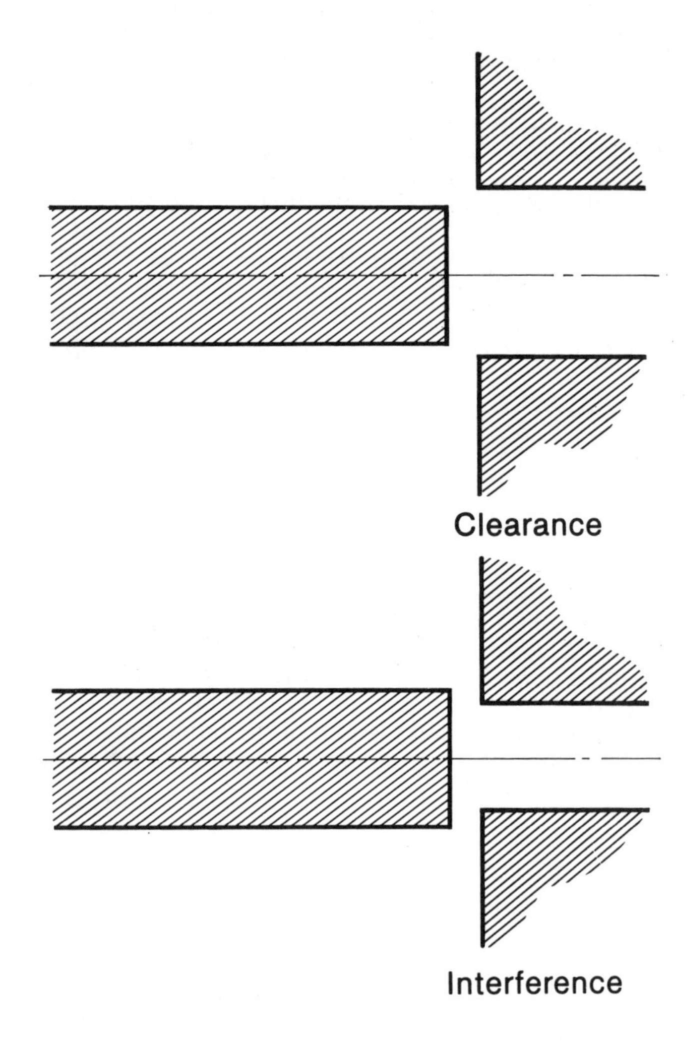

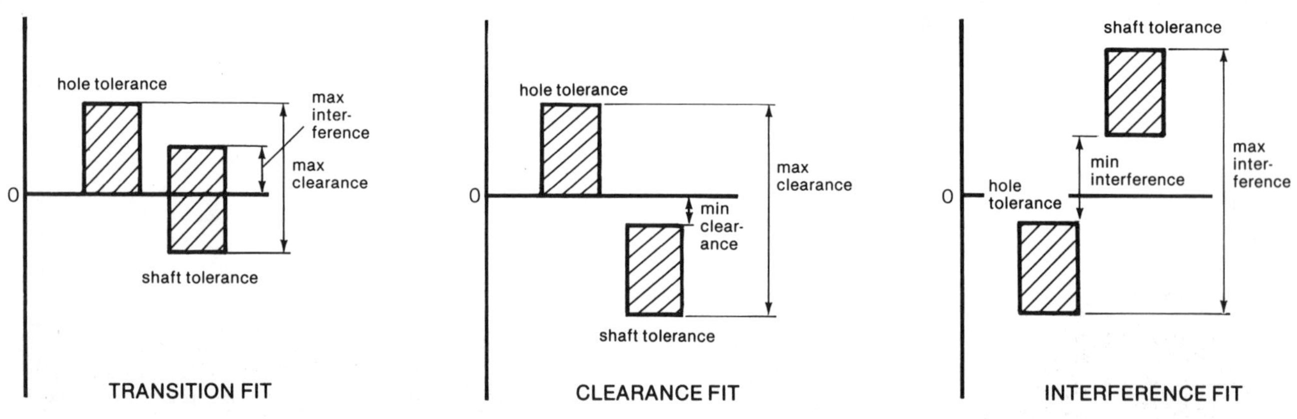

Thread and tool tolerances

The tolerance standards specified by the thread system must be complied with in the manufacture of a screw thread. It is also necessary to choose a tolerance class that can be achieved in production. Furthermore, it is important to check by means of measurement that the thread actually meets the tolerances specified.

In all measurement, it is necessary to have a reference point or a base from which to measure. In the case of screw threads, this reference point is the basic profile. Under the heading "Thread systems", the ISO, SI and Whitworth threads are described with reference to the basic profile. It should be borne in mind that this is a theoretical profile to which the tolerances on external and internal threads are related. The actual thread profiles will vary in appearance, depending on the tolerances.

ISO metric threads

The tolerance system for ISO threads specifies clearance. The figure below shows an example of a diagram of the tolerance positions of the pitch diameter relative to the zero line. The tolerance positions are such that the fundamental deviation of an internal thread (i.e. the difference between the pitch diameter of the basic profile and the lower deviation) is positive or zero, while the fundamental deviation of an external thread (i.e. the difference between the pitch diameter of the basic profile and the upper deviation) is negative or zero.

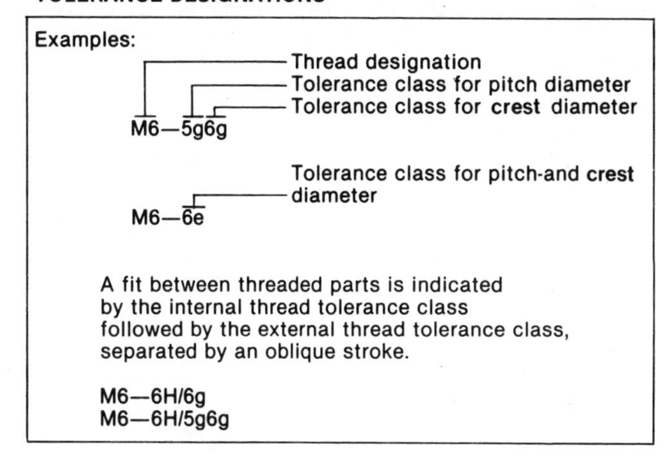
TOLERANCE DESIGNATIONS

Examples:

M6—5g6g
- Thread designation
- Tolerance class for pitch diameter
- Tolerance class for **crest** diameter

M6—6e
- Tolerance class for pitch-and **crest** diameter

A fit between threaded parts is indicated by the internal thread tolerance class followed by the external thread tolerance class, separated by an oblique stroke.

M6—6H/6g
M6—6H/5g6g

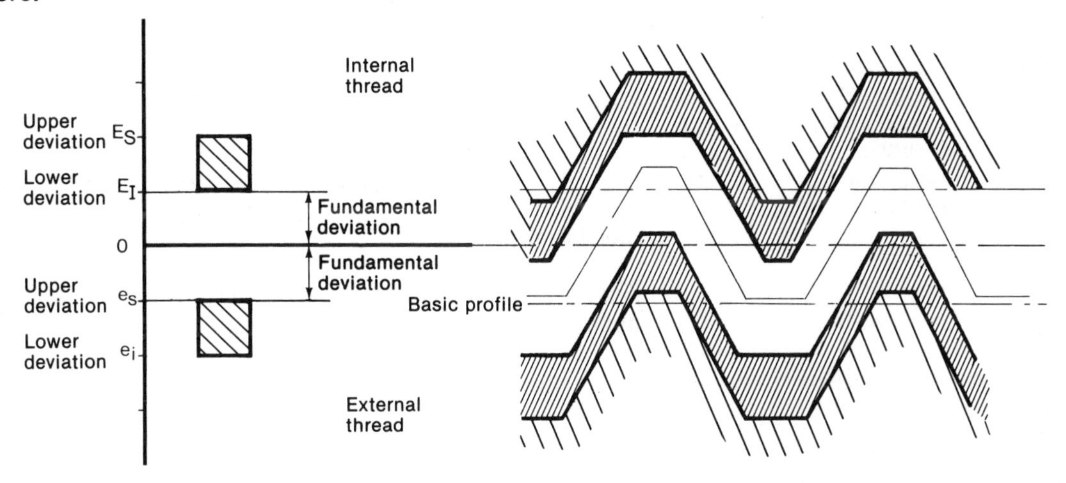

The complete designation for a screw thread comprises both a thread designation and a tolerance designation. The tolerance designation is indicated with a figure for the tolerance grade and letters for the tolerance position.

The tolerance grade is denoted by a number between 3 and 9, where 6 is medium tolerance. The tolerance position denotes the fundamental deviation and is indicated with an upper case letter for internal threads and a lower case letter for external threads. A combination of tolerance grade and tolerance position give the tolerance class.

The tolerance designation for a thread consists of the tolerance class for the pitch diameter followed by the tolerance class for the crest diameter (crest diameter = major diameter on external thread or minor diameter on internal thread). If the tolerance classes for the pitch and crest diameters coincide, only one tolerance class designation is given. The meaning of the tolerance classes in real sizes is given in the standards for the different thread systems.

TOLERANCE GRADES

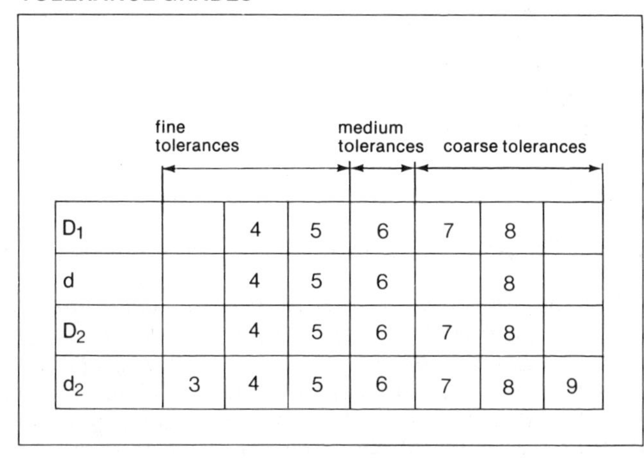

	fine tolerances			medium tolerances	coarse tolerances		
D_1		4	5	6	7	8	
d		4	5	6		8	
D_2		4	5	6	7	8	
d_2	3	4	5	6	7	8	9

TOLERANCE POSITIONS

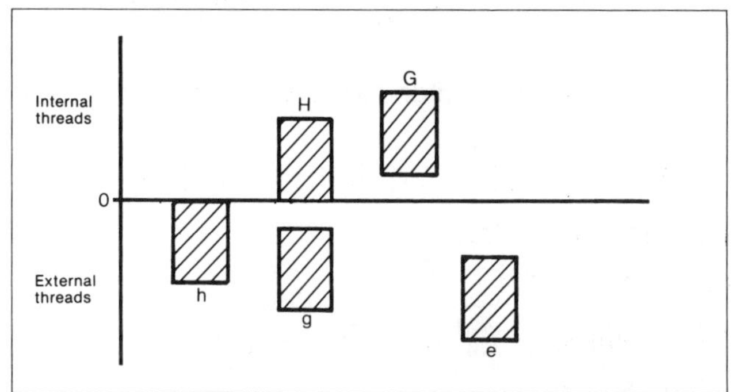

The following table can be used for choosing the tolerance class. The tolerance classes within boxes are intended for first choice; those in bold face are intended for second choice. The shaded tolerance classes are third choice, and those in light should be avoided.

TOLERANCE CLASSES

Tolerance quality	External threads									Internal threads					
	Tolerance position e Large fundamental deviation			Tolerance position g Small fundamental deviation			Tolerance position h No fundamental deviation			Tolerance position G Small fundamental deviation			Tolerance position H No fundamental deviation		
	Length of engagement			Length of engagement			Length of engagement			Length of engagement			Length of engagement		
	S (short)	N (normal)	L (long)	S (short)	N (normal)	L (long)	S (short)	N (normal)	L (long)	S (short)	N (normal)	L (long)	S (short)	N (normal)	L (long)
Fine							3h4h	4h	5h4h				4H	5H	6H
Medium		**6e**	7e6e	5g6g	**6g**	7g6g	5h6h	6h	7h6h	5G	6G	7G	**5H**[1)]	**6H**	**7H**
Coarse					8g	9g8g					7G	8G		7H	8H

When tolerance quality has been determined, tolerance class is selected on the basis of length of engagement. Length of engagement is the axial distance within which the external and internal threads are in contact. Normal lengths of engagement are considered to be 0.5xd — 1.5xd. If length of engagement is not known, use the normal length N.

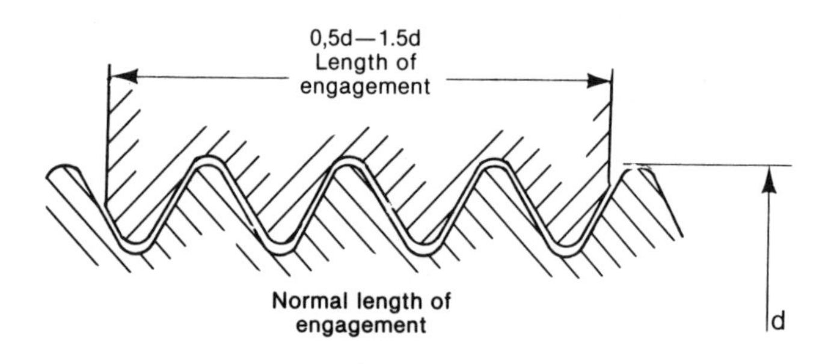

Example:

Assume that we are going to manufacture M30x1.5 bolt and nut threads. The length of engagement between the threaded parts is 20 mm. The tolerance is to be medium. We start by choosing a tolerance class. The tolerance quality is to be medium and therefore lies around tolerance grade 6. The length of engagement, 20 mm, is approximately 0.7xd (0.7x30) and is classified as normal, N. 6e, 6g and 6h can be chosen for the external thread (bolt). Of these, 6g is preferred as a first choice. For the internal thread (nut) we can choose between tolerance classes 6H and 6G, where 6H is preferred as a first choice.

The internal thread then has the following designation:

M30x1.5 — 6g

and the internal thread:

M30x1.5 — 6H

The designation for the joint will be:

M30x1.5 — 6H/6g

For nut threads as well as bolt threads, the actual root contours shall not at any point transgress the basic profile. The tolerance standards permit a root radius for ISO thread profile. For external threads, $R_{min} = 0.125$ x the pitch. (This was changed in 1980. Before the ISO standard permitted $R_{min} = 0.1083$ x the pitch for external threads).

For internal threads, $R_{max} = 0.0722$ x the pitch. R_{min} is not specified.

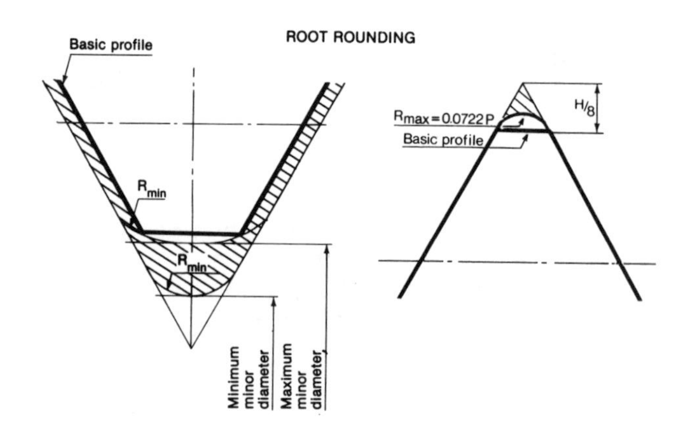

Sandvik Coromant threading inserts are manufactured to tolerance grade 6. If the pitch diameter tolerance is utilized in the manufacture of threads with coarser tolerance grades (7, 8 and 9), the major diameter tolerance cannot always be guaranteed in cases where the pitch diameter tolerance of the thread is larger than the major diameter tolerance of the threading insert. We will show this with an example.

Example:

Assume that a bolt with designation M20x1.25 — 9g8g is to be manufactured with T-MAX threading inserts. According to the table on thread tolerances, the major diameter of the thread may lie between:

20 — 0.363 = 19.637 mm and

20 — 0.028 = 19.972 mm

Major diameter d = D	Pitch P	External threads				
		Tol-class	Deviation, µm Major diameter d		Pitch diameter d_2	
			es	ei	es	ei
		3h4h	0	— 132	0	— 67
		4h	0	— 132	0	— 85
		5g6g	— 28	— 240	— 28	— 134
		5h4h	0	— 132	0	— 106
		5h6h	0	— 212	0	— 106
		6e	— 63	— 275	— 63	— 195
(11,2)—22,4	1,25	6g	— 28	— 240	— 28	— 160
		8h	0	— 212	0	— 132
		7e6e	— 63	— 275	— 63	— 233
		7g6g	— 28	— 240	— 28	— 198
		7h6h	0	— 212	0	— 170
		8g	— 28	— 363	— 28	— 240
		9g8g	— 28	— 363	— 28	— 293

(Taken from SMS 2161)

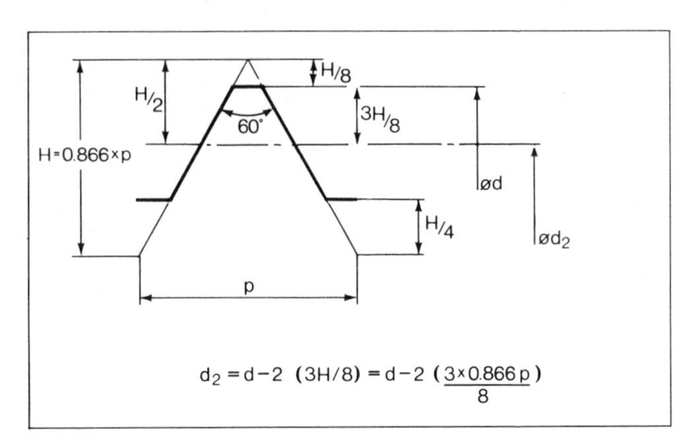

The pitch diameter d_2 has the following basic size (see figure)

$$20 - 2\,\frac{(3 \times 0.866 \times 1.25)}{8} = 19.188 \text{ mm}$$

According to the standards for the thread in question, the tolerance class 9g8g permits the pitch diameter to lie between

19.188 — 0.293 = 18.895 mm and

19.188 — 0.028 = 19.16 mm

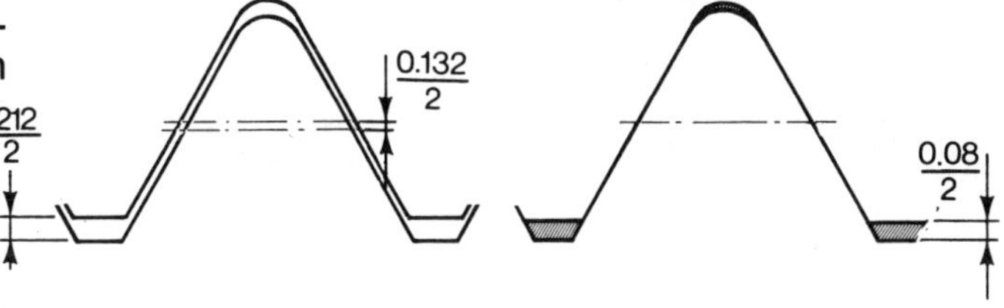

The insert is manufactured to tolerance grade 6, which means that the pitch diameter tolerance is 0.132 and the major diameter tolerance is 0.212. If the limits of tolerance are pulled together so that the pitch diameter tolerance is eliminated, the remaining major diameter tolerance (0.212 — 0.132) = 0.08 mm.

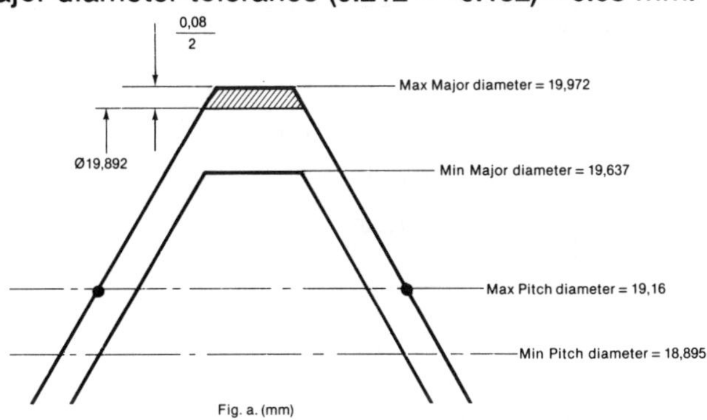

Fig. a. (mm)

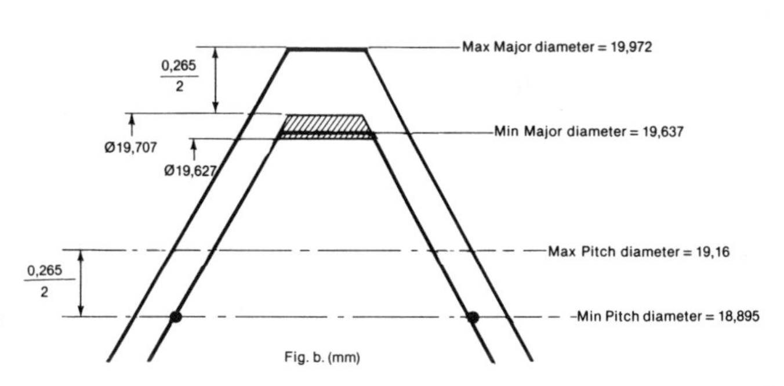

Fig. b. (mm)

If the insert is set to the pitch diameter 19.16 mm, as shown in fig. a, the major diameter of the bolt will lie between 19.892 mm and 19.972 mm.

Assume that the insert is set to the pitch diameter 18.895 mm, as in fig. b. The insert will then be displaced so that the major diameter of the bolt can fall between (19.972 — 0.265) = 19.707 mm and (19.892 — 0.265) = 19.627 mm. Thus, if the pitch diameter tolerance of the bolt is utilized, the major diameter tolerance cannot be guaranteed.

ISO inch threads

The ISO inch thread has three tolerance classes for external threads and three tolerance classes for internal threads.

Classes 1A and 2A have fundamental deviation while class 3 has no fundamental deviation.

Tolerance classes 2A and 2B are used as standard tolerances for nuts and bolts for general usage.

Tolerance classes 1A and 1B are applicable for threads where easy assembly is required and where rough treatment or the presence of foreign matter may prevent the use of closer tolerances.

Tolerance classes 3A and 3B are used where closeness of fit and precision are of importance.

Example:

A nut with designation 7/8" — 14UNF-1B is to be manufactured with T-MAX threading inserts.

According to the applicable thread tolerances, the pitch diameter may be between 21.046 mm and 21.314 mm and the minor diameter may be between 20.262 mm and 20.664 mm.

Sandvik Coromant ISO inch threading inserts are manufactured to tolerance class 2B, which means that the tolerance on the pitch diameter is (21.225 — 21.046) = 0.179 mm and the tolerance on the the minor diameter is (20.664 — 20.262) = 0.402 mm. If the pitch diameter tolerance is eliminated, the remaining "tool tolerance" for the minor diameter (0.402 — 0.179) = 0.223 mm.

If the threading insert is set to the pitch diameter 21.046 mm, the minor diameter of the nut will fall between 20.262 mm and (20.262 + 0.223) = 20.485 mm.

Assume that the threading insert is set to the pitch diameter 21.314 mm, as shown in fig. b. The insert will then give the nut a minor diameter of between 20.53 mm and 20.753 mm, which means that the tolerances on the nut cannot be guaranteed.

Sandvik Coromant threading inserts to ISO inch standard are manufactured to the tolerances for tolerance classes 2A and 2B. If the permissible pitch diameter tolerance is utilized, the minor diameter tolerance cannot be guaranteed for tolerance classes 1A and 1B.

TOLERANCE CLASSES

	EXTERNAL	INTERNAL
COARSE	1A	1B
MEDIUM	2A	2B
FINE	3A	3B

Nut class 1B

Thread designation inkl. tolerance designation	Major diameter		Pitch diameter		Minor diameter	
	$D_{nom} = D_{min}$	D_{max}	$D_{2nom} = D_{2min}$	D_{2max}	$D_{1nom} = D_{1min}$	D_{1max}
1/4"—28 UNF-1B	6,35	Not numerically established	5,761	5,871	5,367	5,581
5/16"—24 UNF-1B	7,938		7,249	7,370	6,792	7,039
3/8"—24 UNF-1B	9,525		8,837	8,963	8,379	8,626
7/16"—20 UNF-1B*	11,112		10,287	10,425	9,738	10,030
1/2"—20 UNF-1B	12,7		11,874	12,016	11,326	11,618
9/16"—18 UNF-1B*	14,288		13,371	13,521	12,761	13,082
5/8"—18 UNF-1B	15,875		14,958	15,112	14,348	14,669
3/4"—16 UNF-1B	19,05		18,019	18,185	17,330	17,687
7/8"—14 UNF-1B	22,225		21,046	21,225	20,262	20,664
1"—12 UNF-1B	25,4		24,026	24,220	23,109	23,568
1 1/8"—12 UNF-1B*	28,575		27,201	27,400	26,284	26,743
1 1/4"—12 UNF-1B	31,75		30,376	30,579	29,459	29,918
1 3/8"—12 UNF-1B*	34,925		33,551	33,759	32,634	33,093
1 1/2"—12 UNF-1B	38,1		36,726	36,938	35,809	36,268

* should be used for special applications only.

(Taken from SMS 1719)

Nut class 2B

Thread designation inkl. tolerance designation	Major diameter		Pitch diameter		Minor diameter	
	$D_{nom} = D_{min}$	D_{max}	$D_{2nom} = D_{2min}$	D_{2max}	$D_{1nom} = D_{1min}$	D_{1max}
1/4"—28 UNF-2B	6,35	Not numerically established	5,761	5,926	5,367	5,581
5/16"—24 UNF-2B	7,938		7,249	7,430	6,792	7,039
3/8"—24 UNF-2B	9,525		8,837	9,025	8,379	8,626
7/16"—20 UNF-2B*	11,112		10,287	10,493	9,738	10,030
1/2"—20 UNF-2B	12,7		11,874	12,086	11,326	11,618
9/16"—18 UNF-2B*	14,288		13,371	13,596	12,761	13,082
5/8"—18 UNF-2B	15,875		14,958	15,188	14,348	14,669
3/4"—16 UNF-2B	19,05		18,019	18,268	17,330	17,687
7/8"—14 UNF-2B	22,225		21,046	21,314	20,262	20,664
1"—12 UNF-2B	25,4		24,026	24,316	23,109	23,568
1 1/8"—12 UNF-2B*	28,575		27,201	27,499	26,284	26,743
1 1/4"—12 UNF-2B	31,75		30,376	30,681	29,459	29,918
1 3/8"—12 UNF-2B*	34,925		33,551	33,863	32,634	33,093
1 1/2"—12 UNF-2B	38,1		36,726	37,044	35,809	36,268

* should be used for special applications only.

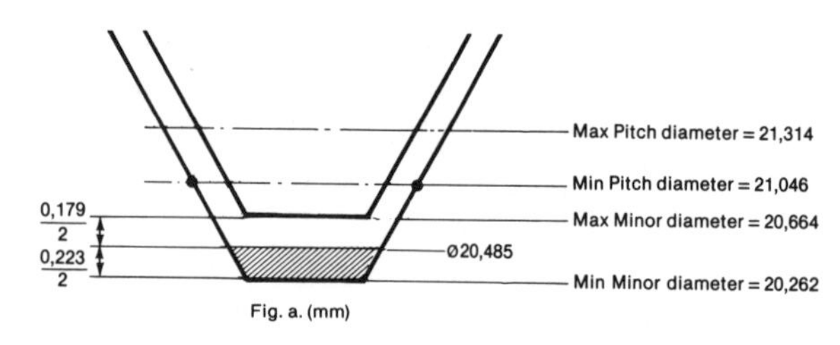

Fig. a. (mm)

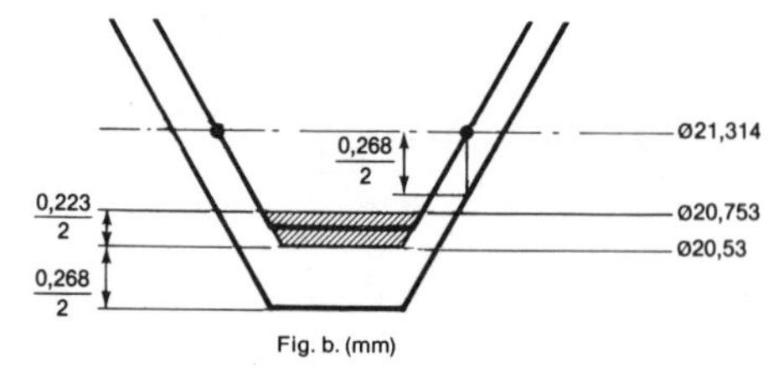

Fig. b. (mm)

SI threads

There are no internationally adopted tolerance standards for the SI profile. Such standards have been established on the national plane, for example in Germany and France. The German DIN standard specifies $R_{max} = 0.1083$ x the pitch for external threads. R_{min} is not a direct function of the pitch. For internal threads, $R_{max} = 0.0722$ x the pitch. R_{min} is not specified.

The SI thread has three tolerance classes: fine, medium and coarse. Sandvik Coromant threading inserts are manufactured to the medium tolerance class (DIN 13 1952). For these threads, the tolerance for the crest diameter (major diameter on external threads and minor diameter on internal threads) is independent of the tolerance class, while the pitch diameter tolerance is smaller in finer tolerance classes. This means that the tolerances on the insert do not guarantee the right crest diameter for the coarse tolerance class.

For the root diameter (minor diameter on external thread and major diameter on internal thread), the fine tolerance class cannot be guaranteed for certain combinations of pitch and major diameter, but in most cases this tolerance can be met as well.

Whitworth threads

This thread system was standardized with different pitch series, namely BSW (coarse thread), BSF (fine thread) and BSP.F (parallel fastening threads).

The Whitworth thread does not yet have an equivalent ISO thread for the finest pitch series, the parallel fastening threads. The parallel fastening threads are available with 11, 14, 19 and 28 threads per inch.

There are two tolerance classes for external Whitworth threads, A and B, where the pitch diameter tolerance for tolerance class B is twice as large as that for tolerance class A. For internal threads there is only one tolerance class.

Sandvik Coromant threading inserts are manufactured to tolerance class A and produce dimensionally accurate threads regardless of pitch series. BSP.F, which gives the closest tolerances, has served as the basis for the insert tolerances.

The crests of the Whitworth thread may be flat or rounded within the stippled area in the figure.

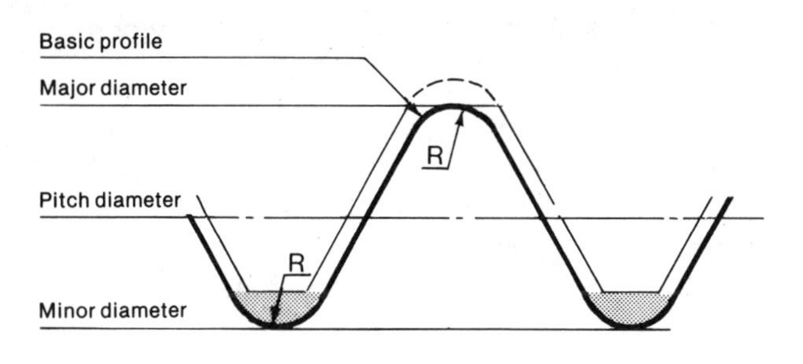

INTERNAL WHITWORTH—THREAD

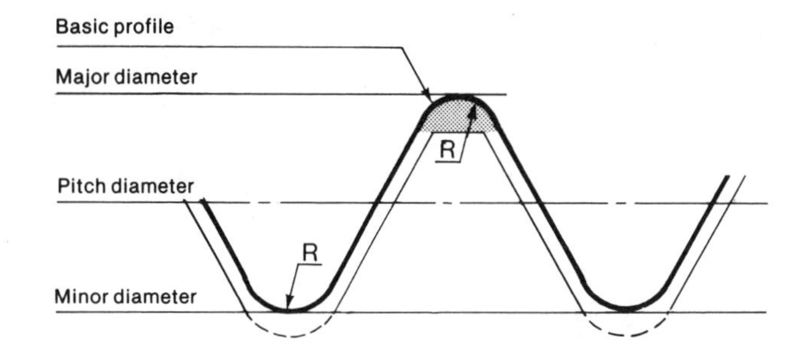

EXTERNAL WHITWORTH—THREAD

Thread inspection

As has already been mentioned, each thread system has its own tolerance system. This tolerance system also includes a gauging system. We shall restrict ourselves here to a general description of different thread gauging methods and gauges. For further information, see the standards for the various thread systems. Gauging procedure is usually described in great detail in the instructions included with the gauges, which should be followed closely.

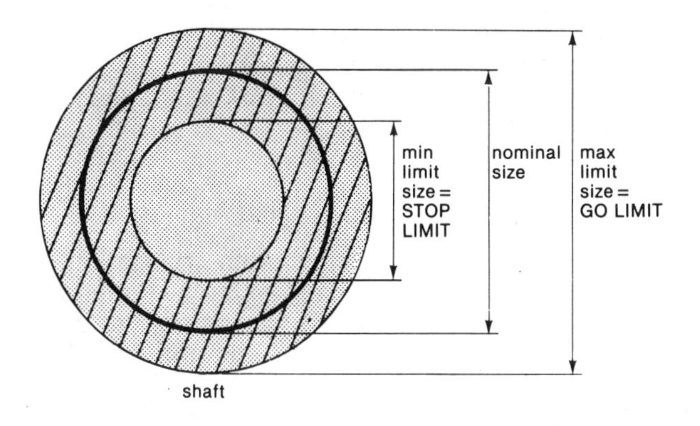

Terminology

A differentiation is made between fixed and indicating gauges. Fixed gauges materialize a given size, while indication gauges have dials that can be read off. The indicating gauges can either permit direct measurement, i.e. when the absolute size can be read off, or indirect measurement, when the deviation from a size to which the gauge is set can be read off.

Tolerance gauges are used for tolerance control. The tolerance gauge consists of two parts: a go gauge and a not go gauge. The go gauge is used for checking the go limit, i.e. the limit that is determined by the maximum limit of size in the case of a shaft tolerance and the minimum limit of size in the case of a hole tolerance (see figure). The not go gauge is used for checking the not go limit, i.e. the limit determined by the minimum limit of size in the case of a shaft tolerance and the maximum limit of size in the case of a hole tolerance (see figure).

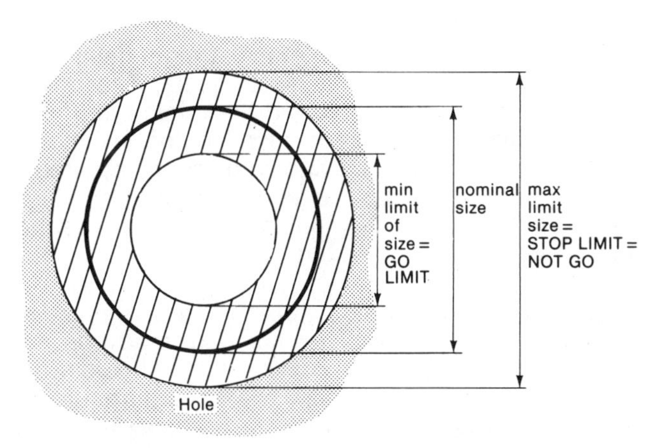

In order to be able to check whether existing deviations of form lie within the tolerance limits, the gauges should be designed in accordance with Taylor's principle. Taylor's principle entails the following:

1) A go gauge shall have such a shape that it permits simultaneous checking of the size and shape of the workpiece along its entire extent.
2) A not go gauge shall have such a shape that each individual size can be checked separately.

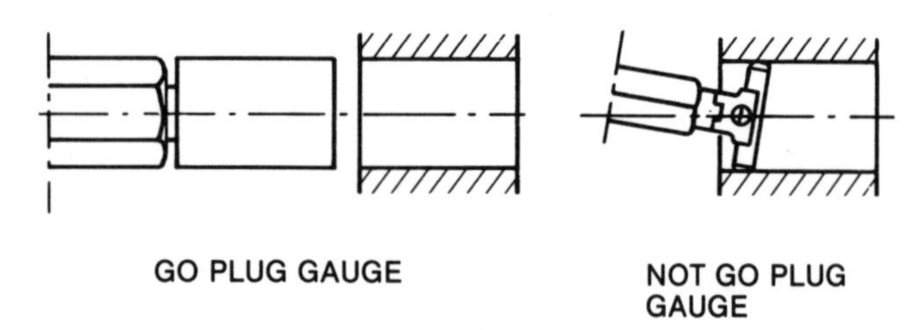

GO PLUG GAUGE

NOT GO PLUG GAUGE

Thread gauging

A thread has five dimensions: major diameter, pitch diameter, minor diameter, pitch and thread angle. A relatively large assortment of measuring apparatuses measuring machines and microscopes is available for measuring these dimensions. The influence of temperature, impurities, wear etc. on the measurement results should be taken into consideration. For example, when pitch is measured, helix variations (i.e. when the crest on a thread turn desribes a crooked line), are often disregarded. Fixed gauges check whether a thread meets the requirements on thread dimensions and tolerances. The deviation of the thread elements cannot, however, be determined. Indicating gauges are required for this.

Fixed gauges for external threads

A thread ring gauge (screw ring gauge) should preferably be used for checking external threads. The thread ring gauge is a fixed or adjustable gauge for checking the pitch diameter of the thread. The go thread ring checks the maximum limit of size simultaneously with respect to deviation from roundness and straightness, pitch errors, helix variations and errors in the flank angles.

It also checks that the length of the straight flank at the minor diameter of the bolt is adequate. The not go thread ring is used for checking the minimum limit of the pitch diameter. If there is suspicion of deviation of form, the thread ring check should be supplemented by checking with a calliper gauge for the not go side or, where great accuracy is required, with wire measurement, to which we shall return.

The go thread ring should be able to pass over the entire length of the thread. The not go thread ring may enter both ends of the thread. However, it should not be able to be screwed on more than two turns of the thread. If the length of thread of the component is three threads or less, the not go thread ring shall not pass completely over it.

Calliper gauges are available for quick checks of external threads. Unlike thread ring gauges, the calliper gauge can be used for both right- and left-hand bolts. The calliper gauge is used for checking the maximum and minimum limits of size of the pitch diameter. It is not suitable for non-rigid components, for example thin-walled components. Since the calliper gauge has limitations when it comes to checking the go side (it does not fulfill Taylor's principle that all dimensions should be able to be checked simultaneously), random checking with a go thread ring is recommended.

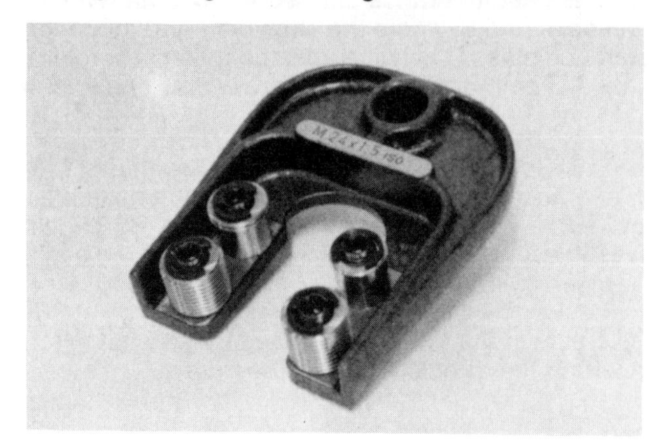

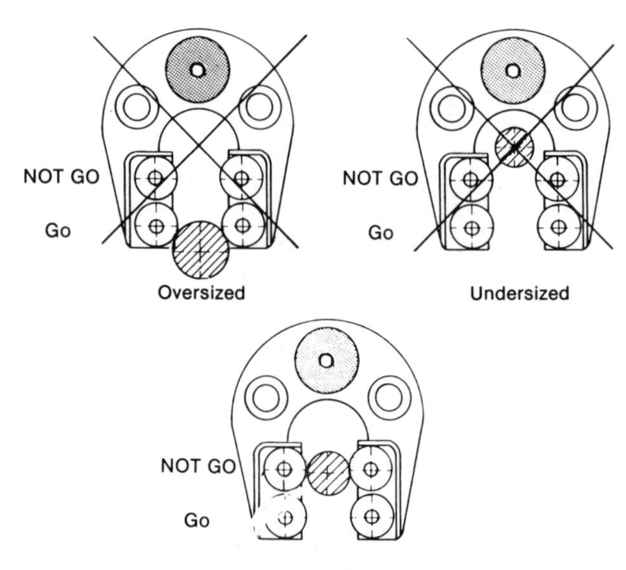

The gauging rolls on the go side shall pass over the diameter of the workpiece being measured. The gauging rolls on the not go side shall not pass over more than the two first turns of thread.

Plain go or not go ring or calliper gauges are used for checking the major diameter of external threads.

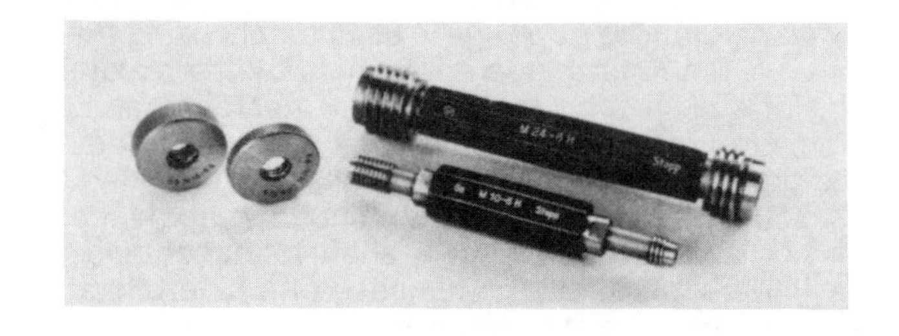

Fixed gauges for internal threads

The tolerance thread gauge is used for checking the minimum and maximum tolerance limits of the pitch diameter of threaded holes. The go thread plug gauge also checks the minimum limit of the major diameter of the threaded hole and the pitch and profile of the thread. For approval of the functional characteristics of a threaded hole, the go thread plug gauge must be able to be screwed in entirely without excessive force. The not go thread plug gauge shall not be able to be screwed in more than two turns from either end of through holes. If the threaded hole has three turns of thread or less, the not go gauge may not pass completely through.

A plain plug gauge is used to check the minor diameter of internal threads.

Indicating gauges

Thread micrometers of various types are available for external and internal screw threads. To meet higher accuracy requirements, three-wire measurement in a frame and a length indicator with sufficient accuracy must be employed. A simple procedure is to mount the component between centres, and since the measuring portion of the thread measuring apparatus is aimed perpendicular to the centre axis, only two wires have to be used. Internal threads are checked with balls instead of wires.

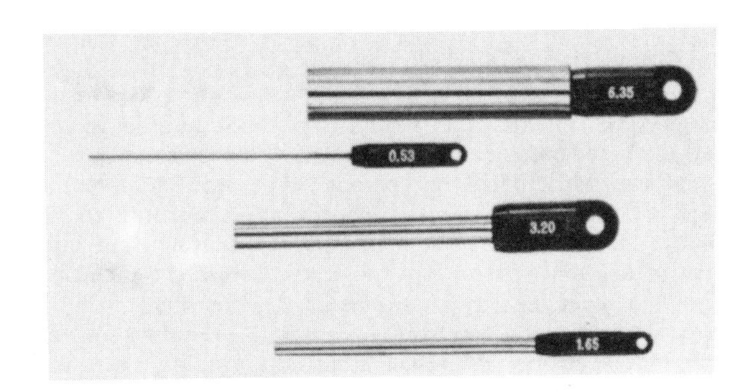

Wire measurement

The most favourable wire diameter is obtained from the equation:

$$W = \frac{P}{2\cos\alpha}$$

The pitch diameter $d_2 = M - B - C$ where

M = width across measuring wires (mm)

B = calculation term (see SMS 1744)

C = correction term for lead angle (see SMS 1744)

Only one measurement need be made to determine the pitch diameter. Two measurements are made to determine the thread angle, one over the wires used for the pitch diameter determination and one over wires of thicker diameter.

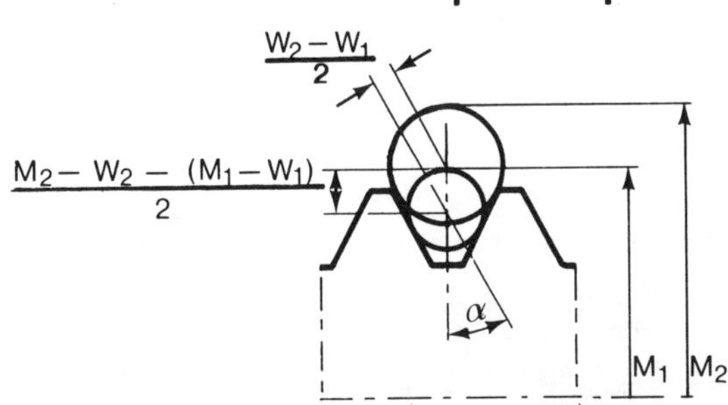

$$\sin\alpha = \frac{W_2 - W_1}{M_2 - W_2 - M_1 + W_1}$$

Thread designation	Thread angle	Wire diameter w	B	C
M, UN, SI	60°	0,5774p	3w—0,86603p	$0{,}076w\ (\frac{p}{d_2})^2$
W	55°	0,5637p	3,1657w—0,9605p	$0{,}0863w\ (\frac{p}{d_2})^2$

(Taken from SMS 1744)

Wire measurement is the most accurate method for measuring the pitch diameter of external screw threads. Where requirements on accuracy are not as high, a shop microscrope can be used.

Reprinted from *Manufacturing Engineering*, January 1980

Thread Cutting in Small ID's

Getting accurate threads and grooves in bores as small as 0.440″ can sometimes present problems, particularly in the harder materials. This system does the job while simultaneously reducing tap inventories

GENERATING THREADS AND GROOVES in small-diameter bores can often present difficulties, particularly in operations where various thread sizes are needed. Usually, the threads are cut with taps, which means that several taps must be stocked to meet the job requirements. But a new development from Kennametal Inc., Latrobe, PA, can cut this inventory problem. It's a line of threading and grooving inserts and boring bars called the #1 Top Notch and is particularly useful for those who have to make regular size changeovers in their small ID work. The system is designed for use on small automatic lathes, chuckers, and NC machines.

The import of this new line is that it facilitates threading and grooving in bores as small as 0.440″ (11.18 mm) in tough-to-thread materials including stainless steel, titanium alloys, and the super alloys. The reason it can handle these materials is that threading and grooving inserts in several carbide grades are available for the bars.

Replaces Taps. Although taps work unquestionably well in most applications, they are generally limited by their very design. That is, a tap is designed to cut a particular thread configuration in materials determined by its construction. So in order to change over to a different thread size or type of material, it is often necessary to use an entirely different tap. According to Kennametal engineers, starting a tap without getting bad threads can be difficult because of a tap's typical four flutes.

The #1 Top Notch system performs single-point cutting, which, according to those engineers, provides considerable cutting flexibility. For example, when used on an NC machine, the operator selects the required diameter bar, insert style, and insert grade, then activates the program to cut the necessary thread configuration. Suppose that the thread was an inch-type and metric threads were needed on the following workpiece. Instead of changing the tooling, the operator would simply change programs. If the material and not the thread style changed, then the operator would change the insert.

High accuracy is achieved with the system. Naturally, if it is used on an NC machine, close control is provided by the program. But even on a conventional machine this accuracy is attained because threads are started and cut with a single point. Very close-tolerance threading can be performed.

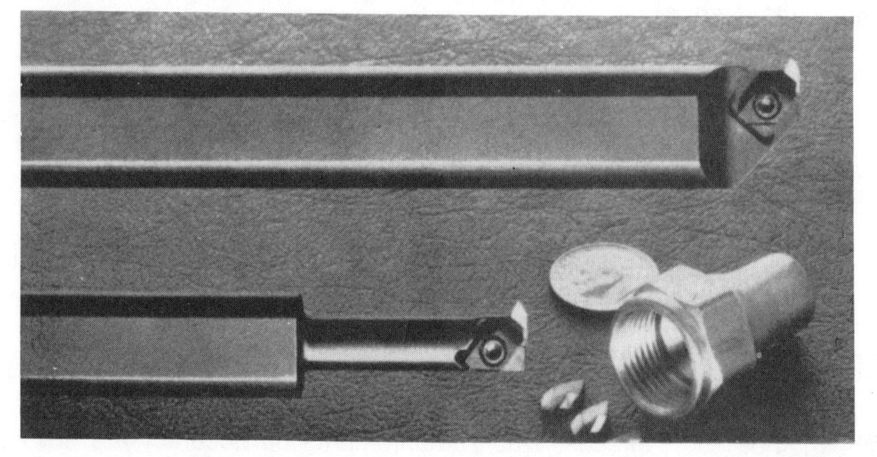

THREADING AND GROOVING bars and inserts provide an alternative to tapping. Work can be performed in bores as small as 0.440″.

System Composition. The system is made up of three steel-shank bars and two types of inserts. The bars have diameters of 3/8, 1/2, and 5/8″ (9.5, 13, and 15.9 mm) and lengths of 6, 8, and 10″ (152, 203, and 254 mm). One insert type is for threading and the other for grooving.

There are four standard insert grades: K1, for threading super and nonferrous alloys; K68, for aluminum alloys, nonferrous alloys, and nonmetals; KC810, for all steels and nonferrous materials requiring excellent edge wear, crater resistance, and material build-up protection; and K420, a general-purpose grade for grooving.

The bars utilize the Top Notch four-way insert locking system, which was developed in the 1960s and which is utilized in the entire Top Notch line. The insert is held down and back in the pocket through the combination of a clamp and lockscrew. This arrangement accurately positions the insert, then holds it securely in place during cutting.

Cutting Capacities/Applications. Grooving to a maximum depth of 0.040″ (1.02 mm) can be performed in a 0.440″ minimum bore. On the other hand, grooves to a maximum 0.070″ (1.78 mm) can be cut in a 0.800″ (20.32-mm) minimum bore. In terms of threading, the system facilitates the cutting of 48 threads per inch and finer when the smallest nominal thread diameter (or root) is 1/2″ (12.7 mm) and the minimum minor thread diameter (or crest) is 0.440″ (11.18 mm). In addition, it can cut as few as 10 threads per inch when the root is 15/16″ (23.8 mm) and the crest is 0.830″ (21.08 mm).

Kennametal engineers say that there are three primary application areas for the new system. Heading the list is aircraft components, in which a large number of threaded holes are needed. Quick insert changing is a real advantage here, particularly since close tolerances must be held in aircraft work. The other two areas are flow-control fittings and miscellaneous new products and prototype work. ■

INDEX

A

B

C

D

E

F

G

H

I

J

K

L

M

N

O

P

Q

R

S

T

U

V

W